Philosophy of Engineering and Technology

Volume 41

D1827087

The *Philosophy of Engineering and Technology* book series provides the multifaceted and rapidly growing discipline of philosophy of technology with a central overarching and integrative platform.

Specifically it publishes edited volumes and monographs in:

- the phenomenology, anthropology and socio-politics of technology and engineering
- the emergent fields of the ontology and epistemology of artifacts, design, knowledge bases, and instrumentation
- engineering ethics and the ethics of specific technologies ranging from nuclear technologies to the converging nano-, bio-, information and cognitive technologies
- written from philosophical and practitioners' perspectives and authored by philosophers and practitioners

The series also welcomes proposals that bring these fields together or advance philosophy of engineering and technology in other integrative ways.

Proposals should include:

- A short synopsis of the work or the introduction chapter
- The proposed Table of Contents
- The CV of the lead author(s)
- If available: one sample chapter

We aim to make a first decision within 1 month of submission. In case of a positive first decision the work will be provisionally contracted: the final decision about publication will depend upon the result of the anonymous peer review of the complete manuscript. We aim to have the complete work peer-reviewed within 3 months of submission.

The series discourages the submission of manuscripts that contain reprints of previous published material and/or manuscripts that are below 150 pages/75,000 words.

For inquiries and submission of proposals authors can contact the editor-in-chief Pieter Vermaas via: p.e.vermaas@tudelft.nl, or contact one of the associate editors.

Darryl Cressman
Editor

The Necessity of Critique

Andrew Feenberg and the Philosophy
of Technology

 Springer

Editor
Darryl Cressman (iD)
Philosophy Department
Maastricht University
Maastricht, The Netherlands

ISSN 1879-7202 ISSN 1879-7210 (electronic)
Philosophy of Engineering and Technology
ISBN 978-3-031-07879-8 ISBN 978-3-031-07877-4 (eBook)
https://doi.org/10.1007/978-3-031-07877-4

This Springer imprint is published by the registered company Springer Nature Switzerland AG
The registered company address is: Gewerbestrasse 11, 6330 Cham, Switzerland

Contents

Chapter 1
Introduction: The Necessity (and Spirit) of Critique in Andrew Feenberg's Philosophy of Technology

Darryl Cressman

Abstract In this introductory chapter I situate the following collection as one that takes as its starting point Andrew Feenberg's philosophical critique of technology. Feenberg's critique of technology is quite wide-ranging and so the chapters in this book fall within three themes that I outline in this introduction. First, Democratic Potentials, which explores how democracy can be used as a standard against which technologies can be critiqued. Trajectories of Contemporary Critique, which looks at how history and hermeneutics can inform the critique of contemporary technologies. And Critical Theories of Technology, which draws together intellectual histories of the critique of technology.

Keywords Andrew Feenberg · Critical theory · Technology · Critique · Democracy · Politics · History · Hermeneutics · Intellectual history · Philosophy of technology

1.1 Introduction

This collection commemorates Andrew Feenberg being awarded the Lifetime Achievement Award in 2019 by the Society for the Philosophy of Technology. It would not be an exaggeration to say that Feenberg is one of the best-known and most-read philosophers of technology of the past 35 years. He is a prolific writer who engages with a range of ideas and so, given the scope of his work, thematizing a volume like this can be a challenge. Other books and journals dedicated to

D. Cressman (✉)
Philosophy Department, Maastricht University, Maastricht, Netherlands
e-mail: darryl.cressman@maastrichtuniversity.nl

Feenberg's work focus on the reform and democratization of technology (Veak, 2006), situating his work as part of the tradition of Frankfurt School critical theory (Arnold & Michel, 2017), and comparing his philosophical project against Don Ihde's project of postphenomenology (*Techné*, 24 [1/2], 2020). To distinguish this collection from these other projects, the following chapters focus on Feenberg's critique of technology.

Feenberg's critique of technology accounts for the sociotechnical world as it is while pointing to concrete potentials of what it could be, "a dialectical critique of technology that is neither irrationalist nor technophobic" (2014, p. 201). The object of his critique is not any one particular technology per se, but rather a type of rationality that normalizes a limited range of possible technical functions, designs, and meanings. Recognizing the inherent potential of sociotechnical rationality to respond to a plurality of experiences and initiatives allows for a shift in thinking about technology, away from neutrality, universal progress, and professional expertise to an idea of technology characterized by ambivalence, local progress, and situated use. This should not be confused with a technical fix. Rather, what Feenberg proposes is a radical re-configuration of social theory itself:

> ...the 20th century, with its world wars, atom bombs, concentration camps, and environmental catastrophes, has made it more and more difficult to ignore the strange aimlessness of modernity. Because we are at such a loss to know where we are going and why, philosophy of technology has emerged in our time as a critique of modernity (Feenberg, 2005, p. 12).

This is a bold claim, but it points to the necessity of critiquing technology. Technology is both a contingent and permanent feature of the human experience such that any critical theory of society that is not also a critical theory of technology is woefully incomplete. Re-considering the concerns of social theory as distinctly sociotechnical concerns can open up trajectories of analysis and change that may prove to be more durable and effective than the law, education, or governance.

The critique of technology is not what it once was. Its apex occurred in the decades following World War II as a response to the carelessness and irresponsibility of the first half of the twentieth century.[1] The triumphant celebration of post-war technological progress, with its biases towards consumerism and militarism, inspired many in the humanities and social sciences to be critical of technology and suspicious of those who were not. The past 40 years has seen this tradition of critique fade.

For critical theorists working in the tradition of the Frankfurt School and Western Marxism, Habermas' influence neutralized the politicization of technology and

---

[1] Of course, the critique of technology did not begin in the twentieth century. In the West, a humanistic style of critique emerged in the wake of the Industrial Revolution. These early expressions took many forms: critiques of technological hubris in literature by writers like Mary Shelley (1797–1851), Henry David Thoreau (1817–1862), and Ralph Waldo Emerson (1803–1882), the humanist political economy of Karl Marx (1818–1883), representations of speed and technology in the paintings of J.M.W. Turner (1775–1851), and the hopeful utopianism of Edward Bellamy's novel *Looking Backwards* (1888). These reactions marked the beginning of a critique that challenged the ideologies of progress that shaped popular ideas about technology.

technological rationality by bounding it to a sphere of functional instrumentality that was independent of social influence and history. For those not convinced by Habermas, the influence of post-colonialism, aesthetics, feminism, and ecology has pushed critical theorists towards many fascinating insights, but without a corresponding critique of technology. Indeed, it is not surprising to find that a recent book on the intellectual history of contemporary critical theory makes only a passing reference to technology as a holdover from Horkheimer & Adorno's remarks on the culture industry (Keucheyan, 2014).[2]

Around the same time that Habermas became the leading voice amongst critical theorists, philosophers of technology began adopting the methodological insights of social scientists and Science and Technology Studies (STS) researchers in what has been termed the empirical turn. Turning to discrete case studies to demonstrate multiple instances of contingency, an explicit methodological a priori is evident in approaches such as actor-network theory, postphenomenology, value-sensitive design, virtue ethics, sociotechnical imaginaries, and many other recent additions to the philosophy of technology. Although these approaches have done much for the philosophical study of technology, they offer little to account for critical concepts like ideology, reason, alienation, and reification.

Despite these changing theories, methods, and objects of critical inquiry, the irresponsibility and carelessness that characterized the first half of the twentieth century persists across our shared technological condition. This demands, I argue, a theoretically coherent, historically conscious, and empirically sound critique of technology.

The following chapters do their best to adhere to this idea of critique. They are not intended to be an elaboration and application of specific concepts found in Feenberg's writings (although many do just that) nor a series of debates regarding the finer points of his philosophy (although this is also present). What connects the contributions to this volume is that they are indebted to Feenberg's insistence that critiquing technology is necessary for any sort of progressive social change. This, then, is the spirit (which can be considered a complement to the necessity) of critique that one can take from Feenberg's work. Whereas other writers can be quick to find fault with methods or theories that are different than their own, reading Feenberg it is clear that his political and philosophical ideals take precedence over any sort of academic tribal allegiances. Borrowing ideas, concepts, and methods that can assist in moving towards a more humane and democratic sociotechnical culture exceeds the value of any one particular approach.

One feature of this collection is that all of the contributors are located in Europe. In one way, this geographical feature distinguishes this collection from others dedicated to Feenberg's work, which tend to draw primarily from the United States and other English-speaking countries. But, it also points to a disparity in the range of perspectives that have come to shape the philosophical critique of technology. The

---

[2] Similarly, as Hans Radder points out in his contribution to this book, "technology" is completely absent from the recent German edited collection *Warum Kritik? Begründungsformen kritischer Theorien*, which takes as its starting point the justification for a critical theory.

object of Feenberg's critique, as noted above, is a form of technological rationality that normalizes a limited range of technical functions while restricting the scope of whose ideas count in relation to design and meaning. Key to this limited notion of technological rationality is the assumption that technology, and technological rationality, is universal. Attached to grand narratives about the inevitability of progress and the inherent benefits of innovation, is the widely assumed belief that technologies provide the same promises to everyone, regardless of who they are, where they come from, or what they desire; a great levelling of expectations determined by functional attributes designed by a small cadre of experts representing a small geographical region. Drawing attention to the voices and perspectives that are not accounted for in the idea of a neutral and universal technological rationality is the basis of a philosophical critique that fits nicely with Feenberg's overall project, "replacing the grand narratives of the past with the many local narratives will free the imagination to explore alternatives to both the existing society and the failed revolutions of the past" (Feenberg, 2017a, p. 204). It is hoped that this book can emphasize the necessity of locality, both geographic and cultural, in order to open up a philosophical critique of technology that recognizes as many experiences and potentials as possible.

Because Feenberg's critique of technology covers a range of ideas, the following chapters are organized around three themes that expand upon his philosophical project. The first theme, **democratic potentials**, is concerned with his work as a political philosopher of technology. The second, **trajectories of contemporary critique**, explores the historical and hermeneutic dimensions of his critique through studies of contemporary digital technologies. And the last theme, **critical theories of technology**, takes as its inspiration the intellectual history of the philosophy of technology. Prior to these themes and chapters, this collection begins with a recent essay by Feenberg that summarizes his approach to technology. Titled "Critical Constructivism: An Exposition and Defense," this is a concise summary of the ways in which his theory draws together insights from Frankfurt School Critical Theory, Heideggerian phenomenology, labour process theory, and Science and Technology Studies (STS) to politicize the rationality manifested in both technology and the ideologies that surround it, arguing that "uncovering the bias of that rationality is the critical task of the study of technology today."

1.2 Democratic Potentials

> What human beings are and will become is decided in the shape of our tools no less than in the action of statesmen and political movements. The design of technology is thus an ontological decision fraught with political consequences. The exclusion of the vast majority from participation in this decision is profoundly undemocratic (Feenberg, 2002, p. 3).

The above quote is used by a few contributors to this collection because it succinctly points to a fundamental problem of modernity: technology remains largely outside of the scope of our democratic expectations. For many, the failure to democratize technology is due to variations of technocracy. At its most imaginative, technocracy has parallels with the totally administered society of Huxley's *Brave New World* while contemporary critics identify it with the top-down decision making associated with nudging or design thinking. Technocracy is problematic because the trajectory of technology should respond to, and not direct, the interests of the people whose lives are mediated through particular designs and functions. Omitting plurality, compromise, and uncertainty from the scope of technical decision-making contributes to a sense of collective disenfranchisement in the face of seemingly autonomous sociotechnical changes, reinforcing the legitimacy of technocratic ideals.

The democratization of technology is a political philosophy that aims to delegitimize technocratic rationality, albeit without predictable dualisms or binary oppositions. For Feenberg, democratization is not a matter of holding an election between different technologies, but accounting for democratic interventions, which he defines as those unexpected and unimagined potentials that emerge from everyday engagements, that, by virtue of their existence and not their intention, destabilize technocratic rationality by demonstrating its limits. Think, for example, of the moment when different social groups transformed turntables, mixers, and LPs into musical instruments (Fikentscher, 2000; Hebdige, 1987) or the processes through which farmers reshaped the design and meaning of the car to better meet goals that were their own, distinct from manufacturers and dealers (Kline & Pinch, 1996). These moments of informal and improvised interjections into formally rational systems make it possible to recognize alternative functions and meanings that reflect values that were not part of the formal design process. There are no appeals to transcendent ideals or evidence of organized resistance to power, just engaged use that leads to unimagined potentials. This is a modest philosophical critique that reminds us that the hubris of planning out technological solutions needs to be balanced alongside the recognition that one cannot know, or plan, the trajectory of technological change.

> …the new politics is neither revolutionary nor reformist…we do not know where these changes lead, but we cannot doubt that they represent a universal advance…critical constructivism gives an account of the process of transcendence without positing a final endpoint the nature of which we do not know (Feenberg, 2017a, p. 119).

Hans Radder's chapter deals with one of the methodological problems that has perplexed critical theorists of technology for the past few decades – how does one balance individual sociotechnical experiences with larger scale concepts, like democracy, that critique demands?[3] On one hand, phenomenological-empirical descriptions of the "thing itself" often reveal fascinating details of small-scale individual experiences and, in doing so, draw out the material and interpretative flexibility of technology. But this attention to micro-level contingencies occurs at the expense of history and an awareness of larger cultural patterns, negating critique by

---

[3] This debate has occurred between empirical researchers of technology and philosophers of technology since the 1990s. See, for example the essays collected in Misa et al. (2003) and Hans Radder's (1992) critique of the methodological and theoretical presuppositions of STS.

affirming the world as it is. On the other hand, reducing the richness of our sociotechnical lifeworld to variations of capitalist political economy or technocracy can
diminish the range of potential by offering little beyond variations on the persistence of one-dimensionality. Radder overcomes this methodological paradox
through a comprehensive critical theory of the common good in which democracy
is used to assess the feasibility and normative desirability of a particular technology.
In his assessment of the Dutch government's attempts to develop a COVID-19
tracking app, Radder demonstrates how a universal concept of the common good
can be used to develop different assessments of particular sociotechnical initiatives.
This particular critique refers back to Marcuse's critical interpretation of theoretical
concepts and Radder reminds readers that, "our concepts contain the seeds of their
own transformation: their nonlocal meaning reflects our potential of seeing the
world in novel ways." The unification of the local and the universal proposed by
Radder fits nicely alongside Feenberg's ideas about local progress. Democratic critique is itself universal, but the character of this resistance is wholly local.

The chapters by Tina Sikka and Roy Bendor complement Feenberg's democratic
proposals by scrutinizing aspects of theory, refining and expanding the processes by
which the democratization of technology can occur. Working from the insights
developed through Feenberg's study of AIDS patients in the United States in the late
1980s and early 1990s, **Tina Sikka** subjects the case of COVID-19 to a similar type
of analysis. Whereas for AIDS activists and patients, increased communication
transformed the hierarchical and paternalistic technical codes that governed experimental drug trials, Sikka highlights how increased communication in a highly politicized and polarized media environment distorts attempts to democratize
sociotechnical rationality, suggesting that Feenberg's democratization thesis falls
short, "not because it is philosophically or empirically suspect, but because of a
number of mitigating factors that have interfered with and distorted the productive
role pro-social democratic values and norms usually play." Sikka overcomes this
problem by drawing on feminist science scholars to imagine discourses around
health care that prioritize an ethic of care that emphasizes the interdependence and
mutual responsibility of all humans alongside (or perhaps even ahead of) their communicative liberty.

Roy Bendor's chapter examines the question of whose sociotechnical agency
counts. One of the more provocative moves that Feenberg makes is that he argues
that it is no longer sufficient to assume that the trajectory of technological change
can be bracketed off as something that only a select class of experts – designers,
engineers, and policy makers – can effectively have agency over. This is not intended
to be antagonistic towards these professions, rather it is a re-consideration of sociotechnical agency based on empirical demonstrations that technical expertise is not
wholly restricted to an idea of technological rationality that only a select few have
mastery over. Bendor, working from a perspective informed by design theory, suggests that professional designers have much to contribute to a democratic politics of
technology while still maintaining practical resonances with Feenberg's politics.
Summarizing different approaches to design that share affinities with Feenberg's
strategies for democratization, Bendor suggests that "a different kind of design

altogether," what he calls (un)design, can open up democratic potentials that go beyond hacking, re-appropriation, or reinvention. Unlike other forms of democratic rationalization, (un)designs are intentionally unfinished, what he calls "ontologically incomplete" designs that invite users to exercise their own agency in directing the function and meaning of design. In Bendor's hands, (un)design recognizes how designers can exercise their professional expertise while maintaining and enhancing the democratic potentials of those who live with design.

Ryan Wittingslow's chapter concludes this section by pointing out the inherently undemocratic processes by which the design and implementation of "smart cities" has been championed by city planners, designers, and philosophers eager to implement a neatly uniform "the good," which he subjects against the standard of democracy. From this perspective, the well-intentioned ambitions of smart city visionaries are deeply problematic. Democracy is pluralistic, containing multiple iterations of "the good" that are excluded by design in the plans of smart cities. Democracy is also slow and sensitive to interests that may delay the goals of competitive efficiency, profit-making, and other positive externalities found in the sociotechnical imaginaries championed by smart city enthusiasts. As Wittingslow demonstrates, democracy is an effective way to think about what smart cities can provide for citizens, concluding that, "democracy plays a foundational role in that it provides the means by which a given polis can collectively conceptualise the smart city…democracy's pluralism makes possible the development of hitherto unimagined potentials for smart urban technologies." Democracy, in this sense, is not about providing another version of a smart city, it is about providing frameworks through which we can evaluate these ambitions, a mode of critique to judge the legitimacy and desirability of technical initiatives and provide a vocabulary to discuss the limitations of top-down and expert-led initiatives.

1.3 Trajectories of Contemporary Critique

> In sum, values are the facts of the future…our world was shaped by the values that presided over its creation. Technologies are the crystallized expression of those values (Feenberg, 2017a, p. 8).

The complex richness of our sociotechnical world is diminished when the meaning of technology is reduced to an ahistorical and decontextualized functionality. Although reducing technology to its functions is a common-sense approach that corresponds with other widely accepted ideas, such as the idea that technology is neutral and that improved functionality is synonymous with progress, this is a misguided effort to discern the essence of technology from a limited idea of how it works. Think of a car: one can understand how a car works by reading the user's manual, but this is an impoverished understanding of what it means for a technology to work. A car also works symbolically as a marker of success (or lack thereof), it works as part of a complex network of roads, traffic laws, manufacturing facilities, gas stations, insurance companies, and a myriad of occupations and institutional arrangements, and it works to mediate debates concerning environmental issues, the expansion of public transportation, and more recently, ethical debates about self-driving cars. Technologies work, in other words, both functionally and hermeneutically.

As the epigraph above hints at, history is one way that the hermeneutic dimension of technology can be revealed as a means of critique. Feenberg draws insights from histories of the competing values and ideas that were present at the origin of technologies, and his work contains numerous references to historians who are sympathetic to these functional and hermeneutic contingencies, such as Carolyn Marvin, Wolfgang Schivelbusch, and Jean-Baptiste Fressoz. The allure of these histories is that they demonstrate that the success or failure of a technology is never wholly technological. The corollary to this insight is that what has come to be assumed as rational and inevitable is dependent on particular historical and social contexts that, if different, would result in a different technology. The assembly line, for example, is often considered to be a model of universal rational efficiency that can be unproblematically used in any context. But this design is dependent upon capitalist notions of labour and power. The functional deskilling, surveillance of labour, and autonomy of management that seems to be an inevitable consequence of productivity and progress are contingent elements that could be otherwise. Forgetting this history, paradoxically, transforms the assembly line into a neutral tool that is also ideologically useful, "The assembly line only appears as technical progress because it extends the kind of administrative rationality on which capitalism already depends" (Feenberg, 2002, p. 78). This is an example of what Feenberg calls the technical code, or the design code, of capitalism, which explains the reification of political and economic biases that influence the design and meaning of technology.

> …the legitimating effectiveness of technology depends on unconsciousness of the cultural-political horizon under which it was designed. A critical theory of technology can uncover that horizon, demystifying the illusion of technical necessity, and expose the relativity of the prevailing technical choices (Feenberg, 1999, pp. 86–87).

Following Feenberg, an attention to history can open up trajectories for critiquing contemporary technologies that go beyond the somewhat restrictive bounds of "new and emerging" technologies. Writing and researching at a pace that attempts to keep up with incessant technological change comes at the cost of an inability to comprehend technological society in its totality, which frustrates attempts at dialectical critique by overlooking the endurance of social and economic relations that are neither new nor emerging. Indeed, the labour processes that characterize technology as imagined by Henry Ford and Fredrick Winslow Taylor in the early twentieth century persist today in both the manufacture and operation of seeming cutting-edge technologies like AI and other innovations (Crawford, 2021; Irani, 2019). Identifying the technical codes that govern the production and consumption of technologies, like the codes that provide the horizon through which AI is imagined, points to historical continuities that tie aspects of technological society together, enabling a critique that transcends any one particular technical object.

The chapters in this section undertake this type of critique through studies of big data, recommender systems, and algorithmic things. **Sally Wyatt's** chapter is a wide-ranging analysis of the many claims that have come to endow big data with particular meanings, such as the idea that data are given, that data are a natural resource, that numbers speak for themselves, and that everything is already data. She easily critiques the assumption that data are neutral and self-evident by pointing out that "data are never simply given, nor are they enough by themselves to make any kind of knowledge claim. Theoretical concepts are always built into the classifications and algorithms used to make sense of large volumes of data." These concepts, rendered invisible, reproduce existing power structures. This is compounded by a myopic reliance on big data as all that truly counts as both knowledge and a way of knowing. Wyatt points to the popular claim that, reliant upon big data, researchers no longer need to know why people do what they do, they only need to know what they do. As she makes clear, this attitude persists in an unthinking reliance on data that "represent the past and not the future, and thus cannot be used to support equal opportunities for groups that have been historically discriminated against in labour markets." She concludes by arguing that it is not enough to critique these moments of reification. If big data are to be harnessed towards socially desirable ends, it is necessary for human intervention to make the biases of big data explicit.

Marit de Jong and **Robert Prey** draw out in fascinating detail the long history of contemporary recommender systems (like the ones used by Spotify, Amazon, and Netflix) in the ideas of behavioral psychologists like B.F. Skinner. What de Jong and Prey call the "behavioral code" of recommender systems, "promotes an impoverished view of what it means to be human." Impoverishment derives from an inherent conservativism, familiar from Wyatt's analysis of big data, that comes from limiting the scope of potential behavior to a simplistic idea of past behavior that avoids questions of intentionality or historical and cultural circumstances – only *what* people do counts, not *why* they do it. For de Jong and Prey, tracing the history of recommender systems to behaviorist psychology opens up a space for critique that looks beyond the software to the ideas upon which it is designed. Considering recommender systems through a critical framework, they trace how a formal bias emerges: a contested intellectual field characterized by multiple debates about human behavior becomes reduced and translated into recommender systems, resulting in a sociotechnical system that effectively aligns both an idea of what counts as behaviour and how behaviour can be measured. This analysis resonates with what Marcuse wrote in his critique of empirical social research, "...the criteria for judging a state of affairs are those offered by...the given state of affairs. The range of judgment is confined within a context of facts which excludes judging the context in which the facts are made" (Marcuse, 1964, pp. 115–116).

Whereas these two chapters open up trajectories for de-reifying the design of contemporary technological systems and processes, **Yoni Van Den Eede** looks to Feenberg to develop an idea of critique appropriate to algorithmic things, those elusive networks of data, algorithms, automated processing, and artifacts that everyone recognizes as both "tech" and "more-than-objects." Algorithmic things have become part of the fabric of our lives, unobtrusive and unquestioned. And so, how does one critique an algorithmic thing? Where are the spaces for resisting and critiquing algorithmic things when there is no thing to resist or critique? As Van Den

Eede writes (with a nod to Heidegger), "Critical constructivism needs break-downs…as a gateway into democratic rationalization…but what if breakdown doesn't take place?" This is not to say that there are not injustices and inequalities mediated through technology, it is just that these biases appear as business as usual and not as issues of concern. Violations of privacy, the extension of surveillance, and the increased commodification of communication processes cannot be addressed via that traditional means of resistance (strikes, protests, and boycotts). Van Den Eede's solution is a re-consideration of the way in which objects should be inter-preted: a new hermeneutic of technology. He proposes this through Object-Oriented Ontology, a theory, he suggests, that allows for a re-conceptualization of technology that in combination with Feenberg's critical constructivism is better suited to mak-ing sense of algorithmic things, and hence, better prepared to critique algorith-mic things.

1.4   Critical Theories of Technology

Capitalism is more than an economic system; it projects a world in the phenomenological sense of the term. The congruence of science, technology, and the form of experience is ultimately rooted in the logic of capitalism. Existing science and technology cannot tran-scend the capitalist world. Rather, they are destined to reproduce it by their very structure. They are inherently conservative, not because they are ideological…because they are intrin-sically adjusted to serving a social order that ignores potentialities and views being as the stuff of domination (Feenberg, 2014, p. 180).

In the quote above, Feenberg argues that existing technology is conservative because it is designed to reproduce the world as it is. Extending this point, philosophical thought about technology can also be conservative by an unquestioned reliance on existing theories or methods that affirm capitalist political economy. Although the object of inquiry may change, conservativism endures when the processes of inquiry remain the same. A key aspect of critique, in this regard, is the development of philosophical critique through theoretical and methodological interrogations of exiting theories and methods, comparative summaries that illuminate resonances and dissonances across different critical approaches to technology, and intellectual histories that draw out different traditions of critique. But, like critique itself, this is a fading dimension of the philosophy of technology. The demands of meritocratic and achievement-oriented inquiry and the lucrative nature of fulfilling the needs of State and Industry favours the safe predictability of theoretically and methodologi-cally similar studies. This is another reason why critique is necessary: if we expect it to effectively challenge the design, function, and meaning of technology and not just reiterate the ambitions of neoliberal capitalism, it is imperative to chart a path that encourages research and writing that is an alternative to these ambitions.

Throughout Feenberg's work is an intellectual history that encourages a theoreti-cal adventurousness familiar from the New Left and journals like *Telos*, *Radical America*, and *New German Critique*. Reading Feenberg is a reminder that the cri-tique of technology has a history. As R.G. Collingwood put it, "the philosophizing mind never simply thinks about an object, it always, while thinking about any object, thinks also about its own thought about that object" (Collingwood, 1946, p. 1). The thought about technology that interests Feenberg begins with Marx's ideas about alienation and machinery and then moves to include the concept of world (or lifeworld) as developed by Heidegger and Husserl; Lukács' theory of reification as a more complete realization of Marx's theory of commodity fetishism; an idea of rationality that begins with Max Weber and extends through Horkheimer, Marcuse, and Habermas; and the recognition of sociotechnical contingency that corresponds with the empirical observations found in Science and Technology Studies (STS) and the labour process theory of Harry Braverman. Along the way, he engages with the ideas of Gilbert Simondon, Michel Foucault, Japanese philosophy, and contemporary critical theorists like Amy Allen, Jodi Dean, Christian Fuchs, and Bernard Stiegler. All of this culminates in a "radical social theory of modernity around the theme of technology" (2018, p. 31), which, given the diminishing place of technology within critical social theory, is a significant contribution.

The chapters in this section re-imagine the intellectual history of critical social theory to emphasize technology and debate the merits of contemporary critical theories of technology. In the first chapter of this section, **Federica Buongiorno** compares Feenberg's critique of technology against Byung-Chul Han's critique of technological modernity. Han is a popular and prominent critical theorist of digital culture whose work is insightful and uncompromising. Buongiorno is a leading scholar of Han's work (she has translated many of his books into Italian) and her analysis of these two philosophers of technology is wide-ranging, exploring how both draw different conclusions from Foucault and Heidegger that correspond with their different attitudes towards essentialism and concluding with a comparison of their critique of digital culture (she also proposes an interesting phenomenological reading of Feenberg that is inspired by Husserl). Buongiorno argues that although both Feenberg and Han can both be considered humanist critics of technology, Feenberg presents a more convincing critique by virtue of his attention to the concrete materiality of design, a reminder that critique needs more than the intellectual gymnastics of Han's poignant and totalizing ideas. Certainly, drawing out the problematic nature of neoliberal society demands uncovering the newly complex biases and consequences of digitally mediated processes and experiences, but this is only one side of critique. The other, as Buongiorno writes, can be found in Feenberg's philosophy, an anti-essentialist corrective to Han's reduction of technology to efficiency and the performance of positivity. Careful consideration of technology as both function and meaning, design and use, allows for critical insights and transformative potentials.

Ivan Landa and Jiří Růžička focus their chapter on the development of Marxist Critical Theory in East-Central Europe, and in particular Marxist humanists writing in 1960s Czechoslovakia, like Karel Kosik and Ivan Sviták. These writers proposed a theory of praxis that differentiated itself from trends in both Western Marxism and existential phenomenology by means of what was termed "onto-creativity," which was neither revolutionary, in the Lukácsian sense, contemplative, in the Heideggerian sense, or fixated on productive labour, as in the Soviet-Marxist sense. Czechoslovak Marxist humanists, instead, proposed a theory of praxis that emphasized how creation, or productive spontaneity, mediates the subject-object relationship through a process of negation and critique. The objective world of things is created by acting subjects whose scope of action is defined by this world, against which the creative desire to re-create this world as something new emerges. This is a theory of praxis that is indebted to a reading of Marx and Hegel and developed as a counter to Stalinist orthodoxy and Western Marxism. Onto-creativity as praxis emphasizes the immediacy of concrete potentials to create something entirely new and distinct from that which is given in social reality. This sits nicely alongside Feenberg's philosophical critique, which begins with our most immediate, and philosophically overlooked, forms of engagement with technology; everyday experiences in which users, through engaged use, imagine concrete potentials that exceed prescribed designs and functions. Interestingly, for philosophers of technology, the critique of existentialists like Heidegger and Sartre by Czech Marxist humanists, that they "were satisfied with a mere description of the given state of affairs, but did not provide any explanation of the possibility of such phenomena," corresponds with contemporary debates in the philosophy of technology between critical theory and phenomenologically-oriented approaches to technology. Beyond this, Landa & Růžička's work points out that the post-war experiences of Marxist theorists in Western Europe and North America were not shared by Marxists working in East-Central Europe, whose experiences demonstrate the importance of locality, and situated knowledge and experience in the critique of technology.

From East-Central Europe to France, **Adeline Barbin's** chapter is a history of French philosophy of technology that emphasizes how technology (or technics) always already is human. Technology, in other words, should be considered a natural and a biological process. Anticipating the seemingly radical ideas of actor-network theorists by decades, Barbin traces this idea across French writers who theorize the relationship between the technical and the natural and how this relationship can inform ideas about technological change. She looks to the work of André Leroi-Gourhan, Gilbert Simondon, and Georges Canguilhem to consider how technological change can open up ways of critically assessing technology by pointing to those aspects of technology that are flexible and contingent and those that are fixed and unchanging. It is not enough, she suggests, to emphasize contingency, as this gives the mistaken impression that the range of technical change is limitless. The challenge is recognizing the limits of what is possible, what she terms "possiblism," to better assess why technologies take the form and function they do. The French tradition, Barbin suggests, provides a more thorough conceptual vocabulary to assess the nature of invention and technological change, opening up modes of critique that can complement Feenberg's philosophical project. This history of the philosophy of technology gives greater insight into Feenberg's reading of Simondon's idea of concretization by providing more conceptual tools to better grasp the continuities and discontinuities of sociotechnical change and potential.

Alberto Romele's chapter takes as its starting point contemporary debates in the philosophy of technology regarding both the "empirical" and "ethical" turns that have shaped the discipline for most of the twenty-first century. Romele, taking up the argument of many political and critical theorists of technology, points out that both of these turns have led to little more than therapeutic philosophies of "living with technology" and consensus building at the expense of real critique. Feenberg's approach, by accounting for both power and resistance, is an alternative to these turns. However, for Romele, his philosophy is not critical enough. Drawing out what he detects to be the consensus-oriented influence of Habermas in Feenberg's work, Romele argues that the way out of this dilemma is by integrating the work of political philosopher Chantal Mouffe (alongside Jacques Rancière and Pierre Bourdieau) whose theories of agonistic politics sees little value in consensus. Although his indictment of Feenberg's theory of democracy as Habermasian in nature can be subjected to debate, Romele's chapter is a valuable contribution to a "critical" turn in the philosophy of technology that offers a real alternative to the emptiness of the ethical and empirical turns by encouraging the subversion of the status quo, not its affirmation.

Taila Picchi focuses her attention on two of the more prominent influences on Feenberg's philosophy, Herbert Marcuse and Martin Heidegger. She traces a genealogy of a dialectical philosophy of technology to Marcuse's early work with Heidegger. Heidegger's concepts of *dasein* and *world* were translated by Marcuse through the lens of Marxism, interjecting a needed dose of political economy to concretize Heidegger's somewhat obscure concepts. In doing so, Marcuse was able to overcome Heidegger's dystopian essentialism regarding technology by arguing that capitalism is a "world" that, although historically contingent, endures through a value-neutral functionalism that is perfectly suited to capitalism. Modern technology may be characterized through an "enframing" that is environmentally and socially disastrous, but this is not the only possible technology. Picchi's intellectual history of Marcuse's dialectics of technology spans decades and culminates in an emphasis of the role that Simondon plays in this history: "We can notice in Marcuse's production the shift from an ontology of Life based on historicity to a critical theory of technology based on the notion of technicity that is influenced by the work of Gilbert Simondon." Picci's chapter is an important contribution to understanding Marcuse's philosophy of technology and an important contribution to the history of a dialectical philosophy of technology.

1.5   Conclusion

Following Feenberg's suggestion that "we are confronted with multiple rationalities instead of a single technocratic rationality" (2017b, p. 5), the chapters in this book demonstrate that critique can take many forms: political philosophies of technology, historical and hermeneutic inquiries, and intellectual histories; critique can be oriented towards specific technical artifacts like smart cities, COVID apps, and big data, or concepts such as reason or ideology; and critique can be developed through any number of disciplines, including STS, media studies, industrial design, urban planning, and philosophy. This range corresponds with a remark from Foucault that Feenberg points out concerning styles of governance. According to Foucault, socialism lacks an art of government that is comparable to liberalism because socialism merely imitates the attempt at universal regulation or attempts to draw limits on the domain of free exchange. From this, Feenberg suggests that a "technological politics" may offer a starting point from which to build a uniquely socialist politics that does not simply reiterate the politics of liberal capitalism. To this, I would like to suggest that a theoretically coherent, historically conscious, and empirically sound critique of technology offers a philosophical approach to technological society that does not merely imitate the ambitions of capitalism. Critique encourages a philosophy in which it is necessary to situate the present moment of technological society with both the past and the wider social contexts that tie together the "thing itself" with its manufacture, use, and meaning, allowing for the development of critical insights that overcome ways of knowing generated and privileged by capitalist technological society. The necessity for this type of critique is wide-ranging and defies headline grabbing technical fixes by focusing on the technical object and the rationality, or to put it differently, the ideas and ways of thinking, that shape our engagements with these technologies.

References

Arnold, D.P. & Michel, A. (2017). *Critical theory and the thought of Andrew Feenberg*. Palgrave Macmillan.

Collingwood, R. G. (1946). *The idea of history*. Oxford University Press.

Crawford, K. (2021). *Atlas of AI: Power, politics, and planetary costs of artificial intelligence*. Yale University Press.

Feenberg, A. (1999). *Questioning technology*. Routledge.

Feenberg, A. (2002). *Transforming technology: A critical theory revisited*. Oxford University Press.

Feenberg, A. (2005). *Heidegger & Marcuse: The catastrophe and redemption of history*. Routledge.

Feenberg, A. (2014). *The philosophy of praxis: Marx, Lukacs and the Frankfurt School*. Verso.

Feenberg, A. (2017a). *Technosystem: The social life of reason*. Harvard University Press.

Feenberg, A. (2017b). Critical theory of technology and STS. *Thesis Eleven, 138*(1), 3–12.

Feenberg, A. (2018). In E. Beira (Ed.), *Technology, modernity, and democracy: Essays by Andrew Feenberg*. Rowman & Littlefield.

Fikentscher, K. (2000). *You better work: Underground dance music in New York City*. University Press of New England.

Hebdige, D. (1987). *Cut 'n' mix: Culture, identity and Caribbean music*. Methuen & Co..

Irani, L. (2019). *Chasing innovation: Making entrepreneurial citizens in modern India*. Yale University Press.

Keucheyan, R. (2014). *The left hemisphere: Mapping critical theory today*. Verso.

Kline, R., & Pinch, T. (1996). Users as agents of technological change: The social construction of the automobile in the rural United States. *Technology and Culture, 37*(4), 763–796.

Marcuse, H. (1964). *One-dimensional man: Studies in the ideology of advanced industrial society*. Beacon Press.

Misa, T., Brey, P., & Feenberg, A. (2003). *Modernity and technology*. MIT Press.

Radder, H. (1992). Normative reflexions on constructivist approaches to science and technology. *Social Studies of Science, 22*(1), 141–173.

Veak, T.J. (2006). *Democratizing Technology: Andrew Feenberg's Critical Theory of Technology*. State University of New York Press.

Chapter 2
Critical Constructivism: An Exposition and Defense

Andrew Feenberg

Abstract In this paper, I explain in detail *Critical Constructivism*, which I situate as a political philosophy of technology that draws from the Frankfurt School, Heideggerian phenomenology, Marxist labour process theory, and Science and Technology Studies (STS). Critical constructivism thus addresses the study of specific designs and the public controversies they provoke, while at the same time reconstructing aspects of the Heideggerian and Frankfurt School critiques of instrumental reason. Critical constructivism "de-ontologizes" these philosophies of technology, capturing their critique of rationality while affirming nevertheless the value of modern science and technology. To clarify my approach to technology and to address a number of misunderstandings regarding my approach to technology, this paper is organized around different aspects of my theory, including Marxism, technology and political theory, operational autonomy, the democratization of technology, formal bias, and instrumentalization theory.

Keywords Critical constructivism · Marx · Democracy · Technology · Labour · Operational autonomy · Technosystem · Technical code · Formal bias · Rationality · Instrumentalization theory

2.1 Introduction

Since publishing *Critical Theory of Technology* (Feenberg, 1991), I have gone on to develop a specifically *political* philosophy of technology which I now call "critical constructivism." That approach is based on a number of intellectual traditions,

---

This essay first appeared in *Logos: A Journal of Modern Society and Culture* 19(2), 2020. http://logosjournal.com/2020-vol-19-no-2/. Thank you to the editors for allowing us to republish it.

A. Feenberg (✉)
School of Communication, Simon Fraser University, Burnaby, Canada
e-mail: feenberg@sfu.ca

including the Frankfurt School, Heideggerian phenomenology, Marxist labour process theory, and Science and Technology Studies (STS).

This eclectic combination of sources recognizes both the empirical specificity of technology and the general crisis of our technological civilization exemplified by such issues as climate change. Critical constructivism thus addresses the study of specific designs and the public controversies they provoke, while at the same time reconstructing aspects of the Heideggerian and Frankfurt School critiques of instrumental reason. The early Frankfurt School is the major influence. It contrasts a "one-dimensional" scientism with the potentialities revealed in everyday experience on the basis of which resistances arise. Critical constructivism "de-ontologizes" these philosophies of technology, capturing their critique of rationality while affirming nevertheless the value of modern science and technology. The task is to conserve their valid insights, made evident by the crisis, without losing modernity itself. Social constructivism plays an essential role in my appropriation of this tradition, but I endeavor to overcome its underestimation of structural features of modern society.

I have presented my approach in many books and articles (see bibliography). Several books and special sections of journals have been devoted to the analysis and critique of my work.[1] Various misunderstandings recur in these commentaries. I will respond here to some of these criticisms in a new way in the hope that a more fruitful discussion can result. In order to achieve maximum clarity, I have refrained from directly addressing criticisms, kept references to a minimum, and reduced my principal positions to schematic arguments. Since I am often accused of favoring one side of a dilemma I attempt to transcend, I have formulated some of the supposedly contradictory propositions explicitly to show how I reconcile them. My goal in this paper is not so much to convince as to clarify my positions on the key issues I have discussed in my work.[2]

2.2 Why Marx?

A Marxist scholar once told me that "Everyone believes 90% of Marxism; the Marxists are those who also believe the other 10%." Who can doubt Marx's most important discovery, the central role of the economy in history and social life? But the other 10%, socialism as democratic control of the economy, is still highly controversial. Two principal arguments challenge it: the inefficiency of planning, and the "imperatives" of modern technology. The latter argument holds that the management of technology is fundamentally incompatible with democracy.

---

[1] For example, see the criticisms and my replies in Veak (2006), Arnold and Michel (2017), *Techné: Research in Philosophy and Technology* 17:1 (Winter, 2013); Thesis Eleven (2017), Vol. 138(1). See also Kirkpatrick (2020).

[2] For a recent summary of the theory, see my contribution to the fourth edition of the *Handbook of Science and Technology Studies* (Feenberg, 2017b), republished as chapter 2 in Feenberg (2017a).

This claim has itself been challenged in recent years. The constructivist notion of "actors" has liberated the study of technology from technological determinism and its vaunted imperatives. (Pinch & Bijker, 1987). No longer is it acceptable to deduce social consequences from a reified notion of technology, presumed to follow a unique track based on strict scientific principles. We now believe that there exist alternative technological choices and designs, and that they may have different social impacts. And we also believe that many different social actors pursue an interest or ideology by attempting to influence those choices and designs. No technological imperatives exclude a more democratic organization of the economy bringing additional actors into the design process.

This is an important methodological advance, but differences in power between actors are not easily explained within the constructivist framework. Foucault challenged the role of power as an explanandum by reducing it to the play of disciplinary techniques. Actor-Network Theory attempts to reduce power to an ever receding list of networked actors. Power would explain nothing but instead would be explained by the number of effective associations actors are able to mobilize in the networks they organize (Latour, 1986).

These strategies aim to avoid positing a substance of power separate from its manifestations. That is convincing as far as it goes but it fails to solve the problem in the most important case. That case is the influence of capitalism on the design and development of technology. Capitalism is a social structure that conditions actors' access to power. As such it is not reducible to techniques or associations. While I cannot argue the case for a structural account of capitalism here, that is the conclusion of most social science since Adam Smith, David Ricardo, and Marx, down to Thomas Piketty. Can structure be incorporated into social theory without reifying power once again?

The reaction against Marxism lies in the background of the constructivist rejection of power as an explanatory factor. This has less to do with Marx's own work than with the dominant interpretation of Marxism which emphasizes political economy and class struggle. Given the extent of government intervention in the market and the disinterest in revolution among workers, that version of Marxism has lost much of the support it once enjoyed.

The "post-Marxists," Ernesto Laclau and Chantal Mouffe, rejected the centrality of class domination (Laclau & Mouffe, 1985). They retained the idea of antagonism but in the context of symbolic differences rather than economic interests. That new approach was congruent with an increasingly diverse radical politics.

But it is becoming increasingly clear that an account of capitalist power is still necessary to address its role in three of the great crises of our time: the climate crisis, the issues surrounding the threats to liberty of current applications of information technology, and the declining faith in science which disarms society in the face of the other two crises. Each of these crises is rooted in the exorbitant exercise of capitalist power over the evolution of the technical system.

- The fossil fuel industry exercises such overwhelming power that so far no amount of scientific evidence of climate change is able to overrule its continuing domination of the energy system.

- The Internet, which once enhanced public debate, has come under the control of a few large corporations and is increasingly incorporated into the propaganda apparatus of corporations and governments.
- Neoliberal deregulation and tax policies, and the extension of simulacra of market rationality into every corner of social life have generated inequalities of wealth and power that have provoked a dangerous reaction against rationality as such. The post-modern decline of the "Grand narratives" of freedom and progress in the face of total rationalization has given way to a return of narrative in the form of dystopian conspiracy theories (Feenberg, 1995, chap. 6).

Despite changes in the economy and much criticism, Marx's *Capital* is still relevant to the understanding of these crises, but it takes some work to extract the useful insight without falling back into class essentialism. Although Marxists often denounce the personal power of capitalists, Marx himself viewed the capitalist as a mere agent of anonymous structures. He situated the capitalist as a dominant technological actor in relation to two determining structures: the irresistible pressures of the market and the resistance of the labour force. Given the relatively untrammeled power of the capitalists of Marx's day, this theory was sufficient to explain the rapid progress of technology accompanied by such negative phenomena as the deskilling of labour and the exhaustion of the soil.

Marx did not need to attribute a particular ideological bias to the capitalist, or personality defects such as greed, nor even a personal economic interest to explain the outcome because these structural factors compelled capitalists to perform in a specific manner or disappear from the scene. In sum, the structures determine *the conditions of possibility of effective action* under capitalism regardless of the motives of an actor situated in the dominant economic position. Hence Marx called capitalism a system of "impersonal" domination. Only other structures would make possible different and more humane forms of effective action associated with what Marx called "socialism."

Today capitalist control of technology faces a far more complex situation than in the nineteenth century. Other actors compete with it. Governments act on behalf of citizens to regulate business, and courts, labour unions, and other organizations within civil society frequently influence business decisions. Furthermore, many of the characteristics of capitalist management have been imported into public institutions, which make technological choices on the basis of quasi-economic criteria that imitate market prices in various ways. The result of all this intermingling and complexity is a situation ripe for the study of the role of a variety of actors and this no doubt explains why the field of technology studies has developed so successfully around this notion.

Nevertheless, Marx's idea that technology is shaped by the distribution of power can still form the basis of a structural approach to current crises. The role of leadership, the split between conception and execution, the control and incentive systems, the forms of psychological manipulation designed to keep subordinates on task, and other similar features of the division of labour have substantive impacts on technological choices and designs. These considerations can be generalized to take account of the harms caused by technologies affecting any social group that, like workers, has no direct access to decisions that affect them.

Marx's contribution for us today is thus his discovery of the political *in* the technology rather than opposing politics and technology as value and fact, ideology and rationality, or any of the other binary oppositions that flow from a discredited technological determinism. In what follows I will show how the Marxian politics of technology plays out in the various aspects of critical constructivism.

2.3 Technology and Political Theory

In the nineteenth and early twentieth centuries bodies of experts were constituted, serving within the administrative structures of modern institutions. Close association with management insured the congruence of technical systems and disciplines and the priorities of capitalist enterprise. Business had the freedom to dismiss hazards suffered by workers and communities, and this dismissive attitude was echoed all too often in science and technical disciplines. Technical disciplines evolved in which the hazards were ignored or minimized. Many scientifically undecidable matters were routinely biased in favor of business (Fressoz, 2012; Bensaude-Vincent, 2013). Government agencies that relied on the science and deployed the technology were little better.

Thus in addition to the unequal access to the wealth of enterprise, subordinates under capitalism suffered an excessive exposure to the discommodities of the technology that produced that wealth. This situation persisted so long as the mass of the population was silenced and in any case too uneducated to engage in a democratic dialogue with experts. That is no longer the case. The heritage of industrialism is now challenged in the public sphere. In this situation political conflict over technology is inevitable.

Political theory has not shown much interest in technology even as the world around us is ever more technified. Exceptionally, for a brief period after World War II, modernization theory and Marxism focused on technology. Technological development was supposed to bring with it democracy or communism, but those predictions have been refuted by history. Since these deterministic theories were discredited, mainstream academic thinking has gone back to its previous indifference to technology.

Surely it is past time to end this state of affairs and to integrate technology into political theory. But the task is not easy. It is necessary to navigate between three apparently contradictory propositions:

- *First, Non-Determinism.* Technological development is not deterministic, characterized by essentialist "imperatives" that inevitably prevail everywhere at every stage, but is socially constructed.

- *Second, System Convergence.* The actual development of technology under both capitalism and (really existing) communism reproduces the power of capitalists and managers.
- *Third, Rationality.* Technical choices are generally, if not always, made on "rational" grounds, that is to say, in response to criteria of efficiency and knowledge held in technical disciplines.

The first and second propositions appear to be incompatible. The convergence between capitalism and communism seems to imply that technological development imperatively requires authoritarian control, otherwise why would such control characterize most development in different social systems over long periods? A whole literature in social science has been based on the notion that technology is responsible for system convergence. In that case, the first anti-deterministic proposition must be false. If, on the contrary, the first proposition is true, if technological development is not deterministic, why do successive waves of technological innovation in different societies yield similar structures of control?

The third proposition holds that most technological decision-making has a rational basis in one or another technical discipline. These disciplines are properly called "rational" because they are evidence based and logically elaborated, in contradistinction to tradition, prejudice, and personal or literal authority. Rationality on these common sense terms has nothing to do with "pure reason." This needs emphasizing since the notion of rationality is confounded in some academic circles with an idealized form that is an easy straw man for relativistic arguments. Critical constructivism is not embroiled in that contentious debate which concerns the truth of natural science rather than the design of technical systems and artifacts. The relativity of rational designs to a social context is self-evident. For example, no one is puzzled that a rational automobile design, rational in the sense of well engineered, is also contingent on a social choice favoring private transportation and must meet aesthetic criteria that enter into the technical design itself.

The third proposition appears to contradict the first since rational technical disciplines are everywhere similar. Perhaps the rationality of technical decisions explains why convergence is observed, as the second proposition claims. To resist this conclusion, Marxists usually condemn the irrationality of capitalism and promises freedom in a future rational society. Not rationality but irrationality–the intrusion of particular capitalist interests–would be responsible for domination. Other Marxists, critical of Soviet communism, attribute convergence to different forms of irrationality, capitalist and bureaucratic, leading to the same result.

Critical constructivism proposes an explanation that preserves the non-deterministic thesis of the first proposition despite system convergence and the rationality of management choices. The similarity between capitalist and communist management methods can be explained by the fact that the Soviets did not elaborate an independent technology and corresponding management methods but simply adopted technologies and methods developed under capitalism in the West. The technology then functioned as a vector of cultural diffusion (Fleron, 1977). The fact that the disciplines and artifacts travel so successfully from one social environment to another is not due to the absence of social influences on design but rather to their transmission *through* the disciplines and artifacts themselves.

This explanation pushes the question back a step to the origins of capitalism and the invention of its systems of control. Labour process theory proposes a convincing explanation for the authoritarian outcome: the deskilling of the workforce. Deskilling is embodied in technology which, as Marx argued, materializes the intellectual content of production formerly held in the minds of craftsmen. The new technologies introduced by the capitalist owners cheapen, pace, and control the work. This insures that the work gets done even by an unmotivated labour force with no stake in the success of the firm. It also deprives workers of the cultural capacity to contest for power (Noble, 1984).

Technological development would thus be contingent on a social cause as proposition one holds, while also following a convergent trajectory wherever capitalist technology is employed, in conformity with proposition two.

But labour process theory cannot quite explain the implications of proposition three. The theory is ambiguous. One could argue, as Marxist often do, that the capitalist preference for deskilling is a product of ideology and interest, in contradiction with technical norms. This is the implication of the claim that socialism can release technology to full development from the limitations placed on it by capitalism. But in fact capitalist technology has been successful *as technology*. Why does it works so well, why do capitalists generally consult neither ideology nor interest but technical considerations to make the choices that get the job done and keep them in control?

This question is raised by my third proposition. It is exemplified by the Weberian "rationalization" thesis, taken up by Lukács in the theory of reification and developed further by the Frankfurt School in the critique of instrumental reason. Heidegger's critique of technology shares this general approach.

Weber describes the "disenchantment" of modern societies. Whereas anthropology finds an indiscriminate blending of symbolic associations and technically valid insights in the artifacts of premodern societies, capitalism relies on formal technical disciplines that eschew all symbolic reference. Industrial machinery is stripped of symbolic associations, unlike the technical means employed in those earlier social formations. Here truly, "form follows function." The only social norm that appears to survive the disenchantment is efficiency, ruthlessly pursued by capitalism under the pressure of competitive markets. Weber extends his thesis to bureaucracy which helps to explain what happened in the Soviet Union. But this approach risks collapsing back into determinism as it seems to have for Weber himself.

The problem now is to explain convergence while recognizing the insights of rationalization theory without contradicting proposition one, non-determinism. A political theory which met these requirements would have to find the cause of convergence in a structural feature of rational technical systems. For that a new concept of power is required.

2.4   Operational Autonomy

Marx's discussion of cooperative labour in the early chapters of *Capital* offers a suggestive starting point. Marx points out that cooperation increases productivity. On the large scales favored by capitalist enterprise, those gains can only be achieved through formal leadership, conscious coordination. Management must now be included in the division of labour and it acquires not only the technical function of coordination but also the power to confiscate the value added by cooperation and to impose machine designs that enhance and perpetuate its control (Marx, 1906 [1867], chap XIII).

Thus Marx does not attribute authoritarian management to technology but rather the reverse. He divides the workforce into a cooperating mass, eventually assembled around machinery, and a directing center, able to exploit its unique position in the division of labour.

Generalizing from Marx's comments, we can conclude that there are different ways of organizing the cooperative activities which characterize modern societies. I distinguish three such ways as ideal types, that is, as distinct theoretical models which clarify the ambiguous realities of actual social life in which mixed modes are more common than pure forms.

- *Control from above* imposes coordination without reciprocal responsibility to subordinates.
- Those subordinates exercise d*emocratic control,* or *control from below* where they select the coordinator of their activity.
- And finally, one must add a third form not discussed by Marx: *collegial control,* in which subordinates in possession of complex skills retain considerable autonomy and the formal right to advise those who coordinate their activities.

These ideal types can be explained in terms of the standard distinction made in organizational theory between policy, operations and control. Policy is the overriding purpose of the organization, while operations involve deployment of the means of action serving that purpose; control consists in the procedures which bring operations into conformity with policy.

In the capitalist enterprise, the operational level, control from above, is exercised under a policy that simply reproduces the operational norm of efficiency in a specific context. This is possible because capitalist property has no obligations, for example to workers or the community, on the basis of which a policy might be formulated other than financial success. In a sense, then, policy is irrelevant, cancels out. The managers need only attend to the tasks associated with the smooth functioning of the means at the disposal of the enterprise to achieve its policy goal. In practical terms this results in shareholders and boards of directors routinely abdicating authority to management, so long as the firm meets financial targets. I call this the "operational autonomy" of management. It is a *formal* characteristic of enterprise in the sense that it specifies procedures without also specifying a separate substantive goal.

In Marxian terms, the point would be that the task of the enterprise is the production of exchange value; the use values it also produces are a byproduct of that activity. The production of exchange value stands under the norm of efficiency as understood under capitalist conditions. It appears neutral insofar as it presupposes no substantive value. The substantive values served by the enterprise are what matters to consumers, not producers. Nevertheless, the neutrality of capitalism is not absolute but is relative to the division of labour. This has consequences for machine design as we will see.

In the case of public institutions a substantive policy must legitimate the activities of the organization. This differs from business which is ruled by the conditions of the market rather than a policy. By "substantive," in contrast, is meant a goal embodying a specific value or effect defined by or for management and realized through the means at its disposal. The substantive policy goal of schools, for example, is education, as defined by standards of some sort, and the school administration and teachers are responsible for achieving that goal to the satisfaction of the government that funds them.

In such cases, a technical discipline or system of rules provides the procedures under which the policy is implemented. The organization is based on explicit values, but it carries out its mission through apparently neutral means. The split between values and means resembles the similar split between use value and exchange value insofar as a certain indifference often characterizes the implementation of the mission of the organization by its agents. Just as the capitalist need not care about the value realized by the product, so bureaucrats, teachers and others working in public institutions may follow the standard procedures while indifferent to the nominal value presiding over their work. The extent of this indifference varies with the system of control.

Where professionals carry out the policy of the institution, collegial forms of control have traditionally prevailed. Schools, for example, used to rely on teachers' expertise. As professionals, teachers were considered competent to judge student outcomes in conformity with the educational mission of the institution. To a greater or lesser extent, periodic consultations between the administration and the leaders of the teaching staff insured consensus around specific policy goals and personnel decisions. A general commitment to the value of education is presupposed by these arrangements.

However, under neoliberalism, the policy goal of education is redefined in terms of quantitative measures such as test scores and rankings that imitate financial objectives in form. Once such measures are in place, collegial administration can be replaced by the operational autonomy of an empowered bureaucracy. Then the work of the institutions exhibits the effects of the division of labour, much like a business enterprise. The consequent demoralization is well known. This is increasingly the pattern not only in schools but throughout the society. Its effect is to submit individuals in nearly every aspect of their lives to the impersonal power of rational systems, against which they rebel in a variety of ways, some constructive, others destructive or even self-destructive where resistance is tainted with irrationality.

In both business and government, specific interests play a role in biasing decision-making, but the structure of control is an even more fundamental, because constant, source of bias. Although the pursuit of efficiency appears to be neutral, it is biased by the mere fact of the independence of management from its subordinates. Managing a cooperative activity without the approval of the participants creates a unique social situation that constrains all parties. Managers must distance themselves from subordinates who are effectively mere resources of the organization, while subordinates tend to be mistrustful of authorities over whom they have no control. Communication processes within the organization are distorted and mistrust has substantive consequences.

The point is not the ill will arising from this situation, which good management attempts to mitigate, but rather the objective constraints on action coordination under these conditions. Management must organize the work with the advantage but also the handicap of autonomy. The advantage is the unity of command, and the handicap is the tendency of subordinates to resist commands emanating from above. Rational decisions made in this situation will have a particular character, for example, privileging what Weber termed "calculation and control" over work satisfaction or morale building.

The concept of operational autonomy focuses attention on the reproduction of management power through technical choices, and the substantive consequences of control from above. That situation determines choices that affect all members of the organization and the community in which it is situated. It is not just unfair from a democratic perspective for managers to act without consulting those whose lives are affected by their decisions, it is also detrimental to concrete concerns of those individuals. From the standpoint of those in control, a limited number of rational technical solutions will appear viable. The excluded solutions might include desirable ones from the standpoint of subordinates that would be viable under the different condition of control from below.

Some economists have studied worker-controlled enterprises in terms of significant effects on technology (Dow, 2003, pp. 136, 162, 288; Pagano & Rowthorn, 1996, chap. 7). Differences in priorities also show up in issues such as health and safety at work, a frequent bone of contention in labour negotiations. Labour process theory argues that the situation of authoritarian management biases machine design toward deskilling, a successful strategy for controlling a reluctant or hostile workforce.

It could be objected that machine design in the post-Fordist economy is no longer based primarily on deskilling. But this does not alter the logic of the argument, which is concerned not only with specific designs such as the assembly line but with an overall pattern of development that centralizes power in a managerial elite. Still, the old Fordist methods are far from irrelevant since most manufacturing is now done on assembly lines in countries such as China and Mexico. The notion that advanced capitalist societies can be analyzed apart from their periphery is fundamentally misguided.

The case for operational autonomy having substantive consequences is obvious with the Internet. Centralization of power in a few large corporations has stimulated a sequence of technical innovations designed to harvest personal data. The prevalence of surveillance and targeted advertising is directly attributable to corporate control. The design of the major sharing platforms is organized around and reinforces operational autonomy despite the democratic potential of the Internet. Where once the free expression of opinion on the Internet counter-balanced the mass media, its successful colonization by these corporations has made much of it into just another propaganda apparatus. Privacy advocates call for a return to the decentralized structure that characterized the system before it was dominated by Google, Amazon and Facebook (Feenberg, 2019).

The substantive impact of the locus of control is also clear in the relations of business to the surrounding community. Where management is free to ignore externalities suffered by the community, efficient technical solutions may have dire consequences for the larger social unit within which the enterprise is situated. If consulted, the community would likely choose different solutions, for example, less polluting ones. Indeed, communities routinely go to court to oblige local businesses to respect the environment.

In sum, decision-making based on criteria of rationality and efficiency is not truly neutral. Furthermore, democratization is not a merely procedural desideratum but is bound up with substantive ends.

2.5 Democratization

The rapid technological change characteristic of modern societies since the nineteenth century has a wide variety of causes and consequences. Because critical constructivism focuses specifically on the role of operational autonomy, it is concerned with those aspects of change related to the power dynamics in the technosystem. The issues involved, for example, environmentalism and privacy, are certainly not the only important ones, but they have had an increasing impact and therefore deserve attention and study.

Struggles over technology have impacts at both the level of substance and procedure. On the one hand, negative experience with technologies and technical systems is articulated critically in the public sphere. The publics excluded in the past seek now to undo the substantive harms of inherited designs. This is most visible today with respect to pollution which business ignored throughout most of its modern history. Today business is constrained by public opinion and regulation to take account of the health of communities affected by its wastes.

On the other hand, each new constraint on business diminishes its operational autonomy to some degree. Extrapolated to the limit, this procedural consequence would yield a truly democratic technical system. Short of that goal, the business system is constrained to better serve substantive public interests in one or another domain.

The question of democratization of technology is contentious. Technical disciplines promise universal benefits and the very notion of rationality includes an appeal to the reason of all members of the human species, without differentiation or qualification. The claims of expertise cannot be ignored but experts, like all human beings, make mistakes and overlook harms and desirable potentials of the artifacts and systems they design and operate. In principle they serve the public in performing their duties in accordance with rational principles, but experts act in a specific context. Where they are employed by capitalist firms, their application of technical knowledge is governed by criteria determined by their employer. The firm itself bears no responsibility not imposed by law and regulation beyond the making of a profit. The exclusion of workers and communities from participation in technical decisions means that some considerations relevant to human wellbeing will be overlooked. As I will argue in the next section, that exclusion also biases technical designs and the technical disciplines themselves.

Technical democratization in capitalist societies involves the recovery of the overlooked considerations under pressure from the public. The demand for democratization is a claim for the extension and formal recognition of these contributions of non-experts, not a rejection of the role of expertise. The widened range of actors consulted in democratic decision-making has the *potential* to introduce considerations overlooked by capitalism, but this is not inevitable. History reveals the many useful interventions of users and victims in the redesign and regulation of technology, but democratic publics have also been known to make bad decisions. Nevertheless, historical experience with dictatorship and democracy inclines one to prefer the latter. The notion of technical democratization simply extends this preference to the technosystem.

Public demands may be articulated in a variety of forms, but almost always a posteriori, after a technology is released into the public domain. For example, consumers may complain about the danger of a product or boycott it. Workers may demand protection from the hazards of technology at the workplace. Hackers may modify a technical system. Communities may hold firms or government agencies legally responsible for pollution. Demonstrators may advocate new technical policies around issues such as energy.

In any case, the interventions have several salient characteristics:

- *First, Punctuality.* They are punctual and occasioned by a specific technical issue which co-constructs a concerned public.
- *Second, Discursivity.* They are formulated discursively in a complex hybrid language.
- *Third, Innovative Dialogue.* They appeal to experts to make design changes in accordance with the relevant technical knowledge.

In sum, technical democratization involves a dialectic of social struggle culminating in an *Aufhebung*, a new technical design that incorporates a wider range of interests. I call this a change in the "technical code," the ideal type that describes the intersection of social and technical requirements.

The first point, registering the punctual nature of public action, indicates an important distinction between the usual democratic procedures and public interventions in technical decisions. Since electoral politics rarely addresses technical issues, discontented citizens often act through other means. Furthermore, electoral politics is organized by local jurisdictions that often do not correspond with the technical networks in which citizens are enrolled. These networks create latent publics that may be activated and become manifest as a result of a scandal, breakdown or change in attitude or opinion in society at large. Once manifest, the new public will exist for a time in the public sphere through punctual actions such as a demonstration, boycott, lawsuit or messaging campaign aimed at political representatives.

The discursive form of these public interventions has a peculiar character that is without precedent in other domains. Typically, technical issues are addressed in a hybrid discourse which combines elements of scientific and technical language with ordinary language. What we might call the "Aristotelian" quality of ordinary language is superimposed on objects properly described in scientific or technical terms.

In Aristotle's world, things have stable essential properties which resist change and those properties include potentialities that are normally realized in the course of development. Modern common sense remains largely Aristotelian in this sense. Values and facts coexist in everyday discourse in a way that is usually inadmissible in scientific and technical languages. That discourse is teleological and so alien to modern natural science and the technical languages of expertise, but useful for articulating everyday experience, including experience with technical artifacts.

The public perception of environmental harm is based on unscientific but intuitively obvious notions such as the "health" of a river or forest, preferred to a fish kill or dead trees. Never mind that from a purely scientific standpoint there is no reason to prefer the one to the other, such indifference is irrelevant in the world of lived experience which operates according to other principles. Those principles include an aesthetic appreciation of the beauty of nature, sympathy for creatures that resemble us humans, prudence when it comes to intervening in natural processes, and concern for the long term future in which our children will live.

In practice, respect for potentialities is translated into scientific-technical language in order to engage the technical systems of a modern society in preservation and repair. The operational significance of the popular notion of harmony with nature is revealed as traditional appeals are de-ontologized and reformulated in terms of scientific-technical solutions. There is no need to defend the idea of harmony on metaphysical grounds; it is an experiential norm that has scientific-technical correlates. Indeed, there is no natural condition to be restored. The way lies forward to nature, not back to it. The ideal of harmony is now based not on return to a pristine past, but on the imagination of a livable world science and technology can help to create.

The third point, which concerns the interaction of lay and expert, shows that the barrier between that common sense and modern science is not absolute. Consider the example of the water crisis in Flint, Michigan. The presence of lead in the municipal water supply became known through a combination of direct observation of color and smell and through testing by a scientist who aided the community. The scientist was able to give the residents a cause and a name for their foul water: "lead." But lead as he tested for it, if not in his personal understanding, was simply an element on the periodic chart with a specific atomic weight, valence, etc. To the members of the community, lead was a threatening invader and a symptom of racial discrimination.

The same object, "lead," crossed the boundary between science and everyday understanding. It had two different lives, a scientific life as an element and an "Aristotelian" life in which it played the role of threat to "normal" human development, that is, to the realization of human potential, specifically the potential of the growing brains of the city's children. This is an extension of the concept of "boundary object" in a new direction (Star & Griesemer, 1989). It has a corresponding formulation in Critical Theory. As Marcuse put it, there are not only "mathematical truths," but also "existential truths" of nature (Marcuse, 1972, p. 69).

The movement for a response to climate change illustrates these conclusions. It depends to an unprecedented extent on science and technology. Individuals do not interpret their experience of the weather in common sense terms alone, but also by reference to scientific studies. This is happening despite the prevalence of industry propaganda that contradicts the science.

The introduction of scientific ideas into the vernacular language of everyday life has occurred gradually throughout the history of modern science. In the case of climate change the scientific facts are immediately interpreted in human terms. We think in terms of potentialities, both favorable and unfavorable, and this introduces radical discontinuities into what for science are continuous quantities. Two degrees centigrade is not simply a number on a graph but signifies specific conditions of life. Climate for us humans is all about the threshold of change compatible with something like a modern way of life. The struggle to preserve that way of life draws on science but informs it with values.

There is thus no absolute opposition between lay and expert; bridges can be built around concepts such as "health" and "security." The effect of those bridges is to translate values into facts, discursively formulated demands into technical specifications. In this way what may appear "irrational" from the narrow standpoint of the prevailing scientific-technical rationality, enters the design of its applications and shapes a rational future.

Bridging concepts enable communication across the boundary between discourses. This is the significance of the third point in the description of public interventions into technology. These interventions aim to communicate the existence and seriousness of a problem by showing it to be a matter of public concern. They are incomplete halves of a dialogue that must proceed between the public and the experts who represent its interests in technical fields.

The concept of "translation" describes this situation. The public translates scientific and technical concepts into everyday language in order to articulate a discontent, and bridging concepts make that discontent comprehensible to technical experts who then translate it into specifications of a modified technical system or artifact. The complete circuit drives technical development forward from one iteration to the next.

This dynamic describes important turning points in the history of modern technology and promises to play an ever increasing role. Concerns with safety in engineering and health care have escalated regularly as experience teaches unpleasant lessons. Those lessons are memorialized in changed technical disciplines, standards and designs. If workers, communities and patients were routinely consulted, it is reasonable to assume that many hazards in production and medical care would be corrected early. Similarly, disastrous administrative decisions in fields such as education might be avoided by wider consultation with those affected.

The picture drawn here of technical democratization may seem overly optimistic. It is often said that the public lacks the qualifications to make judgments about technology. There are in fact cases in which truly dangerous ideas prevail, as for example, the anti-vaccination movement. Rejection of science threatens the very idea of democracy which cannot function in a modern society without a public capable of interacting productively with the experts who operate the systems on which the society depends. For democracy to function, the public need not submit unquestioningly to expertise, but its challenges must uncover actual problems capable of solution, rather than rejecting rationality in favor of wild conspiratorial tales.

While the anti-vaccination movement and similar attacks on science suggest reasons for caution, the story of Flint's drinking water resonates with far more extensive and significant public interventions. The environmental movement has had huge impacts on industry and continues to drive change, especially in the field of energy production. Many of the early changes are now standard technical procedures, their source in public protest forgotten. What might be called a "technological unconscious" covers over the traces of the movements that initiated those changes as their consequences are inscribed in technical specifications.

The successful translation of public demands leaves the impression that more visible phenomena such as the anti-vaccination movement are emblematic of public interventions in general. A proper history of the many engineering, environmental and medical problems corrected in part through public action would provide a different picture. That picture would confirm what we already know in principle, namely, that technical experts and the disciplines on which they base their actions are not perfect, but need periodic correctives.

How important are these considerations on democratization in the larger scheme of things? The argument for democratization of technology has been criticized for over-emphasizing the "binary" distribution of power between managers and experts on one side and subordinates and lay people on the other. Presumably, the fact that they engage in dialogue shows power to be irrelevant, and in any case there are many other sources of technical change. But this argument ignores the inheritance of operational autonomy and its continuing impact. Technical traditions based on an earlier industrial world in which workers and communities were silenced are not easily overthrown. And there are businesses today that profit from their autonomy with dire public consequences. Think Exxon, Purdue, Volkswagen, Boeing…

It is also said that the model of political conflict is inappropriate for technology. It is true that neither a fight to the finish nor compromise describe typical technical developments. Something more interesting and complex occurs where public demands are translated into technically rational designs. I have discussed this in terms of Gilbert Simondon's concept of "concretization," a specific type of technical innovation relevant to democratic intervention (Simondon, 1958; Feenberg, 2017c). Concretizing innovations overcome conflicts over technology between social groups with different agendas by combining their goals in new designs, a new technical code. For example, the electric car combines the seemingly opposed demands of environmentalists and commuters.

But the fact that technical controversies are often resolved by innovations or redesign does not mean politics is irrelevant. Sometimes the dialogue between lay and expert can only get off the ground through compelling political testimony that may appear antagonistic, even though its intent is fundamentally communicative and aims not at victory or compromise but at innovation. The obstacles capitalism places in the way of that communication should not be underestimated. Hence the essential role of politics.

2.6 Formal Bias

The concepts of operational autonomy and technical democratization have implications for the understanding of rationality. These implications can be summed up once again in apparently contradictory propositions.

- *First*, rational artifacts and social arrangements such as technologies, markets and administrative rules are based on technical disciplines that employ evidence and logic, and those disciplines therefore claim universal validity.
- *Second*, under capitalism these artifacts and arrangements are biased, in some cases discriminating by gender or race, and generally to favor the wealth and power of capitalists and their managerial representatives.
- *Third*, other rational artifacts, social arrangements and technical disciplines are conceivable that would favor the wellbeing and power of social groups suffering discrimination and the mass of the population.

It not obvious how to reconcile these three propositions. If the existing systems are rational, then why are they not universal like science and mathematics? How can they favor some rather than serving all? And how can multiple rationalities exist, such that there are rational alternatives to these rational systems?

The important point made by these combined assertions is a negative one: capitalism is not technically irrational; rather its technical decisions are justified by rational claims that are nevertheless biased. The concept of bias I introduce here concerns the different consequences for various social groups of different rational designs, as in the case of pollution discussed above. Democratic interventions into technology respond to systematic harm or injustice resulting from such biased activities.

This concept of bias challenges our common sense understanding of the term. "Bias" usually means prejudice and discrimination. I call this usual sense "substantive bias" because it rests on substantive propositions and myths concerning the object of biased sentiments and actions. Substantive bias in this sense is expressed directly by the biased agents. Their beliefs have been challenged in one domain after another since the Enlightenment by the appeal to evidence. Their untruth is their undoing.

Now we are dealing with a different kind of bias that is, precisely, rational. I call this "formal" bias because it concerns the rational form of artifacts and social arrangements rather than a substantive content of belief. Although less familiar than the usual substantive concept of bias, it is invoked implicitly in public controversy over institutional racism. The red-lining of neighborhoods by mortgage companies depended on formal rules governing the decisions of loan officers. The rules originated in race prejudice but once established those who followed the rules did not have to harbor racist sentiment to achieve the racist outcome, but only to follow procedures.

This sort of bias poses a problem for Enlightenment critique. Formal bias cannot be refuted by an appeal to evidence since it makes no false claims but simply implements rational technical principles or uniform rules. In fact the defenders of formally biased arrangements charge their critics with irrationality! The classic instance is the accusation of "Luddism" aimed at environmentalists.

The conundrum has to do with the connection between rationality, truth, and justice. One expects that in addition to being universally valid, a truth claim will also serve a normative standard such as justice. This is how we understand a fair trial. As rational, science and technology too are "fair" in the sense that they are supposed to serve humanity as a whole. We are so used to this way of envisaging rationality that it is difficult to recognize its different status in the technosystem.

Technical disciplines are surely rational but they do not consist in simple lists of truths, equally useful to all. Rather they are concatenations of many sources of information and multiple propositions coordinated to achieve a practical result. Hidden within that complexity are all the alternative pathways to the result that were eliminated for various reasons in the construction of the discipline. Often no compelling rational account can be given for many of the choices, which sometimes have social implications. This is the "underdetermination" of technical rationality, underdetermined, that is, by technically rational considerations.

An underdetermined rational design may incorporate a bias without losing its claim to rationality. The bias enters through choices that favor those among the many alternative designs that in turn favor the interests of one or another social actor. The double favoritism is essential: formal bias is indirect, achieved through a technical mediation to which a social preference is delegated.

The concept of formal bias complicates the attribution of responsibility for discrimination. Because an original discriminatory intention is materialized in an artifact, its effects may be perpetuated by agents who have no such intention. The substantive effects of formal bias, such as the health problems in Flint discussed above, are due to technical arrangements, not to personal animus, although personal animus might play a role in the original choice of those arrangements. In that case a substantive racial bias was translated into an innocent seeming, i.e. formally biased, technical choice.

Discrimination may also arise where no discriminatory intention is formulated, due simply to the effects of unquestioned cultural presuppositions. For example, when the socially accepted definition of the labour force included children, features of the technology such as the placement of controls were designed for small workers. This was technically rational under the given conditions although today we might consider the whole business of child labour a scandal. In any case, factory owners were not prejudiced against children but acted according to custom.

The case of operational autonomy rests on a custom–the unquestioned assumption of control from above–that has the peculiar feature of structuring capitalist society as a whole. In a society where it is taken for granted that large scale cooperation will be coordinated from above, management will make formally biased choices that preserve its ability to manage and to make such choices in the future. The autonomous power of management enables it to choose similarly biased designs at each stage in the development of the technology. The perpetuation of control from above is due to this social dynamic (Rueschemeyer, 1986, p. 171). It is a system effect rather than a sovereign imposition or a technical imperative.

Under capitalism, management methods and the design of the major technical systems in fact presuppose no intentional discrimination, but simple conformity to the conditions of effective action from above. In such cases the valuative bias may be invisible even to the agents of the discriminatory practices. Their conscious motive is accomplishing the work, but that is only possible *for them* in their position in the capitalist division of labour under conditions that empower them and disempower workers.

The emphasis on the systemic character of operational autonomy gave some commentators the impression that the contrast between substantive and formal bias concerns the *motives* of technical designers rather than the *rational form* of their designs. Supposedly, a formally biased design would not have a substantive motive. My earliest formulations of critical constructivism gave some readers the impression that I considered technical rationality to be free of *any* social influence, "neutral," "value-free." But the distinction was intended to explain rationalized domination not in the absence of motive but regardless of motive.

As the example of red lining shows, a discriminatory motive may be involved in the establishment of a set of rational procedures, but that is not the case with capitalist management and technical design. In that case rational technical designs and disciplines are not an ideological reflection of particular substantive interests. Instead, they translate a more fundamental structural feature of the society. That structure is not neutral but is a social constraint that biases technically rational designs.

The substantive consequences of the systematic exclusion of subordinates from decision-making may motivate social movements to correct particular injustices without challenging the overall pattern of control from above. But a growing awareness of the root of the problem in the formal biases introduced by the system of control could lead to demands for general democratization of decision-making procedures. To be sure, this is a contingent potentiality of social struggle in the technosystem, not an inevitable result. But *only subordinates* are socially positioned to be able to undo the formal bias of control from above.

This understanding of social power is what led Marx to the conclusion that the proletariat could revolutionize capitalist society. It would be able to undo the effects of control from above from its subordinate position in large scale cooperative production. Today subordination is generalized throughout the society and no longer confined to production. This explains the proliferation of resistances of all sorts.

2.7 Instrumentalization Theory

A formally biased device or system cannot be fully explained by an internalist account focused on its causal mechanisms and functions. Such an account leaves out the relation to the social context within which the mechanisms and functions are situated. That relation has consequences for design and a political dimension and so requires an account of the device or system as a social phenomenon.

In some cases the simple presence of an artifact in a specific situation alters behavior or the balance of power. When seventeenth-century Japanese aristocrats observed the effect of guns on the battlefield, they realized that peasants were now their equals. They decided to abolish all firearms in order to maintain a form of martial superiority based on arduous training in swordsmanship which only aristocrats could afford (Perrin, 1979).

One could argue that this has nothing to do with the nature of guns *qua* mechanism, but guns are artifacts and artifact only exists as such in a social context of some sort. Their potentialities are revealed by that context and are thus intrinsic to their nature *qua* artifact. Guns can indeed be "equalizers."

Technology is even more deeply implicated in the social where its design is specifically adapted to customs or the demands of social actors. In this case, the technical object may be conceived as a monad, reflecting in its design the world in which it participates. The design shows the traces of that world in purely technical forms. Thus the technical specification of an artifact, when properly interpreted, reveals its world. Just as we deduce the average height of people in earlier times from the height of the chairs and tables that survive them, so can the many adaptations of technologies to their context be traced in their design.

What I call the "instrumentalization theory" aims to provide a general framework for understanding the sociality of technology. This is problematic because the hard mechanical details of technical specifications do not seem to mix well with the soft stuff of social conventions and values. The engineering department and the philosophy department do not communicate! In devising the instrumentalization theory I attempted to reconcile them.

I represent their supposed incompatibility in two propositions.

- *First*, technical thought, action and artifacts have specific characteristics involving control of nature through causal mechanisms.

- *Second*, technical thought, action and artifacts reflect social and cultural meanings and values.

Once again an understanding of the theory requires a strategy for evading an apparent contradiction.

The instrumentalization theory may remind those familiar with Actor-Network Theory of the concept of enrollment in a network. It reiterates the two-fold operation of "association" through *simplification* of the object to isolate specific causal properties, and *delegation* through which it embodies norms that grant meaning and script users' behavior (Callon, 1987; Latour, 1992). The instrumentalization theory follows this pattern but rather than developing the implications for networks it identifies the principal operations involved in association and delegation. This is a phenomenology of technical action based on what I have called a "double aspect" theory of technology. It addresses the correlated structure of technical objects and subjects at both the causal and cultural levels. This focus has enabled me to recover the critique of instrumental reason, familiar from the Frankfurt School, for the empirical study of technology.

Until recently I called the simplifying characteristics the "primary instrumentalization" and those involved in delegation, the "secondary instrumentalization." Initially I identified two primary instrumentalizations on the side of the object, which I called "decontextualization" and "reduction." These characteristics describe the separation of natural objects from their environment and the stripping away of useless features. Think for example of cutting down a tree and removing its branches and bark to make lumber.

Decontextualizing operations such as these are what is meant by "instrumentality" in the Frankfurt School. They are considered critically insofar as they are assumed to describe a violent world relation. But while these relations are certainly essential to anything we consider technical making, and while they do indeed have a violent potential, they cannot stand alone but only make technical sense when informed by a social content described in the secondary instrumentalization. Again, my initial exposition identified two correlated secondary instrumentalizations, recontextualizing the object in the sociotechnical network to which it belongs and providing it with social meaning. It is at this level that certain trees are singled out to be felled and such things as the width and thickness of the boards are determined, qualifying them as "lumber."

Later expositions complicated this initial picture and also emphasized the *analytic* character of the distinction between the two instrumentalizations. By this I mean that the two instrumentalizations are not separate processes but aspects of a single process. In the lumber example, the "secondary" recontextualizations and meanings are not subsequent to the cutting down of the tree and the stripping of its branches and bark, but rather are present from the very start in the legal and material requirements of the construction system with which the production of lumber is associated. The primary operations are thus determined by the secondary ones and cannot be initiated separately.

This was not clear in my earliest expositions of the theory because I argued that capitalism eliminated secondary instrumentalizations. This formulation implied that the two types of instrumentalization were separate, when in reality capitalism merely eliminates many traditional secondary instrumentalizations while substituting new ones.

The ambiguity in the early versions of the theory led to problems, aggravated by my terminological choice. I never intended the terms primary and secondary to signify a temporal relation but that is exactly how many readers interpreted them. Once this interpretation of the instrumentalization theory was broached, it became extremely difficult to recover my actual meaning. I tried to do so in *Transforming Technology* (2002), a revised version of my first book, but the misinterpretation recurred again and again. Finally in *Technosystem: The Social Life of Reason* (2017a), I abandoned my original terminology and substituted "causal functionalization" and "cultural functionalization." I do not know yet if this will lay the issue to rest.

My mistake was overlooking how my terminology would be read against the background of the common sense instrumentalism prevalent in Western culture. The impression that there is a *real* distinction between causality or function and culture or meaning arises from the notion that causal mechanisms have a basis in nature whereas social meanings are merely conventional. Presumably, this would explain why a machine that works in New York can be made to work similarly in Peking despite the cultural and social differences.

This suggests a resemblance between technology and science, which claims universality beyond merely local differences with more justification. The resemblance gives rise to both technocratic arguments and dystopian critique. If technology is independent of society, it may offer an alternative to ideological contention or an iron cage obliterating human individuality and freedom. A whole series of binaries flow from this original binary of the technical and the social.

That binary itself has a social cause. A totem pole is a communication medium, as is a television, but only the latter can be described as a complex causal mechanism without reference to its social insertion. It is the rise of modern technical disciplines, modeled on natural science, and the complex artifacts they make possible that seems to justify the separation of the technical and the social. The differentiation of these disciplines is an institutional reality with immense consequences, but it is misinterpreted when the role of bias is ignored.

Formal bias enters into technical specifications that reflect the values of the social world in which it is situated. Values enter in two ways. Technical disciplines are bearers of values introduced at earlier times. The values are not present explicitly, *qua* values, but are incorporated into underdetermined aspects of the discipline. Values also appear in contemporary discursive formulations as explicit desiderata for design. Anyone can offer such evaluative suggestions, the marketing department, an engineer, political protesters, even university professors. Some of these discursively formulated values end up in revised designs.

In sum, there is no unbiased design, and hence no intrinsic opposition between technology and values. If the concept of bias is nevertheless useful, that is because it highlights the social forces determining technical codes.

Most criticism of the instrumentalization theory has been motivated by the misunderstanding discussed above. But there is another line of criticism that does not depend on it. It is argued that the instrumentalization theory leaves no room for the agency of things, by which is meant the active contribution of "non-humans."

In fact the instrumentalization theory recognizes the agency of things without using that vocabulary, but it also recognizes the specificity of the technical as the predominance of the human over the non-human. The technical subject stands in a manipulative relation to its objects. That relation is distanced and lacks the prominent role of reciprocal interaction found in human relations. Typically, the actions of the technical subject change the world far more than the world changes the subject. But interaction with the objects does occur, if not in the short term and to the same extent as in other cases such as human relations. The instrumentalization theory recognizes three types of non-human agency.

First, the technical actor's identity is shaped by his or her association with technology. Subject and object co-construct each other, as one says in a later terminology. The carpenter is a carpenter through the practices and tools of the trade. Second, the world of the subject and what it is to be a subject in that world are shaped by the available technology. Third, unintended consequences of technology such as pollution may come back to haunt the actor. In such cases, it is only necessary to enlarge the context in space and time to discover the connection. On the terms of Actor-Network Theory, non-humans are active agents pursuing "anti-programs" countered in turn by public resistance in what I call the "feedback relation" to technology (Feenberg, 2020).

To these three forms of interaction, some critics add the creative power of the material as it reveals itself in the course of technical work. This is the familiar interactive relation to the object in art and craft, but it plays a lesser role in most modern technical systems, organized under bureaucratic controls.

The fact that I do not describe these relations as "co-constructions" is not significant, but it is true that my emphasis has been elsewhere. I reconstruct the culture-critical notion of technological domination in terms of the theory of operational autonomy. Operational autonomy explains the generalization of technical control of human beings in modern societies. Most modern management and technology is designed to suppress the agency of both human beings and things, to reduce them to objects of technical control. The passivity of these manipulated objects is a desired condition rather than a theoretical violation of a supposed universal agency. A realistic look at modern technology does not have to deny agency to recognize that it is purposefully reduced in many technical arrangements and processes. Recognition of the varieties of agency is important for the theory but should not blind us to the realities of the prevailing technical system.

2.8 Consequences

The study of technology in philosophy and the social sciences is politically relevant today as never before. Much discussion in these fields turns on refuting cognitive errors such as the notion that nature can be "conquered," or that a "great divide" isolates human beings from nature. While it is useful to refute erroneous views, a focus on beliefs tends to put the onus on the human species. Cognitive errors do not explain the power structures that are actually responsible for the civilizational crisis we are living today. So long as the plans of powerful organizations are determined by formally biased "standard operating procedures" that privilege existing technology and short term economic growth, nothing will change. The problems can only be solved by political struggles to impose different priorities on the dominant institutions. Critical constructivism proposes an approach to understanding such struggles.

Perhaps the massive entry of young people into the climate change movement will make the role of technical politics clear. That movement has unleashed a radical subjectivity oriented toward potentialities and long-term considerations heavily discounted by orthodox economic thought. This is a subjectivity with a different sense of time and politics from the prevailing one. The protesters demand a new technical code of industry. Their testimony aims ultimately at major structural changes, not the sort of minor reforms compatible with a continuation of neo-liberalism. Capitalism is challenged once again, as it was by the labour movement in the twentieth century, to adapt to a new type and level of social constraint.

At stake in the challenge is the operational autonomy of those in command of technological systems and the technical code of those systems. Business has enjoyed a free ride so far as the environment is concerned throughout the nineteenth and most of the twentieth centuries. Capitalism's margin of maneuver has been gradually eroded by regulation in the public interest. But never before has it confronted a political demand to completely overhaul such basic technology as the energy system. In the past fundamental changes in the technical code of industry took place at a pace governed by business. Today it is the whole system that must change in what, from an economic perspective, is the equivalent of overnight.

In this context philosophical reflection on technology must inform political thought. The old disciplinary divisions in which technology was left to the technologists no longer make sense. Critical constructivism offers theoretical resources for addressing the crisis. The core issue concerns the nature of rationality as it is manifested in the technology that supports the social world. Uncovering the bias of that rationality is the critical task of the study of technology today.

Bibliography

Arnold, D., & Michel, A. (2017). *Critical theory and the thought of Andrew Feenberg*. Palgrave/ Macmillan.

Bensaude-Vincent, B. (2013). *L' Opinion Publique et la Science: à Chacun son Ignorance*. La Découverte.

Callon, M. (1987). Society in the making: The study of technology as a tool for sociological analysis. In T. Pinch, T. Hughes, & W. Bijker (Eds.), *The social construction of technological systems*. MIT Press.

Dow, G. (2003). *Governing the firm: Workers' control in theory and practice*. Cambridge University Press.

Feenberg, A. (1991). *Critical theory of technology*. Oxford University Press.

Feenberg, A. (1995). *Alternative modernity: The technical turn in philosophy and social theory*. University of California Press.

Feenberg, A. (2002). *Transforming technology second edition of critical theory of technology*. Oxford University Press.

Feenberg, A. (2017a). *Technosystem: The social life of reason*. Harvard University Press.

Feenberg, A. (2017b). A critical theory of technology. In U. Felt, R. Fouché, C. A. Miller, & L. Smith-Doerr (Eds.), *Handbook of science and technology studies* (pp. 635–663). MIT Press.

Feenberg, A. (2017c). Concretizing simondon and constructivism: A recursive contribution to the theory of concretization. *Science, Technology and Human Values, 42*(1), 62–85.

Feenberg, A. (2019). The internet as network, world, co-construction, and mode of governance. *The Information Society Journal, 35*, 4.

Feenberg, A. (2020). Critical constructivism, post-phenomenology and the politics of technology. *Techné: Research in Philosophy and Technology, 24*(1–2), 27–40.

Fleron, F. J. (1977). *Technology and communist culture: The socio-cultural impact of technology under socialism*. Praeger.

Fressoz, J.-B. (2012). *L'Apocalypse Joyeuse: Une Histoire du Risque Technologique*. Le Seuil.

Kirkpatrick, G. (2020). *Technical politics: Andrew Feenberg's critical theory of technology*. University of Manchester Press.

Laclau, E., & Mouffe, C. (1985). *Hegemony and socialist strategy: Towards a radical democratic politics*. Verso.

Latour, B. (1986). The powers of association. *The Sociological Review, 32*(1), 264–280.

Latour, B. (1992). Where are the missing masses? The sociology of a few mundane artifacts. In W. Bijker & J. Law (Eds.), *Shaping technology/building society: Studies in sociotechnical change* (p. 1992). MIT Press.

Marcuse, H. (1972). Nature and revolution. In *Counter-revolution and revolt*. Beacon.

Marx, K. (1867). *Capital*. (1906). (trans: Averling E.). Modern Library.

Noble, D. (1984). *Forces of production*. Oxford University Press.

Pagano, U., & Rowthorn, R. (1996). *Democracy and efficiency in the economic enterprise*. Routledge.

Perrin, N. (1979). *Giving up the gun*. David R. Godine.

Pinch, T., & Bijker, W. (1987). The social construction of facts and artefacts. In W. Bijker, T. Hughes, & T. Pinch (Eds.), *The social construction of technological systems*. MIT Press.

Rueschemeyer, D. (1986). *Power and the division of labor*. Stanford University Press.

Simondon, G. (1958). *Du mode existence des objets technique*. Aubier.

Star, S. L., & Griesemer, J. R. (1989). Institutional ecology, 'translations' and boundary objects: Amateurs and professionals in Berkeley's Museum of Vertebrate Zoology, 1907–39. *Social Studies of Science, 19*(3), 387–420.

Techné: Research in Philosophy and Technology, 17, 1 (Winter 2013).

Thesis Eleven, 138(1) (2017).

Veak, T. J. (2006). *Democratizing technology: Andrew Feenberg's critical theory of technology*. SUNY Press.

Part I
Democratic Potentials

Chapter 3
The Critical Theory of the Common Good, Technology, and the Corona Tracking App

Hans Radder

> La multitude qui ne se réduit pas à l'unité est confusion;
> l'unité qui ne dépend pas de la multitude est tyrannie.
> Blaise Pascal (Pensées)*
> Und wir wissen schon, es rettet uns kein höheres Wesen,
> uns bleibt nur dies: Schluß machen
> mit der Herrschaft der Wenigen über die Vielen.
>
> Werner Bräunig (Rummelplatz)**

Abstract This chapter presents a critical theory of the common good and applies it to technology. For this purpose, the notions of theory, critique, and technology are examined in detail. Section 3.1 addresses two major questions concerning (scientific and philosophical) theories: how they relate to reality and how to interpret the meaning of their central concepts. The second section develops a critical theory of the common good. Its two basic ideas are: the nonlocality of the meaning of theoretical and value concepts and a conception of public interests based on a substantive, multidimensional account of democracy. Section 3.3 then defines the notion of technology and examines the implications of the critical theory of the common good for the case of technologies. The critique of technology is illustrated by a detailed analysis and assessment of the Dutch debate on a specific technology, the corona tracking app, during the first phase of the corona crisis in the spring and summer of

---

*Plurality which does not submit itself to unity is confusion; unity which does not depend on plurality is tyranny

**And we already know that no higher being will save us, only one option is left: to put a stop to the rule of the few over the many

---

H. Radder (✉)
Department of Philosophy, VU University Amsterdam, Amsterdam, The Netherlands
e-mail: h.radder@vu.nl

© The Author(s), under exclusive license to Springer Nature Switzerland AG 2022
D. Cressman (ed.), *The Necessity of Critique*, Philosophy of Engineering
and Technology 41, https://doi.org/10.1007/978-3-031-07877-4_3

2020. In developing these ideas and arguments, I engage with the work of other critical theorists, in particular Andrew Feenberg, Jürgen Habermas, Herbert Marcuse and Michel Foucault.

Keywords Critical theory of technology · Andrew Feenberg · Critical theory of the common good · Jürgen Habermas · Herbert Marcuse · Democracy · Michel Foucault · Synthetic philosophy · Covid-19 · The Netherlands

The aim of this chapter is to develop a critical theory of the common good and apply it to technology. For this purpose, I scrutinize the three basic issues that confront any critical theory of technology. A first requirement is to have an appropriate and plausible notion of "theory." Second, there is the question of what to understand by "critique," in particular "philosophical critique." Finally, we need an explicit and appropriate account of the nature and the material, social, moral and political dimensions of technologies. On the basis of an examination of these issues, we can then explain the idea of a critical theory of technology.

The three basic issues are discussed in three sections. Section 3.1 addresses two major questions concerning (scientific and philosophical) theories: how they relate to reality and how to interpret the meaning of their central concepts. I provide an answer to these questions, which draws on recent work in philosophy, in particular the philosophy of the natural and social sciences, and I discuss the relation between theory and the world in the views of Jürgen Habermas and Andrew Feenberg.

The second section starts with a brief review of the history of critical theories, with a focus on the problem of the normative justification of critique. In particular, it critically assesses the views of Habermas and Feenberg concerning this subject. The main aim of this section is to develop a critical theory of the common good. Its two basic ideas are: the nonlocality of the meaning of theoretical and value concepts (in the spirit of Herbert Marcuse) and a conception of public interests based on a substantive, multidimensional account of democracy.

The final section then defines the notion of technology and examines the implications of the critical theory of the common good for the case of technologies. My two-part theory of technology provides both a detailed theoretical characterization of technologies and an analysis of their material and social realizability, including their (actual and potential) moral and political dimensions. Furthermore, I discuss the agreements and disagreements of this theory with Michel Foucault's general approach and with Feenberg's account of primary and secondary instrumentalization. Finally, the critique of technology is illustrated by a detailed analysis and assessment of the Dutch debate on a specific (digital) technology during the first phase of the corona crisis in the spring and summer of 2020. It concerns the use of a so-called tracking app for the purpose of containing the SARS-CoV-2 virus, a central causal factor responsible for the strongly contagious disease COVID-19, and limiting its impact by the early detection of (possibly) infected people.

As I have mentioned, at several points in this chapter I engage with Feenberg's work, which contains important and stimulating material that can be fruitfully used in addressing the issues at stake. More generally, the chapter is also an exercise in "synthetic philosophy" (Radder, 2019, 7–9). It brings together work from different areas of philosophy by a variety of authors; and it synthesizes various strands in my own earlier work published over the past decades. More particularly, with its focus on technology it complements my recent account of science and its relations to democracy and freedom (Radder, 2021).

3.1 Theory

Critical theories are theories of societies and their historical development. As theories, critical theories are usually seen as "scientific" (in the broad European sense that includes the humanities, and hence also philosophy). Yet, debates on critical theories have focused far more on their critical than on their theoretical nature.[1] In contrast, in philosophy of science the nature and function of theories has been discussed in great detail. This chapter addresses two relevant features of theories: the relation between scientific theories and the world and the meaning of theoretical concepts. In this section, I provide a (necessarily concise) review of these features for the purpose of employing it in the subsequent sections on the nature and aims of critical theories, in particular, of technology.

Often, traditional interpretations of scientific theories see them as abstract and universal entities, and still assign them a strong empirical adequacy and explanatory power concerning concrete scientific practices or socio-historical developments. In contrast, more recent interpretations of the nature and function of theories in the natural and social sciences have questioned these claims by scrutinizing the crucial role of specific models and context-dependent assumptions in empirical and explanatory practices. The basic point is that abstract theories are far removed from the complexities of the real world, and because of this distance theories alone cannot provide adequate descriptions or powerful explanations of this world (Cartwright, 1983; Morgan & Morrison, 1999). They can only be connected with this world by adding mediating models. These models often include all kinds of convenient rules of thumb, mathematically feasible approximations, tractable computer methods or pragmatic empirical fitting procedures. An important conclusion is that such modeling assumptions can neither be deductively derived from the theories nor conclusively justified on the basis of the empirical evidence alone. Therefore, modeling practices require relatively independent cognitive work.

Specifically for the social sciences, long ago Max Weber formulated and practiced a social-scientific methodology on the basis of his account of ideal-typical explanation (Weber, 1949[1904]). Because the number of facts that could be relevant to the study of societies and their historical development is almost infinite, all scientific approaches are necessarily selective. In this situation, Weber's

---

[1] See, e.g., the contributions to Ellmers and Hogh (2017).

methodology employs ideal-types, that is, broad theoretical concepts that have to fulfill two general criteria. They should be optimally clear in a conceptual sense and they are chosen because the phenomena they refer to are seen as very significant given the contemporary context of the scientists. Thus, ideal-typical theories, for instance in terms of the concept of bureaucratization, are not universal generalizations of empirical results. Instead, the ideal-types figure centrally in descriptions and analyses of normatively relevant, empirical patterns, and an important aim of the research is the investigation of *the extent* to which these patterns have been realized in actual socio-historical processes. Thus, while Weber is often described as a champion of value-free social science, this is only one side of the coin. The initial choices of specific ideal-types (rather than others) constitute an equally important, but value-laden feature of his methodology.

These results are also relevant for philosophical theories. Against constructivist meta-empiricism, I have argued for an interpretation of philosophy as, in part, a theoretical activity (Radder, 1996, chap. 8). In philosophy, as in the natural and social sciences, explaining the relation of its theories to concrete practices and developments requires substantial additional interpretation and articulation. Furthermore, philosophical theories, for instance Thomas Kuhn's account of the historical development of the sciences, can also be fruitfully interpreted with the help of Weber's methodology of ideal-types (see Mladenović, 2007).

The general lesson from this brief review is that critical theories need to be sharply aware of this gap between their theoretical claims and the relevant empirical practices and socio-historical developments. For instance, in Habermas' critical theory of society this awareness is lacking. This applies in particular to his basic theoretical distinction between the (economic and political) system, where strategic actions are coordinated through the media of money and power, and the lifeworld, which is the exclusive locus of communicative action coordinated by the normative intention to bring about mutual understanding (Habermas, 1984, vol. 2). It has been rightly objected, however, that the relation between this theory and actual empirical practices and socio-cultural developments has been insufficiently conceptualized and can be criticized as empirically inadequate (Kunneman, 1996, sect. 1.5 and chap. 9). The main point is that many actual practices (for instance, the mission-oriented work in schools, in hospitals and even in commercial organizations) include a *mutual* interaction between strategic and communicative action. In these practices, there is not only the one-way colonization of the lifeworld (by the system) but also the "culturalization" of the system (from the lifeworld). Therefore, the basic normative aim of this transformed version of Habermas' critical theory is not only ending the processes of colonizing the lifeworld but also, or even primarily, expanding the processes of culturalizing the system.

In contrast, Feenberg's instrumentalization theory of technology does take account of the complexity of (philosophical) theorizing. This theory analyzes technology at two quite different levels: primary and secondary instrumentalization (Feenberg, 1999, pp.202–208). In primary instrumentalization, we consider technology in terms of the functionality of its objects. This part of the theory provides a general account, which describes a feature of all technological practices in every society. It needs to be complemented, however, by an account of secondary instrumentalization, which describes the actual realization of the functionally constituted objects in concrete socio-technical networks. Thus, "the primary level simplifies objects for incorporation into a device, while the secondary level integrates the simplified objects to a natural and social environment" (Feenberg, 2006, p.186). Although more will be said about the content of the instrumentalization theory, the relevant point in this section is that it, rightly, acknowledges the complexity of the relation between theory and the world. In the terms of Morgan and Morrison (1999), the account of secondary instrumentalization could be seen as a model that mediates between the fundamental theory (of primary instrumentalization) and the concrete technological practices and socio-historical developments.

A second important question concerns the meaning of theoretical concepts, primarily the central concepts that figure in theoretical interpretations and explanations. In earlier work, I have developed a wide-ranging theory of meaning, both in the sciences and in everyday life (Radder, 1996, chap. 4; 2006, chaps. 8–11; 2019, chap. 5). It is related to the uses of language but the focus is on concepts, in particular on the meaning of the concepts underlying linguistic (including theoretical) assertions. In a formula, the claim is that concepts both structure the world and abstract from it. That is to say, this theory distinguishes two major components of the meaning of concepts. The structuring component specifies how we see (relevant parts of) the world. In a broadly Kantian way, it says that our access to the world is mediated, and hence constrained, by the available conceptual resources. In addition, however, concepts also possess an abstracting meaning component. The basic idea is that concepts are *extensible*: existing concepts might be successfully extended to situations that are different, possibly radically different, from the local contexts in which they have been applied thus far. Therefore, concepts refer to an indeterminate set of *potentially* realizable states of affairs. In a broadly hermeneutic spirit, I conclude that concepts possess a nonlocal meaning, that is, their meaning is not exhausted by but rather transcends the structural meanings that derive from their *actual* past and present uses.[2]

In the context of this chapter, the point of this particular form of openness is that our concepts contain the seeds of their own transformation: their nonlocal meaning reflects our potential of seeing the world in novel ways. An example is the concept of labour. Traditionally, this concept was almost invariably applied, and hence limited, to wage labour. However, its meaning may be extended to include unpaid domestic labour as an important type of human activity that substantially contributes to the reproduction or transformation of societies. A second illustration is the concept of social justice. Often, it is limited to the current population of certain nations or regions. Yet, it may also, especially in times of globalization and pandemics, be extended to include the entire world population or, significantly, future

---

[2] Acknowledging the significance of both meaning components makes it possible to avoid the opposition between universalism (as in the case of Habermas' theory) and localism (advocated, for instance, by many proponents of science and technology studies).

generations. In the next section, I will come back to this account of meaning and argue that it points to a crucial dimension of the idea and practices of critique.

3.2   Critical Theory

Critical theories are theories of societies and their historical development. Although invisible in the adjective "critical," there is a relevant distinction between the nouns "criticism" and "critique." Both imply judgement and are thus evaluative, often (but not necessarily) in the negative sense of judging a situation (or specific aspects of it) as not or less desirable. In this sense all academic disciplines are critical. Usually, critique is seen as a specific form of criticism, a form that is comprehensive and fundamental. In this respect Immanuel Kant's critique of pure reason differs from a particular criticism of the empirical support for a hypothesis concerning human cognition.

In the course of human history different forms of critique have been presented. The critical social theories discussed in this chapter are often claimed to have their origin in, or be inspired by, the works of Karl Marx. His critical theory analyzes types of production and the implied social relations, as well as associated systems of ideas. From this analysis, Marx drew far-reaching critical conclusions concerning the actual and future courses of human history.

The kinds of argument used by Marx and the Marxist tradition have been the subject of extensive debate. One major interpretation emphasizes their origin in a scientific approach. In the course of the twentieth century, however, the scientific basis of Marxist critique and the related vision of predestined historical progress have been increasingly questioned. This strengthened another major Marxist interpretation (sometimes called neo-Marxist) which focuses primarily on the normative significance of Marx's work.[3] From an Enlightenment perspective, it claims, social emancipation can be accomplished by consistently raising awareness of, and hence resistance against, the systemic reification, alienation and oppression in capitalist societies. This interpretation has also become a target of sharp criticism. In particular, Max Horkheimer and Theodor Adorno – strongly shaken by the rise of Nazism and the barbarisms of the holocaust – launched a highly pessimistic critique of the emancipatory potential of the Enlightenment vision. Its alleged rationalization was seen to advance an extreme technocratic and dehumanizing instrumentalization of politics rather than a universal emancipation. This state of affairs has constituted a major challenge to later critical theorists. If comprehensive and fundamental social critique can neither be grounded in science nor in rational arguments, how can it be normatively justified? In the remainder of this section, I will focus on this evaluative question.

---

[3] For a recent discussion of these contrasting interpretations, see Kuhne (2017).

An influential answer, based on his interpretation of language and linguistic communication, has been developed by Habermas. In his view, the basic ground of social critique is implicit in the very structure of linguistic communication and understanding.

> The human interest in autonomy and responsibility [in German, *Mündigkeit*, HR] is not mere fancy, for it can be apprehended a priori. What raises us out of nature is the only thing whose nature we can know: *language*. Through its structure autonomy and responsibility are posited for us. (Habermas, 1972, p.314)

In his later works, especially Habermas (1984), he has developed and qualified these early statements (originally published in 1965). First, the normative ground of critique can be found in discursive communicative actions, that is, in ideal forms of argumentative communication fully undistorted by unequal power relations. Furthermore, this ground is not simply empirically given but both operative in actual communications and anticipated as a future development toward undistorted communication. Finally, Habermas has qualified his earlier connection between undistorted communication and universal consensus by acknowledging the common occurrence of dissensus (see also Gabriëls, 2019, p.510). In spite of this, it remains the case that the rationality of mutual understanding (both in occasions of consensus and in cases of dissensus) needs to be judged by the universal criteria of unconstrained discourse.

Given the ubiquity of dissensus, the persistence of systematic asymmetric power relations in linguistic practices and the implausible teleological view of an anticipated power-free future, many authors have criticized Habermas' claim that enlightenment and emancipation can be directly derived from the actual and the historically evolving nature of communicative processes. Therefore, the question of the criterion of critique is still with us.[4] In the remainder of this section, I offer a brief explanation of my answer to this question, based on the interpretation of language and meaning sketched in the preceding section and on the notions of the *common good* and the *public interest*.

In Radder (2019), I have developed an account of the common good of science and technology. The regulative value of the common good is defined and explained, first, on the basis of the nonlocality of the meaning of concepts and the implied nonexhaustibility of (informational and skilled) knowledge and, second, with the help of the normative notion of a public interest with its implied "thick" concept of democracy. Because the central notions of the nonlocality of meaning and the public interest of states of affairs are not limited to science and technology, the original normative criterion of the common good can be generalized. In particular, not only theoretical concepts but also value-concepts possess a nonlocal meaning. Suppose that claims about certain practices and their results are phrased with the help of

---

[4] Annemarije Hagen (2019) rightly states that the current debate is not limited to the Frankfurt School anymore, but includes contributions by a broader group of authors (e.g., Étienne Balibar, Judith Butler, Maeve Cooke, Foucault and James Ingram). In the same spirit, Angela Roothaan (2016) adds the names of Jacques Derrida, Ivan Illich and Joan Tronto, and especially those of post-colonial thinkers like Emmanuel Eze, Frantz Fanon and Achille Mbembe.

extensible concepts, possibly including value-concepts. In that case these activities and their results constitute a common good, if and only if the nonlocal meanings of the relevant concepts have not been structurally curtailed *and* these activities and their results are in the public interest. In this way, the normative claim that human practices should contribute to the common good constitutes a general criterion of critique of (knowledge) claims and socio-historical processes.

Although independently developed in a rather different context, the idea that concepts both structure the world and abstract from it, shares certain features with the view of Herbert Marcuse when he writes:

> However 'man', 'nature', 'justice', 'beauty' or 'freedom' may be defined, they synthesize experiential contents into ideas which transcend their particular realizations as something that is to be surpassed, overcome. Thus, the concept of beauty comprehends all the beauty not *yet* realized; the concept of freedom all the liberty not *yet* attained. (Marcuse, 1968[1964], p.170)

These views provide a general account of social critique as a basic possibility and phenomenon rooted in the nature of human concepts, not only the broad concepts mentioned by Marcuse but also more specific concepts, such as the concept of labour. In other words, adding the abstracting component to the structuring component of the meaning of concepts entails, in the spirit of Marcuse, going beyond a one-dimensional conception of human beings.

This account of the common good differs from the conclusions Habermas draws from the processes of linguistic communication in several significant respects. First, underlying my theory of the meaning of concepts is that "what raises us humans out of nature" is not simply, or not primarily, linguistic communication.[5] More specifically, it is the capability to think and speak about *potential* states of affairs, not in the sense of dreaming but as a part of systematic processes of reflection. This specifically human transcendence of the status quo includes not only what can be readily made present, even if it is temporarily absent. More fundamentally, it is the capability to imagine what has never been present but might come to be realized. Second, the nonlocality of meaning explains the possibility of critique without including some kind of teleological anticipation of future states of affairs.[6] Whether or not current ways of seeing the world will in fact be transcended is a fully contingent matter. Third, the conceptually anticipated possibilities do not entail a fixed, inherent evaluation of these potential future states of affairs. From a normative perspective, they may be seen as neutral, as (more) desirable or as (less) desirable.[7] Therefore, the account of the common good includes a second basic notion, the

---

[5] After all, many (higher) animals possess ways of communicating by means of (sometimes elaborate) languages of signs and sounds.

[6] However, rejecting the philosophical grounding of critique in a teleologically anticipated future should not make us blind for those areas where significant progress has occurred in the course of the past centuries (see Rosling, 2019, for important examples). These developments do foster our hopes and motivate our efforts to work for a better future, even if they cannot provide any guarantee that these efforts will be successful.

[7] This in contrast to what is suggested by the examples in the above quotation of Marcuse.

notion of a public interest, which points out the normative direction of social critique.

A public interest is an interest in realizing and maintaining positive (or in preventing or removing negative) states of affairs of basic significance, that (1) affect (or will probably come to affect) either all members of the public or a specific part of it and that (2) are democratically judged to be of public import. Such valuations of states of affairs can apply both to a variety of human needs and to acknowledged needs of nonhuman species and nature more generally. From this notion, two criteria for being in the public interest can be derived: the degree of inclusion of the (potentially) affected people and the relative quality of the democratic support for ascribing public import to the relevant states of affairs. Elsewhere, I have explained these ideas in detail (Radder, 2019, chap. 6). I cannot repeat this explanation here, but three basic features should be mentioned. First, what is taken to be of public interest is not a matter of fact, because it also requires normative judgement. Second, public interests can be seen as basic regulative values, clearly of a highly ideal nature; yet, the two criteria make it possible to see these values as a matter of degree: we can compare particular states of affairs and judge some as being more (or less) of public interest than others.[8] Third, the criterion of democratic support is based on a multidimensional notion of democracy. Its central elements are: voting, deliberation, a constitution, an inclusive demos, an inclusive electorate, and an inclusive right to stand for election.[9] Because this account of democracy can be applied to all sorts of social institutions, it is not limited to forms of governmental representative democracy but also embraces forms of participatory democracy.[10]

The critical theory of the common good emphasizes the role of democratic voting procedures. This entails a fourth difference with Habermas who makes a "proceduralist conception of deliberative politics the cornerstone of the theory of democracy" (Habermas, 1998, 246). I agree that democratic decision-making should be informed by high-quality deliberation. Yet, fully neutral argumentative deliberation is fundamentally impossible: not only because of unequal power relations, but also because of ineliminable differences in incompatible world views, values, conceptual systems, and epistemological norms. Therefore – in concurrence with Bräunig's statement quoted at the beginning of the chapter – voting (on the basis of an appropriate constitution of the electorate) is an equally ineliminable and independent element of any democracy. In actual decision-making there is *no generally valid* criterion that can tell us what should be the weight of deliberation and

---

[8] For illustrations of the public interest (or lack of it) of a variety of aspects of science and technology, see Radder (2019, chap. 7).

[9] The last element was not explicitly included in my earlier publications on this subject. It is as important as the other elements: countries where this right is structurally violated by excluding, silencing, intimidating, imprisoning, poisoning or killing potential candidates for election cannot count as democracies.

[10] See Radder (2019, pp.201–209; 2021, pp.116–124). See also Mark Brown (2009), who emphasizes the importance of institutionally differentiated forms of democracy.

what of voting. Therefore, democratic decision-making necessarily includes context-dependent judgement of the proportion of these weights.

The inclusion of this substantive notion of democracy provides a solution, or at least a significant mitigation, of a dilemma that has haunted critical theories for a long time. It concerns the question whether or not critical theories need a general, positive conception of "the good society." On the one hand, an affirmative answer seems to be required in order to know that the alternative for the criticized state of affairs is a better one. On the other hand, the pinning down of a specific alternative conception by critical theorists seems to be authoritarian and hence going against the basic principles of critical theories (see, e.g., Freyenhagen, 2017; Hagen, 2019).

At first sight, the critical theory sketched so far might seem to lack positive evaluative content. As I have explained, the guidance it offers derives from the notion of a public interest, while assigning a public interest to some state of affairs requires normative, democratic support. However, in the critical theory of the common good this requirement of democratic support goes far beyond a contentless or formal decision-making procedure. The central elements of this notion of democracy (voting, deliberation, a constitution, an inclusive demos, an inclusive electorate, and an inclusive right to stand for election) presuppose a variety of substantive values: having a vote and a voice in matters of basic significance; enjoying bodily integrity; possessing individual and collective freedom of expression; having a right to education and to participation in deliberation in an independent public sphere; being protected by the separation of the executive, legislative and judicial power of the state. Moreover, and crucially, in a real democracy these rights and arrangements do not exist merely on paper, but are actively brought about and maintained through appropriate institutions, thus fostering a flourishing democratic culture. Yet, in the spirit of Pascal's thought quoted at the start of this chapter, this view also acknowledges the other horn of the dilemma. According to the account of theories discussed in Sect. 3.1, by itself this general theory of democracy does not, and cannot, pin down the specific features of democratic societies. For this purpose, further democratic (but context-dependent) interpretation and specification of its central notions is required. Thus conceived, the proposed critical theory of the common good provides a general guide for the direction to take, but it does not determine the full content of a good society, let alone of a good life.

The nature of social critique and its possible grounding is still a subject of lively debate. Fabian Freyerhagen, for instance, aims to defend Adorno's rejection of any positive conception of the good society and his purely negative social critique of "the inhuman" (Freyenhagen, 2017). Yet, this defense is not convincing, primarily because the aim of critique is not merely to change a situation but to change it for the better. In addition, Freyerhagen's downplaying of the practical motivating power of positive ideas, his use of superficial examples, and his rejection of philosophical claims that have not been "compellingly proven" suggest a conception of philosophy that may have been legitimate in Adorno's time but that looks definitely outdated in the twenty-first century.

From the opposite perspective, Maeve Cook requires that social critique should be guided by "an idea of the good society in which the identified social obstacles to human flourishing would once and for all be overcome" (quoted in Hagen, 2019, p.3). Unfortunately (or, is it fortunately?), such ideas do not exist. The critical theory proposed in this section, with its dependence on democratic practices, will certainly not lead to "once and for all" solutions. But it does possess the advantage of suggesting changes and enabling judgement of developments that may advance the common good.

Annemarie Hagen and Angela Roothaan advocate a view that seems to be more congenial to the one presented here. Hagen emphasizes the significance of open-ended universal concepts, such as equality, democracy or freedom. At the same time, "because the true needs and preferences of agents cannot be decided by critical theorists, critical theory should ... remain sensitive to the claims for inclusion and forms of contestation of those who are denied a political voice" (Hagen, 2019, p.13). Similarly, but more specifically, Roothaan proposes that

> the work done by critical theory, which asked attention for the silence of the oppressed as the starting point of an ethics that can really bear that name, will be increasingly brought into dialogue by the work of postcolonial theorists who have changed the art of reflection by starting from the experiences of those who have suffered from modernity and its racialized discourse of development (Roothaan, 2016, p.64).

As we have seen, my view of critique and critical theory deviates from Habermas' account in significant respects. Yet, it does endorse his emphasis on the necessity of the normative justification of this type of critique. In this respect, I agree with Gerald Doppelt's criticism of Feenberg's critical theory of technology, in particular with his questioning of the notion of democratic rationalization by marginalized interest groups.

> Feenberg presents some intuitively quite attractive contemporary examples of successful lay challenges to established technology. ... But the cases do not provide an explication and defense of a normative standard for judging which interests and challenges might produce a more rational, democratic, morally defensible technology, and which do not. (Doppelt, 2006, p.89)

Along these lines, Doppelt questions, or at least qualifies, the democratic character of some of Feenberg's examples. He argues, for instance, that user challenges for changing one-way information to two-way communication technologies (such as the well-known Minitel example; see Feenberg, 1995, chap. 7) "follow the logic of market rationalization, not democratization" (Doppelt, 2006, p.89). Therefore, he concludes that Feenberg's critical theory needs to be complemented with a basic account of democratic equality, for instance in the form of the political philosophy of liberal democracy.

I agree with Doppelt's general point: challenges to established technology need some kind of democratic support. For instance, while I think that many current websites are fine for getting some information (say, about shopping or hotel accommodation), I am annoyed by the interactive process of being constantly bombarded by requests to review these services. The more so, because these reviews may and will be used to discipline, or even penalize, the people who were concretely involved in providing the relevant service. In these cases, the creation of interactive digital technologies is being used for the purpose of controlling the shop and hotel workers, not for democratic purposes. Although this development was hard to foresee in 1995, in 2022 we should be thoroughly aware of the basic ambivalence of such examples. More generally, although interventions and challenges on social media can be subversive, by no means can all of them be considered as advancing democracy.[11]

In his reply to Doppelt, Feenberg (2006, pp.196–198) makes a number of correct points concerning the limitations of the tradition of political philosophy, in particular its neglect of technology. But he neither addresses nor answers Doppelt's main point of criticism. In contrast, the account of critique developed in this section provides such an answer. Moreover, it does so without appealing to an exclusively liberal and technology-blind theory of democracy (see Radder, 2021).

In sum, critical theories address the normatively relevant issues implied in significant socio-material patterns. The validity of critical theories depends on three crucial conditions. Their approach needs to embrace an openness to alternative interpretations and their claims must be based on an adequate account of (their relation to) the topics criticized. Their normative statements require a cogent explanation and defence, in which the notion of democracy needs to play a central role. Related to this, these statements should not be seen as decrees of a philosopher-king but as contributions to concrete democratic debates and practices in our actual societies. Decisions on the "necessity" of critique can only be made in such debates and practices.

3.3   Critical Theory of Technology

Critical theories are theories of societies and their historical development. Although it is evident that technology strongly influences our societies and their development, even recent critical theories do not, or not adequately, take technologies into account. For instance, as a quick perusal of the bibliographies already suggests, "technology" is completely absent in the almost 400 pages of the volume *Why critique?* (Ellmers & Hogh, 2017). Apparently, technology and society are not seen as important, inherently related entities. More generally, the really disappointing (and even annoying) fact is that the paradigm of two separate cultures, the sciences/technologies and (or even, versus) the humanities, can still be found in all kinds of contemporary debates and policies. Therefore, an important task for a critical theory of technology is a detailed refutation of this dualistic approach.

The theory of technology summarized here consists of two parts. First, technologies are conceptually characterized as "artifactual, functional systems with a certain degree of stability and reproducibility." The second part deals with the issues of

---

[11] See the non-sense (sometimes in the form of "bullshit," that is, without *any* interest in its truth or falsity) that is nowadays being circulated through social media.

their actual material and social realizability and their implied moral and political dimensions (see Radder, 2008, pp.51–60; 2019, chap. 2). The first part of the theory characterizes *potentially* realizable technologies and it explains what *would* be needed for actually realizing particular technologies. Put differently, the theory provides a general characterization of the conditions for successfully realizing technologies. In line with the account of (critical) theories discussed in the preceding sections, in addressing the questions of the actual feasibility and normative desirability the second part involves substantial further interpretation and specification of the core features of technologies. As a whole, the theory is not a generalization of empirical instances of technologies (as is often required in "empirical philosophy of technology"). But neither is it a description of "technology" as a single, philosophically constructed entity or system (as is often the case in what Carl Mitcham (1994) calls "humanities philosophy of technology").

This theory of technology shares some features with the analytic work of the later Foucault. They include, first, an emphasis on the connections between knowledge and power and on the disciplining nature of technologies.[12] A second agreement is a plea for the reflexive character of philosophy, that is, for the intrinsic philosophical significance of developing a history and ontology of our own present (Foucault, 1984; Breitenstein, 2017; Radder, 1996, chaps. 5 and 8). In some other respects, however, the critical theory of the common good differs substantially from Foucault's analyses. The basic problem is the one-dimensionality of these analyses. Foucault himself calls his genealogy "superficial," which he sees as a merit.[13] Although it does suggest or include an (often implicit) normative critique, this critique is exclusively immanent and (like Adorno's) purely negative.

> The critical ontology of ourselves has to be considered not, certainly, as a theory …; it has to be conceived as an attitude, an ethos, a philosophical life in which the critique of what we are is at one and the same time the historical analysis of the limits that are imposed on us and an experiment with the possibility of going beyond them. (Foucault, 1984, p.50).

That is, this critique is fully rooted in specific types of local practices and limited to a refusal of and resistance to those practices. It does not include, and even explicitly rejects, theoretical notions (such as the nonlocality of meaning) that could motivate concrete reflection and action for change or overarching, nonlocal values (as the ones implied in the notion of a public interest) that could provide guidance concerning the direction in which to change the situation for the better.[14]

---

[12] Foucault (1982); Radder (1986, 2019, chap. 2). I have stressed (more strongly than Foucault seems to do) that "discipline" is a condition (rather than a general feature) of realizing a technology. Whether people see the power to discipline as productive rather than repressive (and therefore whether they will co-construct rather than resist the realization of the technology) is a contingent empirical fact, which cannot be settled by means of theoretical analysis.

[13] As Dreyfus and Rabinow (1982, p.106) explain: "Genealogy avoids the search for depth. Instead, it seeks the surfaces of events, small details, minor shifts, and subtle contours."

[14] My position presented in the brief programmatic essay *Wat is progressieve filosofie*? ("What is progressive philosophy?"; see Radder, 1985) was clearly closer to a Foucaultian one.

Two features of my theory of technology are especially relevant for the discussion in this section. The first is the notion of functionality, which presupposes the (possible) existence of a use, a user, or a human context for whom or which the technology is functional. It makes no sense to ascribe a function to a technological artifact in the absence of (actual or potential) use, users or human contexts. A screw driver only possesses the function of driving screws if there are, or there might be, screws and beings that are able to handle it, in the literal sense of having the right form and size of hands. The important implication is that general theoretical analyses of technologies need to take into account the material and social environment in which the technologies are supposed to function.

The second relevant feature, the degree of stability and reproducibility, reinforces this implication. Successfully realizing a technology requires embedding it in an environment that enables and does not disturb its stable and reproducible functioning. And again, because the general characterization of technologies does not assume and cannot guarantee the actuality of this kind of success, a crucial task for a critical theory of technology is to examine and assess the empirical feasibility and normative desirability of current and future technologies.

The connection between the theory of technology and the critical theory of the common good can now be made in a straightforward way. Efforts to advance the common good of a particular technology (or a combination of technologies) require, first, inclusiveness of the people who will, or may, be affected by the technology and a high-quality democratic assessment of its actual feasibility and its normative desirability; second, a genuine openness to seriously explore (possibly radically different) alternatives to this technology[15]; third, a considered judgement of the available options and a choice of the one that best meets the criteria for advancing the common good. Because of the continuing development of new technologies, advancing this good will keep requiring an ongoing effort.

Finally, although it would be "nice" if all technologies were for the common good, my proposal by no means implies that each and every technology should be actually examined by carrying out these three steps. It does not seem very practical, for instance, to go through such extended procedures to examine the common good of specific types of screws used in making a particular type of cupboard. As stated in their definition, public interests are assigned to states of affairs of basic significance. Which states should be qualified as such is a matter of democratic judgement. As citizens, critical theorists are fully entitled (and even have a duty) to take part in relevant deliberations and actions. But it is not up to them to determine by theoretical decree what should count as "of basic significance."

As explained in Sect. 3.1, an important aspect of Feenberg's instrumentalization theory is its two-part nature. In this respect, his account and mine agree. In some other respects, however, I take a different view. This applies to the *content* of primary instrumentalization in particular. According to Feenberg, this aspect of

---

[15] Such a genuine openness requires that realizing relevant alternative technologies has not been blocked by private, financial interests, for instance through strongly constraining patents (see Radder, 2019, pp.113–117 and pp.180–182).

technology describes "our original functional relation to reality" (2006, p.186). More specifically,

> the level of primary instrumentalization demarcates technical action from other modes of action. In specifying the essence of technology, it also suggests technology's appropriate limits. (Feenberg, 2006, p.189)

However, in as far as primary instrumentalization is meant as a general theoretical characterization of technologies, I think we need a more detailed account by including a more developed notion of functionality and by adding the ideas of stability and reproducibility. First, as we have seen a technology can only be functional in a stable and reproducible way if its potential to play some role intended by one or more human agents can be realized by embedding the system in a suitable socio-material environment. Abstracting away this environment, makes it impossible to understand the notion of functionality and the role of the required stability and reproducibility. Second, and related, this more detailed account of functionality entails a general methodological criterion for "technical" success: it specifies what would be needed to make a working technology, that is, a technology that will perform its (intended) function. Adding such a methodological criterion is important for critical theories of technology, since questions about desirability are often intertwined with questions about feasibility. Therefore, they need to be addressed in tandem, as is illustrated by the case of the corona tracking app discussed below.

Feenberg (2008, p.121) writes: "the slippage between abstract functions and concrete devices … seems to me essential to the possibility of modern technical disciplines," and he claims that "there is something obviously right" in Heidegger's substantivist philosophy of technology (2006, p.186). This view concedes too much, not only to Heidegger's philosophy but also to an engineering interpretation of technologies as decontextualized "physical artifacts with a function."[16] My basic objection is that "technical objects" should never be characterized in this way, because the "possibility of technical disciplines," that is, the potential realization of stable and reproducible technologies requires explicit acknowledgment of their systemic nature and the necessity of embedding them in an appropriate environment.[17] As a general feature of any technology, this is a crucial aspect of *primary* instrumentalization. Having said this, I fully agree that we also need to add detailed accounts of the empirical feasibility and normative desirability of realizing (actual or novel) technologies, in the spirit of the idea of secondary instrumentalization.[18]

---

[16] See also the criticism of this "dual nature" interpretation in Vaesen (2011).

[17] Put differently, especially a critical theory of technology needs to emphasize that the specialists of modern technical disciplines are, or should be, "heterogeneous engineers," in the sense described in Law (1987).

[18] Yet, my theory of technology does entail some differences concerning Feenberg's specification of what he calls the "moments" of primary and secondary instrumentalization. In Radder (2008, pp.64–66), I have developed this point by brief comments on the moments of decontextualization, autonomization and systematization. The discussion in this chapter continues, and hopefully clarifies, the issues of our earlier exchange (for Feenberg's response, see his 2008, pp.120–121).

In concluding this part of the section, let me briefly return to Doppelt's criticism of Feenberg's views. In the previous section, I sided with Doppelt's point that critical theories need a normative standard. In addition, Doppelt raised a second objection.

> In the modern world a great deal of technology is private property. Its designers act in the name of the owners, and their rights as owners, to determine the technical code in accordance with their economic interests. ... Dominant technology is provided with powerful rights-based protections. (Doppelt, 2006, 90–91)

He concludes from this that Feenberg's theory is incomplete, because challenging technology would also imply questioning the Lockean moral code of private property rights and free-market exchange. This criticism, however, is unjustified. What can be privately owned are particular tokens of a specific technology. As is clear from its definition as "the dominant social meaning of a technology," a technical code (that is, a meaning) cannot be owned by private owners.[19] For this reason, the existence of private property does not, as Doppelt claims, present an additional problem for Feenberg's critique of technology. On the contrary, claiming that private companies have "the right to determine" social meanings is a form of technocratic politics that is the legitimate target of critical theories.

Like the rest of the world, in the spring and summer of 2020 the Netherlands had to confront a severe healthcare crisis as a consequence of the fast and dangerous spread of the corona virus SARS-CoV-2. Politicians, scientists, healthcare professionals and the population in general struggled to find the best approaches to combat the virus and/or to limit its strongly disrupting impacts on the health of the people and the condition of the economy. My focus here is on one of the many dimensions of this crisis: the plan of developing a so-called tracking app and the ensuing discussion on its prospects and problems in the Netherlands. On April 7, this plan was put forward by Minister Hugo de Jonge of Health, Welfare and Sport (*Volksgezondheid, Welzijn en Sport*, VWS). An extensive debate followed, including expert meetings, open letters to politicians, and a great variety of offline and online contributions.[20] Initially, the tenor of this debate was quite critical. For instance, many experts characterized the attempts at selecting an appropriate app as "rushed and sloppy work." The result of this was that by early May the government seemed to have dropped the plan. However, on May 30 it was reported that the IT unit of the Ministry was working on "a better version" of the app, especially with respect to privacy issues. Next, on June 26 it was announced that this version had been developed and was to be tested soon. Finally, from August 18 on the app (now called *coronamelder*, that is, corona notification app) became available in the app-shops of Google and Apple, on

---

[19] More generally, underlying this argument is a view of technologies as separate artifacts that are supplied for sale on a free market. This view fully misses the point of the comprehensive socio-political value-ladenness of technical codes.

[20] For instance, *De Volkskrant*, one of the major Dutch newspapers, carried many news reports and opinion pieces on this subject. For brief reports on similar initiatives in other countries, see Stupp (2020) and Perry (2020); for an extended, critical analysis of transnational issues in developing competing apps, see Krige and Leonelli (2021).

October 6 its implementation was approved by the Dutch Parliament, and from October 10 the app could be generally used.

For practical reasons, my coverage of the case focuses on the months April through August. Although the tracking app is clearly a moving target and new developments may occur, the analysis of the events during these months captures the basic features of this kind of digital technologies. Furthermore, this analysis makes possible a detailed assessment of the technology on the basis of the critical theory of the common good.

As presented by Minister De Jonge, the idea of the tracking app is to record users' proximity and duration of exposure to known carriers of the corona virus. The app would be installed on people's smartphones. First, if a person is infected, this fact will be registered on his or her smartphone. Then, if people not infected by the virus come close to an infected person (for instance within a critical distance of 1.5 m), this "positive exposure" will be recorded on their phones. After that, these people will either immediately go into quarantine (in the period under discussion usually for 14 days) or will first be tested and, if positive, go into quarantine. Referring to technical expertise and experiences in other countries, De Jonge claimed that this technology is practically realizable. Its main advantage would be the monitoring of the diffusion of the virus at an early stage, which is the best way to contain its spread and limit its broader impact.

My aim here is not to analyze this case in all its detail, which would require at least a full article by itself.[21] What I will do is apply my theory of technology to the case and evaluate the technology on the basis of the critical theory of the common good. At several points, I will mention agreements and disagreements with the views of other contributors to the debate.

First, consider the proposed technology as an artifactual, functional system with a certain degree of stability and reproducibility. For the specific purpose of this chapter, the system in question may be chosen to consist of the collection of the smartphones of people residing in the Netherlands. Furthermore, the relevant parts of the environment, which constitute an equally crucial dimension of the technology, will be chosen to include the people themselves and the quarantine and testing resources available to them.[22] The tracking apps are obviously human-made artifacts, the function of which is to contain the corona virus and limit the number of people infected with it. This can be achieved only if the working of a specific app on a specific phone is stable and if this result can be reproduced on the apps downloaded on a variety of different kinds of smartphones in all kinds of different situations.

---

[21] For an account of several other aspects of the Dutch debate on the corona tracking app, see Siffels (2021).

[22] As explained in Radder (2019, pp.46–53), for other purposes system and environment may be defined in (much) broader ways, depending on the kind of analysis and assessment we aim to make.

So far, a short theoretical sketch of the tracking app technology.[23] Next, consider the socio-material feasibility of the technology, in terms of its false positives, its false negatives, and its expected overall effectivity. False positives are cases where the app of the initially non-infected person wrongly records a critical contact. Here, a major problem is that what is being recorded is a critical distance, a "contact," between two smartphones (Van Straten, 2020). This does not imply that the owners of the smartphones have been within a critical distance.[24] What is more, it tells you nothing about the nature of this bodily contact: did it or didn't it cause a real risk of infection? Phones in purses or on tables may have been close, while the involved people were outside the critical spatial interval; the non-infected person may have borrowed the phone from a friend; the people may have been within the critical distance but the nature of the interaction has been such that infection is less probable (the interaction took place in the open air rather than in a closed space; or the people were wearing face masks or were sitting back-to-back to each other) or even improbable (there was a relevant obstacle, such as a window or a transparent spray shield between the phones). Summed up, such kinds of situations can be expected to abound and many examples have been mentioned in the debate about the case.

False negatives are cases of real infection that are not recorded by the technology. A first source of possible false negatives are the carriers of the virus who cannot record this fact because they do not have a smartphone. The worrying thing is that this group can be expected to include many of the most vulnerable, older people. A further group includes the (primarily younger) people who have been infected but do not know this because they do not or hardly suffer from any of the relevant symptoms. They cannot register as a carrier either, even if they have downloaded the app. In the case of the corona virus, this group is considerable and hence it makes a strong contribution to the set of the false negatives. In addition there are the people who have not installed the app, who do not always take their phones with them, who have turned them off, or whose phones do not work because they need to be recharged or have broken down for any other reason. Furthermore, there are the cases (the opposite of the false-positive situation) where the people have been within the critical distance, while the phones have not. Finally, the app does not detect a possible infection by viruses that diffuse over longer distances (for instance through aerosols) or viruses that are left by infected people on whatever objects they have touched. Summed up, such kinds of situations can again be expected to abound.

---

[23] For reasons of space, I have to leave out several other relevant aspects. For instance, the distance between any two "close by" smartphones needs to be measured accurately; the smartphones beyond the critical distance must be ignored; the time of staying within the critical distance needs to be recorded and all interactions lasting shorter than a certain minimum time (e.g., 10–15 minutes) have to be disregarded. Some more detail about these issues can be found in Stupp (2020) and Perry (2020); for a very detailed discussion, see Van Straten (2020).

[24] In an interview dated May 30, a spokesperson of the Ministry of VWS stated that "a very accurate estimate of distances – about which strong doubts exist – is not necessary" (Wassens, 2020). But of course it is necessary. If the distance is taken too large, the number of false positives will strongly rise. If too small, there will be an increase of false negatives.

Although further quantitative information would be welcome, the total number of false positives and negatives can already be estimated to be substantial. Consider first the false positives. On the one hand, the average number of people infected by a carrier of the corona virus is usually claimed to be at most three and probably even considerably smaller. On the other hand, it is generally assumed that the apps can only achieve their end if at least 60% of the population effectively participates in the tracking procedure. Because it is plausible that the phone of an average carrier of the virus will be in contact with far more than two or three other phones, it follows that the number of false positives must be substantial. Furthermore, if 40% of the population (including the most vulnerable people and the "unaware" carriers) does not participate, the number of false negatives will also be substantial.

A significant number of false positives entails that, in individual cases, a positive notification by the app does not necessarily indicate an infection. Analogously, the occurrence of a substantial number of false negatives implies that "no notification" does not mean "no infection." Yet, in the latter cases the name "corona notification app" may easily induce a mistaken feeling of safety. Therefore, the existence of many false negatives implies that all kinds of other measures (e.g., observing careful hygienic procedures) will still be necessary.

Acknowledging these technical problems is important and highly relevant in assessing the technology. But there is much more to say. The function of this technology is not just to record the number of critical "contacts" but, crucially, to contribute to containing the virus and limiting its impact. In line with my theory of technology, we also have to analyze the feasibility and desirability of the procedures that should follow the recording of a critical "contact."[25] Unfortunately, this crucial phase has been frequently ignored in many, if not most, contributions to the debate. For instance, in many cases "downloading the app" has been simply equated with "effective use of the technology" (see, e.g., Verhagen, 2020).

As we have seen, two follow-up procedures have been proposed. The first is that the people whose phones have recorded a "contact" should immediately go into quarantine.[26] But can we expect that average citizens, especially if they have observed the general protective measures, experience no health problems at all, and know that the functioning of the technology is questionable (because of the large number of false positives), will voluntarily go into a highly consequential, lengthy quarantine? As suggested by several participants in the Dutch debate, the likely answer is that, at the end of the day, a substantial part of the population will simply not do this. This conclusion is supported by the results of an empirical study that finds "a large discord concerning the desirability of the corona-app" (Mouter et al.,

---

[25] In terms of Krige and Leonelli (2021), an adequate implementation of the technology requires it to be embedded in an appropriate, broader healthcare policy.

[26] A complication is that exceptions need to be made. For instance, people working in healthcare institutions (including doctors, nurses, technicians, cleaners, etc.) have a big chance of being permanent members of the critical group. This leads to the question of how to limit the size of this subgroup. Putting all its members in an almost permanent quarantine could entail the collapse of the entire healthcare system.

2020). About one third of the respondents said they would install the app, while another third strongly objected to this. The remaining group was still in doubt: they had many questions and installing the app would depend on convincing answers to these questions. In motivating their views, the respondents took into account both technical issues about the app and issues concerning the follow-up procedures.

The second follow-up procedure involves that people whose phones have recorded a "contact" should be tested for the virus, preferably within a really short term in order to minimize the number of potential people they might infect. If they go for a test and if it is positive, these people may be more motivated to go into quarantine than the other group. This second procedure does presuppose the availability of widespread and reliable testing facilities, a condition that was not at all always met during the period under discussion.

My conclusion from the discussion so far is that the overall effectivity of the technology is questionable.[27] This fact is highly relevant in evaluating its normative desirability. But the Dutch debate has not been limited to this fact, and rightly so. Other major normative issues concerned the impact on people's privacy (including the risks of hacking) and, more generally and more importantly, the dangers of a strong increase of the fine-grained possibilities of surveilling and disciplining entire populations. An additional reason for these worries is that exercising this control will not be limited to democratic governmental organizations but will unavoidably entail a further, invasive dependence on the commercial interests of influential, multinational corporations.

An illustration of these points can be found in an open letter signed by 61 scientists (including me) from a variety of disciplines. It was sent to the most involved members of the Dutch cabinet on April 13. Among several other things, it formulates four crucial necessary conditions for a socially legitimate use of the technology.

> Use of the Apps should be abandoned if: (i) 'contact tracing' … via the Apps does not (or, no longer) achieve its ends or does not employ effective and reliable means; (ii) less invasive solutions are possible; (iii) its social consequences are too troublesome; (iv) no widely supported and justified choice can be made between conflicting (basic) rights and liberties. (Muller et al., 2020, my translation)

By applying the theory of technology and the critical theory of the common good to the case of the tracking app we can check whether this app meets these four necessary conditions. As we have seen, technologies contribute to the common good if they are in the public interest (with a crucial role for sufficient public support and high-quality deliberation) and if the decisions on their implementation have included serious and democratic consideration of alternatives.

---

[27] As explained, for practical reasons the above account has been restricted to the period before the actual launch of the corona notification app. Nonetheless, the concrete uses of the app have amply confirmed the results of this account. See the *Factsheet CoronaMelder* (Rijksoverheid, 2021). For instance, up to January 24, 2021, a mere 25.8% of the Dutch population had downloaded the app; only 12.5% of the positively tested virus carriers had registered this fact on their phones; and the average number of false positives even amounted to 90.5%.

The first point concerns the socio-material feasibility of achieving the aims of the tracking app. In the letter, it is merely mentioned, not really elaborated. However, the straightforward conclusion of the critical analysis in this section is that the corona tracking app does not meet this condition. Point (ii) addresses the openness to alternatives. An important alternative is large-scale testing, which is a feasible approach that was actually taken in several other countries. It is far less invasive and questionable than the use of a tracking app, and hence it constitutes a preferable alternative.[28] More generally, openness to alternatives has two important benefits. It avoids the domination of entrenched interpretations of values and theoretical concepts as the only correct ones. At the same time, this openness enables and motivates people to seek and develop substantially novel interpretations of basic concepts and values relevant to the longer-term protection against this and comparable viruses. For instance, that "a flourishing life" does not need to depend on excessive, global mobility of goods, animals and people, a mobility that constitutes a decisive factor in the worldwide spread of viruses. Or, more specifically, that "employment" is not necessarily dependent on large-scale air traffic, but can be achieved by any activity for which people are willing to pay.[29] The third point of the letter focuses on the more general normative desirability of the app. It refers to the substantial dangers of surveillance, discipline and control that come with the operation and continued use of these types of apps. Although digitalization may advance flexible use (for instance in word processing), the opposite occurs at least as often, especially when the digital devices have been structurally integrated in large technological systems. Once established as a common intervention, such technologies tend to persist. As many people will have experienced, both in their everyday life and in broader socio-political contexts, it is very difficult even to challenge, let alone change, situations where "the computer says no."[30] That is to say, the common use of such tracking apps will severely impede future democratic choices of (far) less disciplining technologies (see Radder, 2019, chaps. 2 and 6). The general point (iv) rejects the use of the app if it lacks broad public support and deliberative justification. In my terms, if it is not in the public interest. On the basis of the discussion in this section, the following conclusions can be drawn with respect to this point. First, the tenor of the public debate and, especially, the empirical survey about people's willingness to install the app suggest that there was no solid electoral support for the app. Second, the serious problems concerning its overall effectivity show that the app really lacks cogent deliberative support. Finally, persisting worries about increasing social control and discipline, which endanger basic rights and liberties, show that its constitu-

---

[28] In addition, of course, to the longer-term search for a working and safe vaccine.

[29] As a consequence of the crisis, governments are now suddenly "forced" to support air companies with huge amounts of tax money for the purpose of "saving employment." The real problem, however, is that they have, for decades, willfully discarded opportunities for a more gradual and therefore less consequential limitation of air traffic by systematically ignoring sustained criticism of its clear drawbacks (and by privileging it instead with special tax advantages and regulating policies).

[30] For many disturbing examples, especially of the impacts on underprivileged social groups, see O'Neil (2016).

tional and legal support may be legitimately questioned. In sum, the claim that the tracking app serves a public interest proves to be highly questionable.

An obvious implication is that this conclusion applies even more to adopting the Chinese approach by legally enforcing the use of the apps and the follow-up procedures. In an informative article, Verhagen and Van Gestel (2020) address the role of cultural differences in Western and Asian approaches to coping with the corona crisis.[31] In contrast to the more individualistic Western culture, Asian culture is characterized as being more collectivistic. The advantage of the latter, it is claimed, is that it serves "the greater good." The problem is that the meaning of this notion remains fully unclear. Does the totalitarianism of the current Chinese government advance the greater good of all the different groups of its population? The notion of the public interest allows a substantial qualification of this problematic argument. It requires democratic support for technological projects of basic significance. And because genuine democratic processes (both voting and deliberation) are essentially collective, the requirement that such projects should be in the public interest goes against an excessive, libertarian individualism.

References

Breitenstein, P. H. (2017). Genealogie als kritische Theorie: Methodologische Überlegungen zur Gesellschaftskritik bei Foucault. In S. Ellmers & P. Hogh (Eds.), *Warum Kritik? Begründungsformen kritischer Theorien* (pp. 258–280). Velbrück Wissenschaft.

Brown, M. B. (2009). *Science in democracy: Expertise, institutions, and representation*. MIT Press.

Cartwright, N. (1983). *How the laws of physics lie*. Clarendon Press.

Doppelt, G. (2006). Democracy and technology. In T. J. Veak (Ed.), *Democratizing technology: Andrew Feenberg's critical theory of technology* (pp. 85–100). State University of New York Press.

Dreyfus, H. L., & Rabinow, P. (1982). *Michel Foucault: Beyond structuralism and hermeneutics* (2nd ed.). University of Chicago Press.

Ellmers, S., & Hogh, P. (Eds.). (2017). *Warum Kritik? Begründungsformen kritischer Theorien*. Velbrück Wissenschaft.

Feenberg, A. (1995). *Alternative modernity: The technical turn in philosophy and social theory*. University of California Press.

Feenberg, A. (1999). *Questioning technology*. Routledge.

Feenberg, A. (2006). Reply to critics. In T. J. Veak (Ed.), *Democratizing technology: Andrew Feenberg's critical theory of technology* (pp. 175–210). State University of New York Press.

Feenberg, A. (2008). Comments. *Social Epistemology, 22*(1), 119–124.

---

[31] Unfortunately, the article carries the misleading heading *Nothing (or something) to hide*. This way of phrasing wrongly suggests that, concerning privacy, the "natural" state is one of complete openness, and that anything less is socially, or even morally, blameworthy. This view, which can frequently be found in popular debates on digital technologies, ignores the unforeseeable but undesirable *interventions* by all kinds of different parties, which will, or may, result from a lack of privacy. As social beings, most people like to share (at least some of) their experiences and information. At the same time, many do not want to be forced to do so. Therefore, an appropriate account of privacy should address what people wish, and do not wish, to share, with whom, through which media, and for how long.

Foucault, M. (1982). The subject and power. In H. L. Dreyfus & P. Rabinow (Eds.), *Michel Foucault: Beyond structuralism and hermeneutics* (2nd ed., pp. 208–226). University of Chicago Press.

Foucault, M. (1984). What is enlightenment? In P. Rabinow (Ed.), *The Foucault reader* (pp. 32–50). Pantheon Books.

Freyenhagen, F. (2017). 'Aber was das Unmenschliche ist, das wissen wir sehr genau': Zur Normativitätsproblematik bei Adorno. In S. Ellmers & P. Hogh (Eds.), *Warum Kritik? Begründungsformen kritischer Theorien* (pp. 229–257). Velbrück Wissenschaft.

Gabriëls, R. (2019). Eine differenzierte Einbettung der kommunikativen Macht: Über die Rezeption von Habermas in den Niederländen. In L. Corchia, S. Müller-Doohm, & W. Outhwaite (Eds.), *Habermas global: Wirkungsgeschichte eines Werks* (pp. 492–519). Suhrkamp.

Habermas, J. (1972). *Knowledge and human interests*. Heinemann.

Habermas, J. (1984). *The theory of communicative action* (Vol. 2). Polity Press.

Habermas, J. (1998). Three normative models of democracy. In J. Habermas (Ed.), *The inclusion of the other: Studies in political theory* (pp. 239–252).

Hagen, A. (2019). How to engage in practices of critique? From a universal conception of the good life to the contestation of universals. *Krisis: Journal for Contemporary Philosophy, 2019*(1), 2–14. Accessible at https://krisis.eu/how-to-engage-in-practices-of-critique-from-a-universal-conception-of-the-good-life-to-the-contestation-of-universals/

Krige, J., & Leonelli, S. (2021). Mobilizing the transnational history of knowledge flows: COVID-19 and the politics of research at the borders. *History and Technology, 37*(1), 125–146.

Kuhne, F. (2017). Moral im 'Kapital'? Hat Marx' Kritik der politischen Ökonomie normative Grundlagen? In S. Ellmers & P. Hogh (Eds.), *Warum Kritik? Begründungsformen kritischer Theorien* (pp. 190–209). Velbrück Wissenschaft.

Kunneman, H. (1996). *Van theemutscultuur naar walkman-ego: Contouren van postmoderne individualiteit*. Boom.

Law, J. (1987). Technology and heterogeneous engineering: The case of Portuguese expansion. In W. E. Bijker, T. P. Hughes, & T. Pinch (Eds.), *The social construction of technological systems* (pp. 111–134). MIT Press.

Marcuse, H. (1968[1964]). *One dimensional man*. Sphere Books.

Mitcham, C. (1994). *Thinking through technology: The path between engineering and philosophy*. University of Chicago Press.

Mladenović, B. (2007). 'Muckraking in history': The role of history of science in Kuhn's philosophy. *Perspectives on Science, 15*(3), 261–294.

Morgan, M. S., & Morrison, M. (Eds.). (1999). *Models as mediators: Perspectives on the natural and social sciences*. Cambridge University Press.

Mouter, N., Kessels, R., De Wit, A., Rotteveel, A., Lambooij, M., & Collewet, M. (2020). Grote verdeeldheid over wenselijkheid van de corona-app. *Economisch Statistische Berichten, 105*(4788), 394–397.

Muller, C., et al. (2020). *Inzake: COVID-19 tracking- en tracingapp en gezondheidsapp*. Accessible at https://tinyurl.com/uasrb3h; for an English translation, see https://tinyurl.com/urj2tor

O'Neil, C. (2016). *Weapons of math destruction: How big data increases inequality and threatens democracy*. Penguin Books.

Perry, S. (2020). Coronavirus app has changed the way the Isle of Wight sees itself. *The Guardian* (May 9). Accessible at https://www.theguardian.com/world/2020/may/09/coronavirus-app-has-changed-the-way-the-isle-of-wight-sees-itself

Radder, H. (1985). Wat is progressieve filosofie? *Krisis, nr.18*, 101–106.

Radder, H. (1986). Experiment, technology and the intrinsic connection between knowledge and power. *Social Studies of Science, 16*(4), 663–683.

Radder, H. (1996). *In and about the world: Philosophical studies of science and technology*. State University of New York Press.

Radder, H. (2006). *The world observed/The world conceived*. University of Pittsburgh Press.

Radder, H. (2008). Critical philosophy of technology: The basic issues. *Social Epistemology,* *22*(1), 51–70.

Radder, H. (2019). *From commodification to the common good: Reconstructing science, technology, and society.* University of Pittsburgh Press.

Radder, H. (2021). Which science, which democracy, and which freedom? In P. Hartl & Á. T. Tuboly (Eds.), *Science, freedom, democracy* (pp. 113–134). Routledge.

Rijksoverheid. (2021). *Factsheet CoronaMelder.* Accessible at https://www.rijksoverheid.nl/onderwerpen/coronavirus-app/resultaten-praktijktest-en-uitvoeringstoets-coronamelder. Accessed February 4, 2021.

Roothaan, A. C. M. (2016). Ethics, critical theory, and postcolonial criticism of development. In M. Masaeli (Ed.), *Globality, uneven development, and ethics of duty* (pp. 47–65). Cambridge Scholars Publishing.

Rosling, H. (2019). *Factfulness: Ten reasons we're wrong about the world – And why things are better than you think.* Flatiron Books.

Siffels, L. E. (2021). Beyond privacy vs. health: A justification analysis of the contact-tracing apps debate in the Netherlands. *Ethics and Information Technology, 23*(supplement issue 1*)*, 99-103. Accessible at https://doi.org/10.1007/s10676-020-09555-x

Stupp, C. (2020). Coronavirus tracking apps raise questions about Bluetooth security. *Wall Street Journal* (April 30). Accessible at https://www.wsj.com/articles/coronavirus-tracking-apps-raise-questions-about-bluetooth-security-11588239000

Vaesen, K. (2011). The functional bias of the dual nature of technical artefacts program. *Studies in History and Philosophy of Science, 42*(1), 190–197.

Van Straten, E. (2020). *Risico's Corona tracking app.* Accessible at https://www.security.nl/posting/652884/Risico%27s+Corona+tracking+app

Verhagen, L. (2020). Corona-app is er, maar wat moeten we met de kritiek? *De Volkskrant* (August 18).

Verhagen, L., & Van Gestel, M. (2020). (N)iets te verbergen. *De Volkskrant* (May 23).

Wassens, R. (2020). Ministerie test 'notificatie-app' regionaal. *NRC* (May 30–31).

Weber, M. (1949[1904]). 'Objectivity' in social science and social policy. In M. Weber, *The methodology of the social sciences*, translated and edited by E.A. Shils and H.A. Finch (pp. 49–112). Free Press.

Chapter 4
Andrew Feenberg and the Distorted Democratization of Technology: Covid-19 and the Case of Hydroxychloroquine

Tina Sikka

Abstract This chapter examines the contentious Covid therapeutic, hydroxychloroquine, using Andrew Feenberg's theory of technological democratization. I explore whether the use of this experimental medicine is suitable, fit for, or reflective of a process of technological democratization in a manner that is similar to that of HIV/AIDS medicines and trials. In answering this, I draw on Feenberg's technological democratization thesis and extend his conception of care, bodily integrity, and communication in medicine using a reconstructed concept of care as expressed by feminist ethics. My central argument is that technological democratization of Covid-19 treatments and the underlying science has been made extremely difficult because hydroxychloroquine has become emblematic of polarized and polarizing political battles. In doing so, I articulate a model of "distorted technological democratization" to explain this phenomenon.

Keywords Andrew Feenberg · Democratization · Communication · Disinformation · Covid-19 · Hydroxychloroquine · Feminism · Science studies · Helen Longino · AIDS

4.1 Introduction

Socio-cultural understandings of Covid vaccines have been discussed in detail by a number of scholars, journalists, and commentators (Woods et al., 2020; Ali, 2020; Coconel Group, 2020). However, there is little analysis in the way of the myriad therapeutics that have been tested, used, rejected, and argued over ad nauseum. These include the much hyped, and then derided, hydroxychloroquine, remdesivir (another contentious medicine), convalescent plasma, and the steroid dexamethasone. These are technological artifacts that have come to stand as significant

T. Sikka (✉)
Department of Media, Culture, Heritage, Newcastle University, Newcastle upon Tyne, UK
e-mail: tina.sikka@newcastle.ac.uk

© The Author(s), under exclusive license to Springer Nature Switzerland AG 2022
D. Cressman (ed.), *The Necessity of Critique*, Philosophy of Engineering
and Technology 41, https://doi.org/10.1007/978-3-031-07877-4_4

markers of power, privilege, scientific rigour, and risk in our contemporary environment.

This chapter aims to address this lacuna by examining the category of the contentious Covid therapeutic, hydroxychloroquine, using Andrew Feenberg's theory of technological democratization. I ask whether the use of this experimental medicine is suitable, fit for, or reflective of a process of technological democratization in a manner similar to that of HIV/AIDS medicines and trials. In answering this, I draw on Feenberg's technological democratization thesis and extend his conception of care, bodily integrity, and communication in medicine using a reconstructed concept of care as expressed by feminist ethics. Bringing these two frameworks together, I argue, creates room for citizen engagement and the cultivation of prosocial technical formations that are currently lacking. My central argument is that the technological democratization of Covid-19 treatments and the underlying science has been made extremely difficult because hydroxychloroquine has become indicative of polarized and polarizing political battles. In doing so, I articulate a model of "distorted technological democratization" to explain this phenomenon.

I begin this chapter with a basic introduction to Andrew Feenberg's technological democratization thesis and his examination of HIV/AIDS medications as an exemplar of successful technological democratization. I then engage in a case study of the use of the anti-malaria drug hydroxychloroquine to treat Covid-19 which was transformed from a medicinal cause celebre to the butt of jokes in three short months. I argue that this medicine exemplifies a process "distorted technological democratization"which I define as a process of *nominal* democratization that is attenuated by one or more of the following: (1) Disinformation and political polarization; (2) Scientific and technological opacity; (3) Reified economic structures and imperatives; and (4) An ossified communication and media landscape. Using hydroxychloroquine as a case study, I argue, allows for the rigorous examination of this therapeutic in situ while also making room for a deeper understanding of larger problems and issues. This approach is also helpful as a method of theory development, particularly around the "distorted technological democratization"thesis, by permitting the examination of problems connected to "social interaction, historical processes, and organizational structures" (Feagin et al., 1991, p. 68; Gomm et al., 2000; Hyett et al., 2014).

In addition to the case study, I also draw on a combination of traditional/legacy and social media thematic analysis which Feenberg does not use in his own work but is important vis-à-vis conflicts and conversations around hydroxychloroquine wherein "distorted democratization" is taking place less through official fora and insurgent activism (e.g. ACT UP) than online via news media and Twitter. It is from this analysis that the themes of disinformation, political polarization, scientific and technological opacity, reified economic structures and imperatives, and an ossified communication and media landscape were first identified. Methodologically, I gathered data using keyword searches focusing on hydroxychloroquine in conjunction with terms like "trials," "use," "harm," "prophylactic," "prevention," "risk," "danger," "WHO," "FDA," "Trump," "hoarding," "capitalism," "profit," and "politicization" using Twitter's Analytics Platform, googlenews, and Nexis which were searched between the dates of March 1st and September 1st 2020. Media articles that fit the parameters of this study (i.e. focusing on the support or rejection of hydroxychloroquine) were then further categorized according to their use of polarizing language and the changing discourse around its use as a possible treatment. This multimodal approach was chosen because it allows for the excavation and discussion of the "intentions that lie behind the production of mass media documents…The usual strategy," as Pawson puts it, "is to pick on a specific area of reportage and subject it to a very detailed analysis in the hope of unearthing the underlying purposes and intentions of the authors of the communication" (Pawson, 1995, p. 107; Anderson, 2007; Vaismoradi et al., 2013). This mixture of inductive and deductive analysis also permits the identification of pertinent meta-themes around health, risk, and fear applicable to the discussion of technological democratization (Creswell & Plano Clark, 2011; Attride-Stirling, 2001).

It is important to point out that this analysis represents a probative and preliminary bit of research, rather than a systematic one. It aims to encourage further study not only of Covid-19 therapeutics, but other technologies and systems that have become denatured and tainted by political and economic distortions. Close content and discursive media analysis as well as perspectives from political economy and Science and Technology Studies (STS) would round out this prefatory research.

I close this chapter with the argument that intensifying Feenberg's framework and supplementing it with the tools proffered by a feminist ethic of care is ideal. This approach, I contend, offers a way in which to "rescue" the democratization thesis from the distorting effects of our contemporary milieu.

4.2 Technological Democratization

Andrew Feenberg's democratization thesis begins with the central premise that for technologies that claim rational neutrality under capitalism to be transformed toward pro-social ends three central principles need to be acknowledged. First, is that their current instantiation is neither universal nor immutable; second, that modern technologies contain formal biases such that they can claim rationality but still be biased towards certain ends. This explains, for example, how an algorithm used for hiring CEOs can reflect a formal bias but also be distorted since recommendations are based on the profiles of successful past hires (i.e. white men). Third, technologies tend to be underdetermined – meaning that more than one potential design can be adopted to realize a particular end. It is often the case that the systems and designs that are chosen, whether by intention or inertia, reflect existing capitalist social arrangements. The technical apparatuses governing medical and therapeutic decision making, according to Feenberg, tend to reflect this as well (Feenberg, 2008, 2020).

Proprietary technologies that are constituted by values and norms that reify profit and the application of "rational principles" are likely to be resistant to democratization and discount the importance of social wellbeing and human health. Feenberg proposes the theory of "technical codes" as a means to assess systems that operate at the nexus of ideology and technique such that the "technically coherent solution of a general type of problem," i.e. one that fits with existing technologies of control and exploitation, becomes the default choice (Feenberg, 2002, p. 20). In the case of medical decision making specific to medicines, it is an opaque, expert led, and control-oriented apparatus that dominates trials and experimentation that activism by patients suffering from HIV/AIDS, for example, were able to transform.

It is important to point out, however, that Feenberg retains a strong role for technical expertise in any process of democratization thereby eschewing charges of relativism. He writes,

> Technical democratization in capitalist societies involves the recovery of the overlooked considerations under pressure from the public. The demand for democratization is a claim for the extension and formal recognition of these *contributions of non-experts, not a rejection of the role of expertise* (Feenberg, 2020, emphasis added).

Feminist scientific ethics, of the kind I discuss towards the end of this chapter, are similarly attendant to practices that are necessary for empirical rigour in addition to social justice. This means that empirical success (i.e. reliability) remains an important part of scientific practice but, like Feenberg, feminist scholars like Helen Longino also maintain that the aims of research are often achievable by more than one theory (undetermination) and thus that alternative kinds of scientific practice are possible (Longino, 2002; Intemann, 2010).

Before delving into the case study, a short note on the practice of social struggle is needed. Practically, social struggle involving legal, physical, technical, and dialogic modes of intervention aim at the transformation of technocracy to democracy. Technologies can serve as "boundary objects," or what Feenberg refers to as bridging objects, that mean different things to different actors but which work to challenge cultural boundaries and hegemonic norms. Collective groups of actors are thus able to push against bureaucracy's legitimacy and "disrupt the networks and even reconfigure them in new form" (Feenberg, 2002, p. 34). In the next section, I turn to an example of successful technological democratization in action as a way to demonstrate what is possible before moving to an examination of the myriad political, bureaucratic, symbolic, and technological impediments that have made a similar process vis-à-vis experimental Covid-19 drugs like hydroxychloroquine fail so completely.

4.3   Case Study: HIV/AIDS

The most prominent case in the domain of medicine in which this kind of democratization has worked is HIV/AIDS medicines and trials in the late 1980s and 1990s in the US. Feenberg discusses this case in a number of books and articles in which he demonstrates how a group of highly motivated patients were able to pressure insular and expert-dominated groups of decision makers to work with them and reorient their practices in ways that better reflected the "symbolic dimension and caring functions" of medicine (Feenberg, 1992a, p. 319, 2020). The gay rights movement propelled the communicative power of these groups who had been disproportionately impacted by the AIDS epidemic and who argued the system needed to change. Feenberg explains how the formalization of the patient-provider relationship from one of care and communication to that of side-effects, efficacy, and binary logic had harmed HIV/AIDS patients. The gatekeeping of medical knowledge, which had been publicly justified as necessary to safeguard patient welfare, was revealed to be flawed and harmful.

In the end, groups like ACT UP (AIDS Coalition to Unleash Power) were able to persuade the government, The Food and Drug Administration (FDA), physician groups, and pharmaceutical companies, often using disruptive political tactics, that experimental medicines should be given to patients in line with "the ethical obligation to the patient" which they argued was "better fulfilled by extraordinary efforts to achieve a higher quality of consent rather than by restricting opportunities to participate in research" (Feenberg, 1992b, p. 213; Thorgaard, 2014). Paternalistic treatment, a singular focus on side-effects, and a narrow understanding of "efficacy" were critiqued so persuasively that the FDA ultimately modified some of its rules and regulations around access to experimental treatments. In the end, AIDS activists were able to form a kind of "counter hegemony" that democratized how medicines were tested and accessed (Thomson, 2000). What is most important here, and will be made clear further on, is the moral content of Feenberg's framework wherein he provides a "moral justification for taking AIDS victims" desire/demand for access as a good ethical reason for revising the technical code of clinical trials" and for treating medical care as a "non-commodity good" (Veak, 2012, pp. 97 & 129). This activism was mirrored by the Treatment Action Committee (TAC), who built on the work of gay rights and anti-apartheid advocates in post-apartheid South Africa by using the discourse of human rights to push for access to antiretroviral drug treatment as part of their newfound right to "gender equality and non-discrimination on the grounds of sexual orientation" (Mbali, 2005, p. 216).

Notably, the media in the early 1990s, after being roundly criticised by social scientists, activists, patients, and the medical community for perpetuating misinformation and homophobia, began to take on a more informative role as well (Berridge, 1992; Flowers, 2001). The Kaiser Family Foundation, in an exhaustive review of national coverage of HIV/AIDS between 1981 and 2002, found a marked change in coverage with a growing focus on prevention, the state of research (including the introduction of AZT), stigma, and transmission rather than a sensationalist focus on HIV/AIDS as a death sentence, as sinful, and as endemic to the gay community (Kaiser Family Foundation, 2004). This change in media discourse was supplemented by early Internet organizing and instances of activist-press cooperation, which James Gillett chronicles in his article "Media Activism and Internet Use by People with HIV/AIDS" (Gillett, 2003). It should be noted, however, that the mainstream press, while less sensational and stigmatizing – e.g. by no longer calling AIDS the "gay pneumonia," which 83% of stories did in 1981 (Brodie et al., 2004) – still did not engage with and cover many of the aspects of the disease that members of ACT UP tried desperately to get them to pay attention to including that of poverty, marginalization, racism, and inequality (Swain, 2005; Hertog et al., 1994).

The alignment of communicative structures, activism, relative media support, a more responsive government, and a medical community open to change created the conditions under which the hierarchical and generally opaque technical codes that governed clinical trials and access to experimental drugs could be infused with a discourse of mutuality, collective rights, human dignity, and an ethic of care (Doppelt, 2001). According to Annas (1989), the "code" that ultimately won out is one of therapy over research typified by one of ACT Up's slogans: "A Drug Trial is Health Care Too," and exemplified by the government's position, put forth by US Undersecretary of Health and Human Services, S. Jay Plager who stated that the new rules would give AIDS patients the ability to decide "whether they would rather take an experimental drug or die of the disease untreated" (Pear, 1987).

Taken together, this assemblage of relational factors produced the democratization of technology through the intromission of meaning and human relations into technical disciplines. For Feenberg, cases like this tell us what successful resistance consists of – namely, "the struggle for democratization of the institutions of a society based on technical rationality" (Feenberg, 2018, p. 125). While there were structures to support this struggle in the case of HIV/AIDS treatments, I argue that they are in a state of complete chaos as it relates to the technical systems governing therapeutic medications for Covid-19 as I demonstrate in the next section.

4.4   Hydroxychloroquine

Hydroxychloroquine is a drug that has been used to treat malaria, rheumatoid arthritis, and lupus since the mid 1950s. What propelled its initial use in 2020 was a small scale, non-randomised/controlled trial conducted by Didier Raould, the director of the Research Unit in Infectious and Tropical Emergent Diseases (URMITE) in Marseille, France. Raould claimed that a combination of hydroxychloroquine and an antibiotic had "cured" patients of Covid-19 (this was the research President Donald Trump later favourably tweeted about) (Schneider, 2020).

Using hydroxychloroquine to treat Covid-19 was subsequently supported by several governments and bodies including the Indian government and the FDA. After a Brazilian trial was stopped due to participants developing heart complications, other trials of the drug were also terminated and a critical eye was cast as to the rigour of past published research – some of which was retracted (BHF, 2020; Tleyjeh et al., 2020; Mahase, 2020). The WHO, the Journal of the American Medical Association (JAMA), and the FDA have all officially made statements underlining that there is no evidence to suggest hydroxychloroquine is effective in preventing or treating Covid-19 and have warned against its use due to potential side effects (Howard, 2020; Gander, 2020).

Contra to the case of HIV/AIDS, Feenberg's democratization of technology thesis vis-à-vis hydroxychloroquine falls short not because it is philosophically or empirically suspect or unrealizable, but because of a number of mitigating factors that have interfered with and distorted the productive role pro-social democratic values and norms usually play. These structural impediments have led to a distortion of democratic processes that are unique to this particular moment and require concerted action to rectify. Nolen Gertz (2020) makes a similar argument in his analysis of digital behaviours in which he calls for the addition of a theory of technological mediation to make up for Feenberg's focus on the internet's *potentialities* rather than on its *actualities*. In what follows, I discuss four of these impediments, namely: disinformation and political polarization, scientific and technological opacity, reified economic structures and imperatives, and an ossified communication and media landscape. These structures were identified out of the thematic and interpretative analysis of media texts as described above and integrated into the analysis below.

Before diving into that analysis, however, a short clarification is needed around the kind of comparison that is being presented. My position, and thus the position of this piece, is three-fold: first, that I agree with Feenberg that the transformation of clinical trials and medical practice surrounding HIV/AIDS treatment represents an example of the democratization of technology realized in practice; second, that any discussion of democratic access to or public participation in research for Covid-19 has been impeded by disinformation and political polarization, scientific and technological opacity, reified economic structures and imperatives, and an ossified communication and media landscape; and third, that for the treatment of this virus to be pulled out of the realm of differentiated access, opaque science, profit imperatives, and venal politicization, an *even more robust* conception of technological democratization is needed. I contend that this can be found in work being done by feminist scientists and particularly in feminist conceptions of care (Bentley & Watts, 1986; de La Bellacasa, 2011). In the next sections I discuss the central distortions preventing democratization before presenting a way forward.

4.5 Disinformation and Political Polarization

Disinformation and political polarization have been placed together here because of the way in which they operate synergistically to produce a distinctive kind of democratic distortion. Disinformation can be defined quite simply as the spreading of content that is false, misleading, or both, in a manner that has negative effects on the public (Wardle & Derekshan, 2017; Gelfert, 2018). The most harmful forms of disinformation, or what has now come to be called "fake news," are those that aim to manipulate, exploit, and/or profit. I maintain that political polarization, specific to ones' engagement with media, and particularly in the form informational echo chambers, has politicized information in deleterious ways (Bail et al., 2018; Tucker et al., 2018).

Yet former President Trump, even having recovered from the disease, continues to tout the combination of hydroxychloroquine as "one of the biggest game changers in the history of medicine" (Trump, 2020). In July of 2020, Jair Bolsonaro,

President of Brazil, posted a "good morning" tweet accompanied by the message "RT-PCR for Sars-Cov 2: negative" and a picture of him giving a thumbs up while holding a box of hydroxychloroquine (Bolsonaro, 2020). El Salvador's President told reporters in Late May that he takes the drug to avoid getting the virus stating, "I use it as a prophylaxis, President Trump uses it as a prophylaxis, most of the world's leaders use it as a prophylaxis" (National Post, 2020).

Early missteps regarding the communication of risk by scientific bodies helped to create the conditions for the politicization of treatments and trials which was reflected in the media. This disinformation piggybacks on a citizenry that has low trust in institutions and for whom information tends to be siloed. The amplificatory effects of social media, which I discuss further below, exacerbated the effects of this disinformation. As such, the more fervent public statements from doctors, the WHO, and the FDA who warned against taking hydroxychloroquine in light of it having been shown to not only not work and having serious side effects including heart arrythmia, had less of an impact (FDA, 2020; DW, 2020).

News outlets on the right and *far right* in particular also spread this disinformation. *The Washington Times* (not to be mistaken with *The Washington Post*), for example, wrote a scathing and widely cited article criticizing the Governor of Michigan for moves to limit access to hydroxychloroquine while right wing pundits like Ben Shapiro, Charlie Kirk and even Rudy Giuliani touted its effectiveness on Twitter and through posts online (Richardson, 2020; Walker & Hagle, 2020). Criticism of such coverage and of the misplaced faith in the efficacy of the drug was also prevalent from left and centre left media. Vox's *Recode*, for example, wrote a critical piece calling this a manufactured "infodemic" and criticising Trump's "endless stream of false and misleading claims" citing a Cornell University study which found that "the president is the single largest driver of misinformation around Covid" (Clark Estes, 2020). *Scientific American*, in an article debunking Covid myths, linked to articles demonstrating hydroxychloroquine's potentially dangerous side effects and which they described as functioning as an anchoring bias wherein "his [Trump's] supporters may be more likely to believe reports that confirm their views rather than those that challenge them" (Lewis, 2020). Australian University expert on drug resistance Gaetan Burgio tweeted, "This is insane!" in response to Trump's support for hydroxychloroquine (MSNBC, 2020), while MSNBC tweeted Governor Andrew Cuomo's comments calling the president's push for hydroxychloroquine "one of the greatest charades" that "turned out to be a total failure" (MSNBC, 2020).

In contrast to HIV/AIDS trials and medicine, polarized discourse around hydroxychloroquine transformed it into a symbol of fealty or loyalty to the political right. While maintaining a common identity across political boundaries (i.e. we can all agree it is a medicine), hydroxychloroquine has taken on different kinds of meaning for diverse groups while also exerting an agential power of its own in networks that privilege some kinds of knowledge over others (Martin, 2010; Barad, 2003; Foucault, 1980). For Feenberg, it is this agonistic struggle that is a marker of "progress" wherein a technology's "cultural horizon" is validated. "A recontextualizing critique of technology," he argues, "can uncover that horizon, demystify the illusion of technical necessity, and expose the relativity of the prevailing technical choices" (Feenberg as cited by Winston & Edelbach, 2011, p. 86). My argument is that the kinds of disinformation and polarized discourse that exists in our current context makes this a distant prospect.

In political terms, as Amanda Marcotte (2020) in an article for *Salon* argues, this miracle cure discourse is indicative of a kind of social Darwinism that is constitutive of modern conservatism in which a shared sense of mutuality and sociality is seen as morally suspect. One specific drug, made available to the deserving citizenry of an authentic (white) America, is thus seen as ideologically preferable to community quarantine, mask wearing, social distancing, or public health care. This is in keeping with the harsh social libertarianism that has become inextricably linked to Trump's base as well as other technological controversies in which high-tech solutions are preferred over easier and less costly ones (e.g. climate change and education) (Liao et al., 2012; Ames, 2019).

To be sure, similar kinds of rumour mixed with pseudoscience were also rife in the case of HIV/AIDS – albeit on an attenuated scale. Myths that HIV does not cause AIDS, that both are a racist colonial plot, and that AIDS could be cured by alternative medicines were initially ubiquitous but by 1985 much less potent (Mbali, 2004; Kalichman, 2018). After 5 years of denial, disregard, and the pathologization, in 1985 US President Ronald Reagan delivered a speech on AIDS to the WHO announcing new investments in medication as well as a nation-wide HIV prevention campaign (Fourie & Meyer, 2016). The AIDS activist group ACT UP formed in the same year.

I contend that it is the persistence of the denialism and strength of the rumours and innuendo around Covid-19 that is of particular interest here. What little hope there was for public actors to step in and push for citizen engagement, access to decision making structures, and the ability to see the science has been negated by a unique form of disinformation on the internet that includes hedging (and outright lying) on the part of political leaders.

Sattui et al. (2020) characterizes the public discourse about hydroxychloroquine, inclusive of social media, as a pendulum which moved from early adopters and rising interest, to scepticism and rejection upon increasing evidence that hydroxychloroquine was harmful (Mahevas et al., 2020). The intersection of unconstrained public discourse, dis- and misinformation, and political polarization, such that if you supported Trump you supported access to hydroxychloroquine (even while denying the severity and even existence of the virus – an odd contradiction), created conditions quite unlike that of HIV/AIDS drugs and the ethos of equality, justice, and care ACT UP and other groups were able to command.

4.6   Scientific and Technological Opacity

The lack of scientific clarity around Covid-19 played out through journalism wherein on-the-spot reporting of "science under pressure" reflected generalized medical uncertainty, retracted findings, the lack of double-blind protocols, and questions concerning methodological rigour. These uncertainties were not success-fully communicated to a terrified public by politicians, medical advisors, or the media. Malecki, Keating, and Safdar, in their article "Crisis Communication and Public Perception of COVID-19 Risk in the Era of Social Media," write about haz-ard and outrage as the primary responses of a public ill-served by experts and cri-tique the media for whom social media became a key resource for Covid-19-related information (Malecki et al., 2021). If we compare the headlines towards the end of June and July 2020 between *The New York Times* and *Fox News,* for example, the implications of this conflicting "information tornado" becomes patently clear. On the one hand, reportage coming out of *The New York Times* included articles like, "Misleading Virus Video, Pushed by the Trumps, Spreads Online," "The Doctor Behind the Disputed Covid Data," and "Malaria Drug Promoted by Trump Did Not Prevent Covid Infections, Study Finds" (Frenkel & Alba, 2020; Gabler & Caryn Rabin, 2020; Grady, 2020), while followers of *Fox News*, were given such articles as, "Hydroxychloroquine Helped Save Coronavirus Patients Study Shows," "Hydroxychloroquine Could Save Up To 100,000 Lives If Used For COVID-19," and "Twitter Deletes Video Promoted by Trump on Hydroxychloroquine Use For Coronavirus" (Re, 2020a, b; DeMarche, 2020). This conflicting coverage subse-quently spread through social media and made the science a central site of conflict.

Another important element in the case of Covid-19 is that much of the reporting on the science, and even some treatment decision making, had drawn on research that had not yet been peer reviewed (often called pre-print publications). Again, the institutional pressure to reach dispositive finding, coupled with emergency-like con-ditions, meant that misinformation proliferated and was picked up and transmitted by experts and non-experts alike. This created an informational tornado around Covid-19 such that the integrity of science became seen as suspect which has had serious implications as it relates to public trust (Brainard, 2020; Bagdasarian et al., 2020). Sattui et al. demonstrate this concretely in their study of hydroxychloroquine and public discourse in which they found that the chaotic dissemination of results had showed a significant amount of public scepticism about who to trust and what constitutes valid, actionable advice (Sattui et al., 2020).

Moreover, science journalists, often tasked with communicating medical infor-mation to the public, have had to contend with new agendas and norms that are resistant to complexity while working on a schedule that makes extensive research difficult (Bennett, 1996; Waisbord, 2018). While most scientists bemoan this dis-juncture, some have "become facile at guiding coverage. For example, if an impor-tant paper is about to be published in a journal, scientists may hire consultants to help them" market "their discovery to the press by appealing to the demand for news pegs" (Dunwoody, 2014). A group of scientists from Oxford University, for example, took to the media to criticize the politicization of the drug and are hoping to begin a trial testing the drug's preventative potential. One of the investigators, in a particularly compelling pull quote, declared, "I don't think there's been a more politicised and controversial medicine than hydroxychloroquine" (Giles, 2020).

It is also the case that there is generally poor public understanding of the scien-tific process specifically in terms of the determination of validity, the status of truth, and scientific change – all of which have been exacerbated in the digital age (Ward, 2008; Vos, 2012; Latour, 2000). Added to this is the fact that journalists tend to be reticent to take on the role of educator preferring, instead, to cover startling findings and innovations or two-siding scientific controversies (as in the case climate change) rather than engaging in complex discussions of truth and the vagaries of scientific practice (Boykoff & Boykoff, 2004). These uncertainties and misunderstandings have played out in the media around hydroxychloroquine, often with politics playing a disproportionate role. In May, *The New York Times* put out an article criticizing *Fox News* for their contradictory coverage of hydroxychloroquine in an article titled "At Fox News, Mixed Message on Malaria Drug: 'Very Safe' vs. 'It Will Kill You'" (Grynbaum, 2020). The article itself, while critical of the network for its inconsistent messaging, also told a story of the drug as a contested object while also clarifying that the best scientific evidence has not shown there to be any benefits vis-à-vis treatment for Covid-19. It is difficult, however, for measured language to counter the inflammatory rhetoric from the right with media pundits like Laura Ingram tweeting patently false statements like, "Why does CNN not recognize the COVID patients all over the world who have walked out of hospitals testing negative after being treated with hydroxychloroquine?" (Barr, 2020).

In an odd turn of events, *The Lancet*, in June, made a commitment to strengthen its peer review process after having to retract research connecting hydroxychloroquine to an increase Covid-19 deaths due to problems with how the data was gathered and interpreted. These kind of lapses are not unexpected given how much of the reporting relies on "real world data" consisting of "huge registers where you can collect data from multiple sources" (Furlong & Manancourt, 2020). Overall, hydroxychloroquine represents a paradigmatic case of these failings in that it has suffered from a convergence of fast changing science, much of it in preprint, journalistic failings in the form of mixed messages and lack of context, and social media dis- and misinformation. And yet, as we know from cases like that of HIV/AIDS, "laypersons" are perfectly capable of understanding complex science if given the opportunity and access. Democratization of information and citizen engagement forms a central plank of Feenberg's approach and has been a part of citizen science initiatives around environmental change, public planning, and technologies (Wachelder, 2003; Mahajan et al., 2020).

4.7   Reified Economic Structures

The third central difference between the cases articulated by Feenberg which further exemplifies the myriad barriers preventing a similar kind of democratization of Covid-19 are that medicines in this category reflect the exploitative economic imperatives and structures of the neoliberal responsibilization of health. In the first instance, it is the existent model of proprietary and privatized, profit-driven medical science that I contend is ill-suited to the practice of democratization. What is needed for this to occur is the implementation of a publicly funded, collective, and engaged process of therapeutic testing and reporting wherein patient knowledge is seen as valid. This could prove to be beneficial both on the level of practice and as it relates to knowledge generation (Fiske et al., 2019).

The critique of big-pharma, which often relies on initial public funding, is replete with stories of profit-driven decision-making, secrecy, and exploitative pricing (Calnan, 2020; Owen, 2014). Discussion of these reified economic structures, imperatives, and interests are reflected in the media – albeit rather obliquely when it comes to mainstream media. What is required is a clear discussion of the ways in which capitalism operates to incentivise particular forms of exploitation and profiteering in the context of medicines, including that of hydroxychloroquine. In April 2020, an article that bucked the trend in *Quartz,* traced the economic "scramble" for companies to land on "the drug" that will make contracting Covi-19 much less potentially lethal (McDonnell, 2020), while a piece in *Vanity Fair* took on "Big Pharma" for cravenly "meddling in trials, manipulating patents, and inflating prices." The author cites a prominent Mississippi doctor who castigated the industry for "trying to intimidate research and corner a market on potential COVID-19 therapies" for profit (Falzon, 2020).

On a socio-psychological level, it is necessary to consider the unique pressures Covid-19 continues to place on existing economies and how this has multiplied the burdens on individual people who have turned to populism and angry denialism rather than the kinds of democratic organization Feenberg calls for. As a result of Covid-19, the IMF estimates that the economic output of the global economy will have suffered from $28 trillion or 4.4% in lost output, increased poverty, "with close to 90 million people expected to fall into extreme deprivation this year," and skyrocketing unemployment (Elliot, 2020). According to the Congressional Budget Office and the Census Bureau, the US has suffered from skyrocketing unemployment levels (close to 15%), a year to year loss in GDP (which is not expected to reach 2019 levels again until after 2022), food scarcity, delays in receiving medical care, and housing insecurity (CBO, 2020; Census Bureau, 2020). This despair has been funnelled into anti-mask protests, Covid-19 scepticism, and the embrace of one-off and unsubstantiated "cures" supported by a President who has his own agenda including personal reputation, national prestige, and partisan one-upmanship. It also creates the conditions for the responsibility for health to be placed on individual self-regulation rather than collective provisioning. This is in line with the neoliberal focus on health as an individual responsibility which, "seeks to reshape the sensemaking, even subjectivity, of individuals in such a manner as to shift their explanations for problems or concerns from external agents or forces to the self" (Pyysiäinen et al., 2017, p. 215). This has resulted in the minimising of economic structures and institutions as responsible and the disproportionate funnelling of communicative energies to countering disinformation rather than finding ways for individuals to become involved in citizen-led initiatives such that their "social influences" are "exercised and institutionalised in…rationalised domains" as exemplified by the HIV/AIDS case (Feenberg, 2010, p. 48).

As a result of the capriciousness of the virus and the attendant uncertainty it cultivates, there has also been considerable pressure on the medical and pharmaceutical community to find that "magic bullet" from governments who control much of the funding. All of this is occurring removed from public scrutiny inline with economic imperatives. Having already demonstrated the media's initial hyping of the drug, it would behoove us to look more closely at the ways in which hydroxychloroquine has been discursively tied to the economy by the media in ways that make patient-led democratization significantly more difficult. First, is the lack of a clear articulation of the connection between economic imperatives and medical approvals. One of the few clear discussions of profits by the media was in an article printed by *The Irish Times* who noted that Novartis, a Swiss pharmaceutical company, saw sales of this generic drug jump with their core net income increasing by a third to $3.55 billion and sales rising by 13 per cent to $12.3 billion in the immediate aftermath (The Irish Times, 2020). This highly commercialised and profit driven model of treatment normalizes a model of health and health provision which does not meet public needs or the principles of democratisation. As Heled, Santos Rutschman, and Vertisnsky argue, the "consequences of relying on profit-driven entities for pandemic preparedness and response are playing out in real time, as companies that are inadequately prepared search their existing portfolios for any candidates that may be used in the treatment of COVID-19 with an eye on stock price, government emergency funds, and the lure of lucrative pricing for any resulting treatments" (Heled et al., 2020, p. 22), rather than focusing on other incentives (i.e. new modes of drug discovery), compulsory licensing, public drug manufacturing, patient input etc.

Taken together, these economic structures and imperatives, often aided by the media and supported by disinformation as well as scientific and technological opacity, denudes Feenberg's democratizing project.

4.8   Ossified Internet Space

Another central attenuating factor that decreases the possibility for the democratization of technology is the influence of the internet. Specifically, it is the fracturing and polarization of social media that has created the conditions for the proliferation of conspiracy theories, myths, and disinformation. Much has been written about how the affordances of digital technologies have facilitated instantaneous, anonymous, and collaborative communication by "allow[ing] people to interpret events in ways that align well with their worldviews..." which, unfortunately, is also "an important component of maintaining conspiracy beliefs" (Wood, 2013, p. 31; Mosinzova et al., 2019; Vosoughi et al., 2018). In the case of Covid-19 treatments, discursive struggle online are rife. Moreover, much of the research in this area supports critiques of the internet as exemplary of capitalist exploitation, mundane "clicktivism," harmful disinformation, meaningless frivolity, and consumerism (Fuchs, 2010, 2019; Dean, 2013; McChesney, 2013). In the case of Covid-19 drugs, the early headlines around hydroxychloroquine are illuminating. In March of 2020, the headlines from major media outlets touting hydroxychloroquine were shared most online included, "WHO officials enrol first patients from Norway and Spain in 'historic' coronavirus drug trial" (Lovelace Jr. & Feuer, 2020), from *CNBC*, "India bans export of malaria drug Trump touted as coronavirus treatment" (Bloomberg, 2020), from *Bloomberg*, and, from *Forbes*, "New York to begin clinical trials for coronavirus treatment Tuesday, Cuomo says" (Voyttko, 2020). Demand for hydroxychloroquine surged, supported by the President of the United States who claimed to have taken it as a prophylactic, leading to fears of shortages and even reported hoarding by doctors (Sander et al., 2020; McCarthy & Greve, 2020).

Social media discourse reflected this with almost half a million tweets containing "hydroxychloroquine" or "hcq" being tweeted between the end of February and May 22nd and which and cleaved along partisan lines (with Republican identified individuals being particularly active) (Hamamsy & Bonneau, 2020). Much of this was sparked by a viral video, shared by Trump and viewed over 17 million times on Facebook, from the dubious group "America's Frontline Doctors," which promoted hydroxychloroquine's use as a treatment for Covid-19. The video was subsequently taken down by social media companies including Twitter and Facebook for spreading false health information (Giles et al., 2020). Similar tweets from Trump supported Dr. Stella Immanuel, a Houston doctor, who has vocally denied the efficacy of masks and promoted hydroxychloroquine while also making statements linking gynaecological issues to dreams about demons. These were also taken down (Sommer, 2020).

However, while my own belief is that the internet poses more of a problem, particularly at this point, I would be remiss if I did not emphasise Feenberg's own account of the online spaces as potential sites of democratization and asserted human agency. The fact that the internet is socially constructed by us, Feenberg argues, is manifested in the myriad ways in which participants have been able to, one, "overcome the narrowness of the communication channel; two, "actively appropriate what is available…in unexpected ways;" three, "create dynamic and rich communities by inventing new forms of expression and through interactive negotiation of meanings, norms and values;" and four, produce "distinctive normative orientations established and maintained through written ethical codes" (think memes and gifs) (Feenberg, 2002, p. 187; Inefuku, 2017; Mößner & Kitcher, 2017; Beck et al., 1994). As noted earlier, it was through conversations that took place in early forms of the internet that helped propel ACT UP and other forms HIV/AIDS activism. For this to occur vis-à-vis Covid-19 and hydroxychloroquine, it would require a significant amount of action by social media platforms. This might include removing falsehoods, reconfiguring their click-driven algorithms, and working with communities to cultivate communicative norms that take advantages of open information in ways that break down silos and encourage "shared civic goals" which "offer opportunities for citizens" (Nelson et al., 2017, p. 4; Smith, 2013; Allen & Light, 2015).

It is also worth pointing out the need to mediate between Feenberg's (qualified) optimism about the democratic potential of digital spaces and more critical perspectives. This tension has persisted for decades. In a recent 2019 article, Feenberg acknowledges the troubling rise of disinformation, propaganda, and populism but argues that "the network still functions very much like a common carrier for online communities of all sorts" and continues to allow "Billions of people communicate more or less freely on the network" (Feenberg, 2019, p. 236). This perspective sits in sharp contrast to the work of scholars like Jodi Dean and Christian Fuchs who have been particularly critical of the internet asserting that the "the Internet has emerged as the zero institution of communicative capitalism…that materialize[s] specific fantasies of unity and wholeness as the global" (Dean, 2005, p. 67; Fuchs, 2009). My approach aims to find a middle ground between the two in light of novel impediments to democratization and the possibilities offered by feminist science.

4.9 Ways Forward and Concluding Thoughts

As I have argued thus far, Andrew Feenberg's democratization of technology thesis represents an ideal and practicable approach to transforming exploitative technologies and systems into ones that, in a non-essentialist fashion, are able to establish a new governing technical code. Participation by diverse publics motivated to present their interests irrespective of social position and/or specific technical expertise is critical. Moreover, these transformations must go beyond formal changes to "alter actual power relations in the technical sphere" in pursuit of a new democratic dispensation (Feenberg, 2001, p. 192). The case of HIV/AIDS demonstrates, in clear terms, that this is possible. AIDS patients, many from the gay community, were able to transform the technocratic regulatory system constituted by a seemingly entrenched and inflexible system through direct action, organization, protest, and lobbying. While successful in this case, similar kinds of medical democratization (e.g. Covid-19) have failed due to structural and political distortions including that of disinformation and political polarization, scientific and technological opacity, reified economic structures and imperatives, and an ossified communication and media landscape. The controversy over hydroxychloroquine represents a limit case, one that speaks to the deterioration of democratic norms such that a possible treatment for Covid-19 becomes a tool of political partisanship (e.g. disinformation) wherein interested parties are stymied by scientific gatekeeping and polarized by (social) media misinformation and economic interests.

The case of hydroxychloroquine is thus a prime example of "distorted technological democratization" which, to be clear, does not imply a failure of Feenberg's thesis but a call for its intensification. Through concerted activism, AIDS patients were able to draw ethical norms back into the public realm, something Steven Epstein (1996) examines in even more detail in his book *Impure Science: AIDS Activism and the Politics of Knowledge*. For the technical codes governing Covid-19 treatment to be reconstructed, I contend that we must double down on the recuperative ethos Feenberg articulates throughout his work. Namely, via the reclamation of medicine's ethic of care which avoids the inevitability of substantivism and the naïve neutrality of instrumentalism for an approach that underscores "communicative interaction," healing, and patient agency such that "ethical, technical, and epistemic procedures" are able to "merge into a single complex that supplies knowledge and protects human dignity" (Beira & Feenberg, 2018, pp. 21–22).

For such an ethic to be in place for hydroxychloroquine and other Covid-19 drugs, what would be required is a well-defined process – something feminist scientists have focused on for decades. I argue that Helen Longino's feminist model of scientific community is sufficiently capacious and structured to mitigate most of the worst discursive and socio-material excesses surrounding the treatment of Covid-19. The principles governing these communities include the "design and constitution of a [in this case scientific] community" or communities; an openness to "transformative criticism;" participatory public forums for debate; "publicly recognized standards" of discourse and validity; openness to criticism, and "equality of intellectual authority" (i.e. layperson *and* expert) (Longino, 1994, pp. 144–145; Wray, 1999). Were these principles implemented via, for example, citizen science panels wherein the fallibility and constructedness of science is established as a priori, perhaps the most egregious forms of disinformation, scientific confusion, corporate self-interest, and media disorder around hydroxychloroquine could have been avoided. Moreover, this approach would institutionalise, in a flexible manner, the lessons of HIV/AIDS treatment such that democratization is able to take place precisely along these lines. Moreover, the competing interests that exploited Covid-19 and our fractured media environment for political gain could have been identified, heard, and dismissed for violating established public standards.

The other resource that feminist science offers which might help unravel and delegitimize some of the more egregious manifestations of fake news and moderate the impacts of a sensationalist media ecosystem is an ethic of care. While Feenberg gestures to this, I think highlighting the significance of care in the history of medicine, particularly when read through a feminist lens, offers a way to strengthen democratic principles and processes. These norms, to be clear, are not biological or reflective of a female essence, but a product of gendered experiences, structural conditions, and concomitant convictions about what society should look like. They include interconnected relationality, concern for the vulnerable, mutual trust, and care for the other (Koehn, 1998; Larrabee, 2016; de La Bellacasa, 2011). Medicine and other healing occupations are entangled with care as defined by caring for, looking after, protecting, and, sustaining others (Murphy, 2015). As Ben-Porath argues, care is "grounded in those practices of human life that are reflective of our dependence on each other" rather than the solipsistic neoliberalism that characterises the case of hydroxychloroquine (Ben-Porath, 2008, p. 65; Spivak, 1999).

This approach also cultivates an understanding of action, inclusive of what takes place in a laboratory or medical office, as a search for what "may allow the other to exist beyond the grasp of the present" and thus protect them from harm (Ahmed, 2007, pp. 139–140, 2013). Applications of Sara Ahmed's framework to medical ethics has produced movements that see patients as situated agents, as part of the healing process, imbricated with larger networks of people, institutions, and things (Gregg et al., 2010; Harbin et al., 2012). This ethic is precisely what AIDS activists were able to recuperate and what I contend needs to be recaptured vis-à-vis Covid-19 treatment. The scarcity of democratization, communication, and care are what have produced the very distortions that were then exacerbated by media biases, political polarisation, neoliberal profiteering, and unclear science. Transforming these processes should be the priority of scholars, scientists, politicians, media actors, and the public going forward.

References

Ahmed, S. (2007). A phenomenology of whiteness. *Feminist Theory, 8*(2), 149–168.

Ahmed, S. (2013). *The cultural politics of emotion*. Oxfordshire Routledge.

Ali, I. (2020). The COVID-19 pandemic: Making sense of rumor and fear: Op-Ed. *Medical Anthropology*, 1–4.

Allen, D., & Light, J. (2015). In D. Allen & J. S. Light (Eds.), *From voice to influence: Understanding citizenship in a digital age*. The University of Chicago Press.

Ames, M. G. (2019). *The charisma machine: The life, death, and legacy of one laptop per child*. MIT Press.

Anderson, R. (2007). *Thematic content analysis (TCA): Descriptive presentation of qualitative data*. http://www.wellknowingconsulting.org/publications/pdfs/ThematicContentAnalysis.pdf

Annas, G. J. (1989). Faith (healing), hope and charity at the FDA: The politics of AIDS drug trials. *Villanova Law Review, 34*, 771.

Attride-Stirling, J. (2001). Thematic networks: An analytic tool for qualitative research. *Qualitative Research, 1*, 385–405.

Bagdasarian, N., Cross, G. B., & Fisher, D. (2020). Rapid publications risk the integrity of science in the era of COVID-19. *BMC Medicine, 18*(1), 1–5.

Bail, C. A., Argyle, L. P., Brown, T. W., Bumpus, J. P., Chen, H., Hunzaker, M. F., et al. (2018). Exposure to opposing views on social media can increase political polarization. *Proceedings of the National Academy of Sciences, 115*(37), 9216–9221.

Barad, K., (2003). Posthumanist performativity: Toward an understanding of how matter comes to matter. Signs: *Journal of women in culture and society, 28*(3), pp.801–831.

Barr, J. (2020, March 30). Twitter pulls down Laura Ingraham's coronavirus tweet for violating policy. *Hollywood Reporter*. Retrieved from https://www.hollywoodreporter.com/news/twitter-pulls-down-laura-ingrahams-coronavirus-tweet-violating-policy-1287441

Beira, E., & Feenberg, A. (Eds.). (2018). *Technology, modernity, and democracy: Essays by Andrew Feenberg*. Rowman & Littlefield.

Bennett, W. L. (1996). An introduction to journalism norms and representations of politics. *Political Communication, 13*, 373–384.

Ben-Porath, S. (2008). Care ethics and dependence—Rethinking jus post bellum. *Hypatia, 23*(2), 61–71.

Beck, U., Giddens, A., & Lash, S. (1994). Reflexive modernization: Politics, tradition and aesthetics in the modern social order. California: Stanford University Press

Bentley, D., & Watts, D. M. (1986). Courting the positive virtues: A case for feminist science. *European Journal of Science Education, 8*(2), 121–134.

Berridge, V. (1992). AIDS: History and contemporary history. In G. Herdt & S. Lindenbaum (Eds.), *The time of AIDS: Social analysis, theory and method*. Sage.

BHF. (2020). Why hydroxychloroquine isn't a "miracle cure" for coronavirus. *BHF*. Retrieved from https://www.bhf.org.uk/informationsupport/heart-matters-magazine/news/behind-the-headlines/coronavirus/coronavirus-and-hydroxychloroquine

Bloomberg. (2020, March 25). India bans export of malaria drug Trump touted as coronavirus treatment. *Bloomberg*. Retrieved from https://fortune.com/2020/03/25/coronavirus-hydroxychloroquine-trump-india-export/

Bolsonaro, J. M. [jairbolsonaro]. (2020, May 25). RT-PCR for Sars-Cov 2: negative. – GOOD MORNING EVERYONE [Tweet]. Retrieved from https://twitter.com/jairbolsonaro/status/1286994557440348160

Boykoff, M. T., & Boykoff, J. M. (2004). Balance as bias: Global warming and the US prestige press. *Global Environmental Change, 14*, 125–136.

Brainard, J. (2020, October 8). Researchers face hurdles to evaluate, synthesize COVID-19 evidence at top speed. *Science.* Retrieved from https://www.sciencemag.org/news/2020/10/researchers-face-hurdles-evaluate-synthesize-covid-19-evidence-top-speed

Brodie, M., Hamel, E. B., Kates, L. A., Altman, J., & Drew, E. (2004). AIDS at 21: Media coverage of the HIV epidemic 1981–2002. *Columbia Journalism Review, 42*(6), A1–A8.

Calnan, M. (2020). Rationing, regulating and Big Pharma. In *Health policy, power and politics: Sociological insights.* Emerald Publishing Limited.

CBO. (2020, May 19). Interim economic projectins for 2020 and 2021. *Congressional Budget Office.* Retrieved from https://www.cbo.gov/publication/56351

Census Bureau. (2020, August 12). Household pulse survey. *US Census Bureau.* Retrieved from https://www.census.gov/data-tools/demo/hhp/#/?measures=HIR

Clark Estes. (2020, October 7). Hydroxychloroquine conspiracies are back, but Trump's the patient now. *Vox's Recode.* Retrieved from https://www.vox.com/recode/2020/10/7/21504748/hydroxychloroquine-trump-covid-treatment-misinformation

Coconel Group. (2020). A future vaccination campaign against COVID-19 at risk of vaccine hesitancy and politicisation. *The Lancet. Infectious diseases, 20*(7), 769.

Creswell, J. W., & Plano Clark, V. L. (2011). *Designing and con- ducting mixed methods research.* Sage.

de La Bellacasa, M. P. (2011). Matters of care in technoscience: Assembling neglected things. *Social Studies of Science, 41*(1), 85–106.

Dean, J. (2005). Communicative capitalism: Circulation and the foreclosure of politics. *Cultural Politics, 1*(1), 51–74.

Dean, J. (2013). Whatever blogging. *Digital Labor: The Internet as Playground and Factory, 127*, 46.

DeMarche, E. (2020). Twitter deletes video promoted by Trump on hydroxychloroquine use for coronavirus. *Fox News.* Retrieved from https://www.foxnews.com/politics/twitter-post-that-seems-to-show-doctors-praising-hydroxychloroquine-use-for-coronavirus.

Doppelt, G. (2001). What sort of ethics does technology require? *The Journal of Ethics, 5*(2), 155–175.

Dunwoody, S. (2014). Science journalism: Prospects in the digital age. In M. Bucchi & B. Trench (Eds.), *Routledge handbook of public communication of science and technology* (2nd ed., pp. 27–39). Routledge.

DW. (2020, May 25). WHO stops clinical test for malaria drug hydroxychloroquine. *DW.* Retrieved from https://www.dw.com/en/who-stops-clinical-test-for-malaria-drug-hydroxychloroquine/a-53564772

Elliot, L. (2020, October 13). IMF estimates global Covid cost at $28tn in lost output. *The Guardian.* Retrieved from https://www.theguardian.com/business/2020/oct/13/imf-covid-cost-world-economic-outlook

Epstein, S. (1996). *Impure science: AIDS, activism, and the politics of knowledge* (Vol. 7). University of California Press.

Falzon, D. (2020, April 24). "I take that as a threat": Big Pharma is meddling in the race for a COVID-19 treatment. *Vanity Fair.* Retrieved from https://www.vanityfair.com/news/2020/04/big-pharma-meddling-in-race-for-covid-19-treatment.

FDA. (2020, July 1). FDA cautions against use of hydroxychloroquine or chloroquine for COVID-19 outside of the hospital setting or a clinical trial due to risk of heart rhythm problems. *FDA.* Retrieved from https://www.fda.gov/drugs/drug-safety-and-availability/fda-cautions-against-use-hydroxychloroquine-or-chloroquine-covid-19-outside-hospital-setting-or

Feagin, J. R., Orum, A. M., & Sjoberg, G. (Eds.). (1991). *A case for the case study*. UNC Press Books.

Feenberg, A. (1992a). Subversive rationalization: Technology, power, and democracy. *Inquiry, 35*(3–4), 301–322.

Feenberg, A. (1992b). On being a human subject: Interest and obligation in the experimental treatment of incurable disease. *Philosophical Forum., 23*(3), 213–230.

Feenberg, A. (2018). Marcuse: Reason, imagination, and utopia. Radical Philosophy Review. https://doi.org/10.5840/radphilrev201891190

Feenberg, A. (2001). Democratizing technology: Interests, codes, rights. *The Journal of Ethics, 5*, 177–195. https://doi.org/10.1023/A:1011908323811

Feenberg, A. (2002). *Transforming technology: A critical theory revisited*. Oxford University Press.

Feenberg, A. (2008). From critical theory of technology to the rational critique of rationality. *Social Epistemology, 22*(1), 5–28.

Feenberg, A. (2010). Marxism and the critique of social rationality: From surplus value to the politics of technology. *Cambridge Journal of Economics, 34*(1), 37–49.

Feenberg, A. (2019). The internet as network, world, co-construction, and mode of governance. *The Information Society, 35*(4), 229–243.

Feenberg, A. (2020). Critical Constructivism, Postphenomenology and the Politics of Technology. Techné: Research in Philosophy and Technology. https://doi.org/10.5840/techne2020210116

Fiske, A., Prainsack, B., & Buyx, A. (2019). Meeting the needs of underserved populations: Setting the agenda for more inclusive citizen science of medicine. *Journal of Medical Ethics, 45*(9), 617–622.

Flowers, P. (2001). Gay men and HIV/AIDS risk management. *Health, 5*(1), 50–75.

Foucault, M. (1980). Power/knowledge: Selected interviews and other writings, 1972–1977. Vintage.

Fourie, P., & Meyer, M. (2016). *The politics of AIDS denialism: South Africa's failure to respond*. Routledge.

Frenkel, S., & Alba, D. (2020, July 29). Misleading virus video, pushed by the Trumps, spreads online. *The New York Times*. Retrieved from: https://www.nytimes.com/2020/07/28/technology/virus-video-trump.html

Fuchs, C., (2009). Information and communication technologies and society: A contribution to the critique of the political economy of the Internet. *European journal of communication, 24*(1), pp.69–87.

Fuchs, C. (2010). Labor in informational capitalism and on the Internet. *The Information Society, 26*(3), 179–196.

Fuchs, C. (2019). *Nationalism on the Internet: Critical theory and ideology in the age of social media and fake news*. Routledge.

Furlong, Á., & Manancourt, V. (2020, June 30). Lancet to change peer review process following COVID-19 retraction. *Politico*. Retrieved from https://www.politico.eu/article/lancet-review-process-following-covid-19-saga-coronavirus/

Gabler, E., & Caryn Rabin, R. (2020, July 27). The doctor behind the disputed covid data. *The New York Times*. Retrieved from https://www.nytimes.com/2020/07/27/science/coronavirus-retracted-studies-data.html

Gander, K. (2020, August 4). CDC hydroxychloroquine prescription advice removed after 'unusual' move to issue guidelines without strong evidence. *Newsweek*. Retrieved from https://www.newsweek.com/cdc-hydroxychloroquine-prescription-advice-removed-after-unusual-move-issue-guidelines-without-1496736

Gelfert, A. (2018). *Fake news: a definition. Informal Logic, 38*(1), 84–117.

Gertz, N. (2020). Democratic potentialities and toxic actualities. *Techné: Research in Philosophy and Technology, 24*(1/2), 178–194.

Giles, C. (2020, August 8). Hydroxychloroquine being 'discarded prematurely', say scientists. *BBC News*. Retrieved from https://www.bbc.co.uk/news/health-53679498

Giles, C., Saradarizadeh, G., & Goodman, J. (2020, July 28). Hydroxychloroquine: Why a video promoted by Trump was pulled on social media. *BBC News*. Retrieved from https://www.bbc.co.uk/news/53559938

Gillett, J. (2003). Media activism and internet use by people with HIV/AIDS. *Sociology of Health & Illness, 25*(6), 608–624.

Gomm, R., Hammersley, M., & Foster, P. (Eds.). (2000). *Case study method: Key issues, key texts*. Sage.

Grady. (2020, June 3). Malaria drug promoted by Trump did not prevent covid infections, study finds. *The New York Times*. Retrieved from https://www.nytimes.com/2020/06/03/health/hydroxychloroquine-coronavirus-trump.html

Gregg, M., Seigworth, G. J., & Ahmed, S. (Eds.). (2010). *The affect theory reader*. Duke University Press.

Grynbaum, M. M. (2020, May 19). At Fox News, mixed message on malaria drug: 'Very safe' vs. 'it will kill you.' *The New York Times*. Retrieved from https://www.nytimes.com/2020/05/19/business/media/coronavirus-hydroxychloroquine-fox-news.html

Hamamsy, T. & Bonneau, R. (2020). Twitter activity about treatments during the COVID-19 pandemic: case studies of remdesivir, hydroxychloroquine, and convalescent plasma. medRxiv.

Harbin, A., Beagan, B., & Goldberg, L. (2012). Discomfort, judgment, and health care for queers. *Journal of Bioethical Inquiry, 9*(2), 149–160.

Heled, Y., Rutschman, A. S., & Vertinsky, L. (2020). The problem with relying on profit-driven models to produce pandemic drugs. *Journal of Law and the Biosciences, 7*(1), lsaa060.

Hertog, J. K., Finnegan, J. R., Jr., & Kahn, E., Jr. (1994). Media coverage of AIDS, cancer, and sexually transmitted diseases: A test of the public arenas model. *Journalism Quarterly, 71*(2), 291–304.

Howard, J. (2020, September 30). Hydroxychloroquine didn't prevent Covid-19 among health care workers in new study. *CNN*. Retrieved from https://edition.cnn.com/2020/09/30/health/hydroxychloroquine-covid-prevention-jama-study/index.html

Hyett, N., Kenny, A., & Dickson-Swift, V. (2014). Methodology or method? A critical review of qualitative case study reports. *International Journal of Qualitative Studies on Health and Well-Being, 9*(1), 23606.

Inefuku, H. W. (2017). Globalization, open access, and the democratization of knowledge. *Educause Review, 52*(4), 62.

Intemann, K. (2010). 25 years of feminist empiricism and standpoint theory: Where are we now? *Hypatia, 25*(4), 778–796.

Kaiser Family Foundation. (2004). *AIDS at 21: Media coverage of the HIV epidemic 1981–2002*. Retrieved from www.cjr.org/issues

Kalichman, S. C. (2018). 19 "HIV does not cause AIDS": A journey into AIDS denialism. *Pseudoscience: The Conspiracy Against Science*, 419.

Koehn, D. (1998). *Rethinking feminist ethics: Care, trust and empathy*. Psychology Press.

Larrabee, M. J. (Ed.). (2016). *An ethic of care: Feminist and interdisciplinary perspectives*. Routledge.

Latour, B. (2000). When things strike back: A possible contribution of 'science studies' to the social sciences. *The British Journal of Sociology, 51*(1), 107–123.

Lewis, T (2020, October 12). Eight persistent COVID-19 myths and why people believe them. *Scientific American*. Retrieved from https://www.scientificamerican.com/article/eight-persistent-covid-19-myths-and-why-people-believe-them/

Liao, S. M., Sandberg, A., & Roache, R. (2012). Human engineering and climate change. *Ethics, Policy & Environment, 15*(2), 206–221.

Longino, H. (1994). The fate of knowledge in social theories of science. In F. F. Schmitt (Ed.), *Socializing epistemology: The social dimensions of knowledge* (pp. 135–158). Rowman and Littlefield.

Longino, H. E. (2002). *The fate of knowledge*. Princeton University Press.

Lovelace Jr., B., & Feuer, W. (2020, March 20). WHO officials enroll first patients from Norway and Spain in 'historic' coronavirus drug trial. *CNBC*. Retrieved from https://www.cnbc.com/2020/03/27/who-officials-enroll-first-patients-from-norway-and-spain-in-historic-coronavirus-drug-trial.html.

Mahajan, S., Kumar, P., Pinto, J. A., Riccetti, A., Schaaf, K., Camprodon, G., et al. (2020). A citizen science approach for enhancing public understanding of air pollution. *Sustainable Cities and Society, 52*, 101800.

Mahase, E. (2020). Covid-19: WHO halts hydroxychloroquine trial to review links with increased mortality risk. *BMJ (Clinical Research ed.), 369*, m2126–m2126.

Mahevas, M., Tran, V. T., Roumier, M., et al. (2020). Clinical efficacy of hydroxychloroquine in patients with covid-19 pneumonia who require oxygen: Observational comparative study using routine care data. *BMJ., 14*(369), m1844.

Malecki, K., Keating, J. A., & Safdar, N. (2021). Crisis communication and public perception of COVID-19 risk in the era of social media. *Clinical Infectious Diseases, 72*, 697–702.

Marcotte, A. (2020). Behind the right's obsession with a miracle cure for coronavirus: It's not just about Trump. *Salon*. Retrieved from https://www.salon.com/2020/04/03/behind-the-rights-obsession-with-a-miracle-cure-for-coronavirus-its-not-just-about-trump/

Martin, B. R. (2010). The origins of the concept of 'foresight' in science and technology: An insider's perspective. *Technological Forecasting and Social Change, 77*(9), pp.1438–1447.

Mbali, M. (2004). AIDS discourses and the South African state: Government denialism and post-apartheid AIDS policy-making. *Transformation: Critical Perspectives on Southern Africa, 54*(1), 104–122.

Mbali, M. (2005). The treatment action campaign and the history of rights-based, patient-driven HIV/AIDS activism in South Africa. In *Democratising development: The politics of socio-economic rights in South Africa* (pp. 213–243). Martinus Nijhoff.

McCarthy, T., & Greve, J. E. (2020, May 19). Trump is taking tydroxychloroquine, White House confirms. *The Guardian*. Retreived from https://www.theguardian.com/us-news/2020/may/19/trump-hydroxychloroquine-covid-19-white-house

McChesney, R. W. (2013). *Digital disconnect: How capitalism is turning the internet against democracy*. The New Press.

McDonnell, T. (2020, April). A 'bridge to a vaccine': The race to roll out antibody-based Covid-19 drugs. *Quartz*. Retrieved from https://qz.com/1835197/pharma-companies-race-to-roll-out-antibody-based-covid-19-drugs/

Mosinzova, V., Fabian, B., Ermakova, T., & Baumann, A. (2019, February 3). Fake news, conspiracies and myth debunking in social media-a literature survey across disciplines. *Conspiracies and myth debunking in social media-a literature survey across disciplines.*

Mößner, N., & Kitcher, P. (2017). Knowledge, democracy, and the internet. *Minerva, 55*(1), 1–24.

MSNBC [MSNBC]. (2020, October 14). NY Gov. Cuomo says the Trump admin.'s push of hydroxychloroquine was "one of the greatest charades" and "turned out to be a total failure" [Twitter moment]. Retrieved from https://twitter.com/MSNBC/status/1316445946239234050

Murphy, M. (2015). Unsettling care: Troubling transnational itineraries of care in feminist health practices. *Social Studies of Science, 45*(5), 717–737.

National Post. (2020, May 27). 'Why?' El Salvador's Bukele says world leaders being told to take hydroxychloroquine, while public is warned away from it. *National Post*. Retrieved from https://nationalpost.com/news/world/why-el-salvadors-bukele-says-world-leaders-being-told-to-take-hydroxychloroquine-while-public-is-warned-away-from-it

Nelson, J. L., Lewis, D. A., & Lei, R. (2017). Digital democracy in America: A look at civic engagement in an internet age. *Journalism & Mass Communication Quarterly, 94*(1), 318–334.

Owen, T. (2014). The 'access to medicines campaign' vs Big Pharma. *Critical Discourse Studies, 11*(3), 288–304.

Pawson, R. (1995). Methods of content/document/media analysis. *Developments in Sociology, 11*, 107–107.

Pear, R. (1987, May 21). U.S. to allow use of trial drugs for AIDS and other terminal ills. *The New York Times*. Retrieved from https://www.nytimes.com/1987/05/21/us/us-to-allow-use-of-trial-drugs-for-aids-and-other-terminal-ills.html.practice?

Pyysiäinen, J., Halpin, D., & Guilfoyle, A. (2017). Neoliberal governance and 'responsibilization' of agents: Reassessing the mechanisms of responsibility-shift in neoliberal discursive environments. *Distinktion: Journal of Social Theory, 18*(2), 215–235.

Re, G. (2020a). Hydroxychloroquine helped save coronavirus patients, study show. *The New York Times*. Retrieved from https://www.foxnews.com/politics/hydroxychloroquine-helped-save-coronavirus-study

Re, G. (2020b). Hydroxychloroquine could save up to 100,000 lives if used for COVID-19. *Fox New*. Retrieved from https://www.foxnews.com/politics/hydroxychloroquine-helped-save-coronavirus-study

Richardson, V. (2020, March 29). Michigan Gov. Gretchen Whitmer hit for hydroxychloroquine crackdown as debate escalates. *Washington Examiner*. Retrieved from https://www.washingtontimes.com/news/2020/mar/29/gretchen-whitmer-michigan-governor-hydroxychloroqu/

Sander, T., Armstrong, D., & Kofman, A. (2020, March 24). Doctors are hoarding unproven coronavirus medicine by writing prescriptions for themselves and their families. *Propublica*. Retrieved from https://www.propublica.org/article/doctors-are-hoarding-unproven-coronavirus-medicine-by-writing-prescriptions-for-themselves-and-their-families

Sattui, S. E., Liew, J. W., Graef, E. R., Coler-Reilly, A., Berenbaum, F., Duarte-García, A., et al. (2020). Swinging the pendulum: Lessons learned from public discourse concerning hydroxychloroquine and COVID-19. *Expert Review of Clinical Immunology, 16*(7), 659–666.

Schneider, L. (2020, March 26). Chloroquine genius Didier Raoult to save the world from COVID-19. *For Better Science*. Retrieved from https://forbetterscience.com/2020/03/26/chloroquine-genius-didier-raoult-to-save-the-world-from-covid-19/

Smith, A. (2013). *Civic engagement in the digital age*. Retrieved from http://www.pewinternet.org/2013/04/25/civic-engagement-in-the-digital-age/

Sommer, W. (2020, July 28). Trump's new favorite COVID doctor believes in alien DNA, demon sperm, and hydroxychloroquine. *Daily Beast*. Retrieved from https://www.thedailybeast.com/stella-immanuel-trumps-new-covid-doctor-believes-in-alien-dna-demon-sperm-and-hydroxychloroquine

Spivak, G. C. (1999). *A critique of postcolonial reason: Toward a history of the vanishing present*. Harvard University Press.

Swain, K. A. (2005). Approaching the quarter-century mark: AIDS coverage and research decline as infection spreads. *Critical Studies in Media Communication, 22*(3), 258–262.

The Irish Times. (2020, April 28). Covid stockpiling of medicines boosts profit at Swiss group Novartis. *The Irish Times*. Retrieved from https://www.irishtimes.com/business/health-pharma/covid-stockpiling-of-medicines-boosts-profit-at-swiss-group-novartis-1.4239880

Thomson, I. (2000). From the question concerning technology to the quest for a democratic technology: *Heidegger, Marcuse, Feenberg*. *Inquiry, 43*(2), pp.203–215.

Thorgaard, K. (2014). Is evidence-based medicine about democratizing medical practice. *Outlines. Critical Practice Studies, 15*(1), 49–62.

Tleyjeh, I., Kashour, Z., AlDosary, O., Riaz, M., Tlayjeh, H., Garbati, M. A., et al. (2020). The cardiac toxicity of chloroquine or hydroxychloroquine in COVID-19 patients: a systematic review and meta-regression analysis. *medRxiv*.

Trump, D. J. [realDonaldTrump] (2020, March 21). HYDROXYCHLOROQUINE & AZITHROMYCIN, taken together, have a real chance to be one of the biggest game changers in the history of medicine. The FDA has moved mountains - Thank You! Hopefully they will BOTH (H works better with A, International Journal of Antimicrobial Agents) [Tweet]. Retrieved from https://twitter.com/realDonaldTrump/status/1241367239900778501

Tucker, J. A., Guess, A., Barberá, P., Vaccari, C., Siegel, A., Sanovich, S., et al. (2018, March 19). Social media, political polarization, and political disinformation: A review of the scientific literature. *Political polarization, and political disinformation: a review of the scientific literature*.

Vaismoradi, M., Turunen, H., & Bondas, T. (2013). Content analysis and thematic analysis: Implications for conducting a qualitative descriptive study. *Nursing & Health Sciences, 15*(3), 398–405.

Veak, T. J. (Ed.). (2012). *Democratizing technology: Andrew Feenberg's critical theory of technology*. Suny Press.

Vos, T. P. (2012). 'Homo journalisticus': Journalism education's role in articulating the objectivity norm. *Journalism, 13*(4), 435–449.

Vosoughi, S., Roy, D., & Aral, S. (2018). The spread of true and false news online. *Science, 359*(6380), 1146–1151.

Voyttko, L. (2020, March 22). New York to begin clinical trials for coronavirus treatment Tuesday, Cuomo says. *Forbes*. Retrieved from https://www.forbes.com/sites/lisette-voytko/2020/03/22/new-york-to-begin-clinical-trials-for-coronavirus-treatment-tuesday-cuomo-says/#4fd8878f4203

Wachelder, J. (2003). Democratizing science: Various routes and visions of Dutch science shops. *Science, Technology, & Human Values, 28*(2), 244–273.

Waisbord, S. (2018). Truth is what happens to news: On journalism, fake news, and post-truth. *Journalism Studies, 19*(13), 1866–1878.

Walker, A., & Hagle, C. (2020, March 30). Emergency authorization of antimalarial drugs for COVID-19 follows weeks of promotion on Fox News. *Media Matters*. Retrieved from https://www.mediamatters.org/coronavirus-covid-19/emergency-authorization-antimalarial-drugs-covid-19-follows-weeks-promotion

Ward, S. J. A. (2008). Truth and objectivity. In L. Wilkins & C. G. Christians (Eds.), *The handbook of mass media ethics* (pp. 71–83). Routledge.

Wardle, C., & Derekshan, H. (2017). *Information disorder: Toward an interdisciplinary framework for research and policy making*. Council of Europe report, DGI(2017)09.

Winston, M., & Edelbach, R. (2011). *Society, ethics, and technology*. Cengage Learning.

Wood, M. (2013). Has the internet been good for conspiracy theorising. *PsyPAG Quarterly, 88*, 31–34.

Woods, E. T., Schertzer, R., Greenfeld, L., Hughes, C., & Miller-Idriss, C. (2020). COVID-19, nationalism, and the politics of crisis: A scholarly exchange. *Nations and Nationalism, 26*, 807–825.

Wray, K. B. (1999). A defense of Longino's social epistemology. *Philosophy of Science, 66*, S538–S552.

Chapter 5
Beyond the Design Code: Critical Design and Democratic Rationalizations

Roy Bendor

Abstract This chapter suggests a few touchpoints between Andrew Feenberg's critical constructivism and current work in critical design, with the hope of initiating a productive dialogue between the two. Taking Feenberg's notion of the design code as an invitation for such an encounter, the chapter illustrates how designers can take an active, meaningful role in facilitating what Feenberg calls "democratic rationalizations" in three different ways: in (co)design processes designers facilitate collaboration and nurture collective creativity while maintaining as much autonomy for participants to make the design process and its outcomes their own. In (re)design processes designers may be forced to respond to their users' needs post-hoc, but they may also anticipate users' needs and even embrace opportunities to provoke users to reflect on the social, cultural and political implications of the technologies they use. Lastly, and most radically, (un)design processes ask designers to let go of (some of) their intentionality and seed potentials for users to appropriate the technologies they design. In Feenberg's terms, this equals an almost complete relinquishing of "operational autonomy" and the valorization of technical rationality that characterizes it.

Keywords Critical constructivism · Design theory · Design code · Andrew Feenberg · Designers · (co)Design · (re)Design · (un)Design · Double aspect theory · Operational autonomy

R. Bendor (✉)
Department of Human-Centred Design, Delft University of Technology,
Delft, The Netherlands
e-mail: r.bendor@tudelft.nl

5.1   A Missed Encounter

The point of departure for this chapter is the observation that critical constructivism has yet to figure significantly in design theory. This is rather strange given the emergence of critical forms of design that reaffirm design's political essence while aiming to leverage design's particular strengths into envisioning and pursuing "ways of being alert to the reconstructive possibilities and potentialities that may exist in the present" (White, 2015).[1] One would think that the shared prefix alone would be enough to trigger the curiosity of critical designers. More substantively, Feenberg's view that a critical deconstruction of the power relations that shape technology should lead to generative accounts of democratized technology, in essence pivoting "from demystification to a positive account of the political implications of the social contingency of the technosystem" (Feenberg, 2017, p. 162), is well aligned with critical design, while his analysis of technology and social rationality could help critical design further develop as a coherent programme. Furthermore, Feenberg's incorporation of social constructivist accounts of technology-in-the-making into the core of his theory – what Achterhuis (2001, p. 77) describes as his "internalist" account of technological change – would indicate a philosophical project much more attuned to the micropolitics of design. All this leads me to suggest that Feenberg's critical constructivism would be welcome by design theorists seeking a sophisticated, comprehensive account of technology and ideology.

And yet, for a variety of reasons on which I will not speculate here, this has not been the case. But I believe that to some extent, the missed encounter between critical constructivism and critical design results from the fact that Feenberg has yet to explicitly engage with designers as potential allies or partners in democratizing technology. This is not just a matter of publishing in design journals (something other philosophers of technology such as Peter-Paul Verbeek and Bruno Latour are more wont to do),[2] but has to do with the unmistakable impression one gets when reading Feenberg that the "technical disciplines" are not key actors in the kind of democratic transformation he envisions. I will say more about this below, but suffice it to say here that despite stating that "technosystem can only be changed *from within* through a gestalt switch" (Feenberg, 2017, p. 167; emphasis added), the real protagonist in Feenberg's account are the public and not engineers, technicians or

---

[1] The question of what counts as critical design is quite contentious, as is the question of how it relates to other, more or less congruent design approaches. The term was first suggested by Anthony Dunne in his dissertation (1997), submitted at Royal College of Art and later published as *Hertzian Tales: electronic products, aesthetic experience, and critical design* (2005 [1999]). Haylock (2018) suggests that the contours of critical design can be drawn along similar lines to those used by Horkheimer to distinguish "critical" theory from the affirmative role of "traditional" theory. In contrast, Bardzell and Bardzell (2013, p. 3300) put forth a more generative role for critical design as a set of practices that move beyond merely providing a social critique of design, as does Malpass (2017, p. 6) who approaches critical design as "a form of socially and politically engaged activity and creative activism."

[2] See for instance Verbeek (2015) and Latour (2008).

designers. This is despite the fact that, as I will argue here, the latter have much to offer the democratization of technology.

Accordingly, the aim of this chapter is not so much to provoke a rapprochement – after all, Feenberg's *oeuvre* is in no way antithetical to critical approaches to design – but more to bring critical constructivism and critical design into a productive dialogue. I am prompted in this by a recent change of terminology. Beginning with his *Questioning Technology* (1999), Feenberg referred to the way political interests are inscribed in technological artifacts as the negotiation of "technical codes," but in his more recent *Technosystem: The Social Life of Reason* (2017), technical codes were replaced by "design codes."[3] While this may seem at first unimportant or even pedantic, I take it as an indication of Feenberg's willingness to engage with designers. Such an engagement, I believe, may not only help bring Feenberg's thinking about design up to date with contemporary work in design theory, but it may help expand Feenberg's framework to include new designerly practices with democratic potentials. What I will call here (un)design denotes such new forms of democratic-potentiality-by-design, pointing to the limitations of Feenberg's theory while charting a path beyond the design code as it is currently articulated.

5.2   The Design Code

Feenberg's analysis of technological development and change is premised on what he identifies as a dialectical relation between the interests of those powerful enough to shape technology (by controlling the means of technological innovation, so to speak), and the needs of those pushed to the periphery of design, relegated to merely using the technology developed for them by others. The former pursue what Feenberg calls "operational autonomy", that is, the capacity to shape technology in ways that produce political legitimation and maintain existing social and economic hierarchies, while the latter aim to democratize technology by consolidating and leveraging their "participant interests" to "undermine the existing social hierarchy or to force it to meet needs it has ignored" (Feenberg, 1999, p. 76). Much in the tradition of the Frankfurt School, this dialectic is described in terms of contrasting rationalities. Whereas most technology conforms to the imperatives of efficiency and control upheld by "technological rationality" (a term borrowed from Marcuse), counterhegemonic forces pursue "democratic rationalizations" that seek to integrate the wellbeing, interests and desires of the public into the technical sphere as core criteria for technical innovation and progress.

The dialectic of operational autonomy and participant interests plays out and sediments in what Feenberg calls the "design code:"

---

[3] On first mentioning the design code in *Technosystem*, Feenberg clarifies that it is the same as what he called technical code "in some earlier presentations" (2017, p. 31).

> Design codes translate worldviews and interests between the everyday language of social actors and the technical languages of engineers or managers. The translation hides the social significance of the codes behind a veil of technical necessity. Codes that achieve unquestioned authority constitute a technical culture. The task of critique from Marx down to the present is to reverse the process of translation and reveal the code's human significance. (Feenberg, 2017, p. 57)

Because design codes are where function and meaning – the "double aspect" of technology – meet, they are key to understanding technology as "a site of social struggle" (Feenberg, 1999, p. 83). It is here, in other words, that different social, cultural and political understandings of what a technology is, what it does, and what it should do, are contested. It should be clear, however, that the design code is not an object per se, nor a 'thing' in the way a software code or program is; it is a fluid space or, more accurately, a marker of a dynamic process through which design is materialized. In this sense, Feenberg foreshadows what Björgvinsson et al. (2012, p. 102) describe as a move from "designing 'things' (objects) to designing Things (socio-material assemblies)."

5.3 What Designers Do

Feenberg's analysis of what we may call a generic design process foregrounds design as a series of ontological operations through which artifacts/systems are first created by bringing together material and functions ("objectification"), and then embedded in everyday life ("subjectivation"). On the surface, the main task of designers is to select among different material and functional possibilities based on the availability of resources and the anticipation of future use. However, in a more fundamental way, designers mediate between competing values, worldviews, interests and needs. This is done by what Feenberg recognizes as the material layering of function and meaning (again, corresponding to the "double aspect" of technology):

> The design process brings together meaning and matter. It is a terrain on which social groups express their worldview materially and advance their perspectives and their interests. Design proceeds through bringing together layers of function corresponding to the various meanings actors attribute to the artifact. (Feenberg, 2017, p. 32)

This view of design clearly brings out its political and ethical dimensions. These, Feenberg insists, are inseparable from functional and aesthetic considerations. The designer's role, it follows, is to select among competing requirements in ways that agree with the interests of the ideologies, organizations or individuals they serve. The outcome is overdetermined by the interests of those actors invested in the current system, yet tends to be concealed as universal requirements derived from neutral, technical rationality: "engineers and other technical workers act in accordance with technical disciplines. Those disciplines codify the impersonal power of technology in conformity with hegemonic social demands" (Feenberg, 2017, p. 194). Design codes of technologies developed under neocolonial, neoliberal capitalism ineluctably promote the values and survival of neocolonial, neoliberal capitalism.

The standard way of designing may be disrupted or reoriented by what Feenberg calls "democratic rationalizations" or "democratic interventions." Here, "public demands are translated into technically rational designs" (Feenberg, 2020), and may do so in three different modes: (1) *during* the design process design codes may be co-created by teams of designers and members of the public. *After* the conclusion of the design process, (2) design codes may be reluctantly renegotiated following public interventions in the environment within which technology functions (through "hearings, lawsuits, and boycotts"), or (3) may be forced to change (or 'hacked') by knowledgeable users who find that the technology does not fit their needs and desires in the real world (Feenberg, 2017, pp. 53–54). In Feenberg's work it is predominantly the second mode that is used to exemplify democratic rationalizations, and it in this sense that the public is the protagonist in the struggle to democratize technology. That said, designers play important roles in all three modes. This is often obscured by Feenberg's treatment of the technosphere in a rather undifferentiated manner despite the fact that software and hardware engineers, repair technicians, product managers and interaction designers (to name a few more recognizable technical professions) approach their work in significantly different ways that open up to considerably different democratic potentials. Without going into detail, and while acknowledging that the following remarks may themselves require some qualification,[4] some technical disciplines maintain a high-resolution view of the design process and its outcomes while others are immersed in the smallest of details. Some technical disciplines come into close contact with end-users while others not at all. And in some technical disciplines questions of ethics and politics are asked more frequently and vigorously than in others. Painting "engineers and other technical workers" (Feenberg, 2017, p. 194) with the same colours, therefore, obscures the fact that some are "closer to the machine" (Ullman, 1997) and may be better positioned to acquire and hold onto operational autonomy, and that within the technical disciplines themselves roles and responsibilities change over time (see for instance Sanders and Stappers (2012, pp. 23–25) in the context of design). In response, the remainder of this chapter discusses the three modes of democratic rationalization identified by Feenberg, albeit from the perspective of design. By doing so I will illustrate several touchpoints between critical constructivism and critical design, and argue that Feenberg's framework is missing an additional form of democratic rationalization that can only become apparent once the analysis allows for a more variegated view of design.

---

[4] A fully fleshed account of the differences between all technical disciplines is beyond the scope of this paper, but for a few influential views of what is specific about design as a form of thinking and doing see Attfield (2000), Cross (2006), Jones (2002), Margolin (2002), Schön (1988), and Simon (1969).

5.4   (co)Design

There are clear advantages to involving the public in design processes prior to the release of a technology (what Feenberg describes as "reflexive design" in earlier work, and what I will refer to here as (co)design – not to be confused with co-design which is a specific approach to engaging users in design). This allows designers to address user (or public) needs in a more comprehensive and meaningful manner and to prevent unanticipated (and potentially harmful) consequences. This has been one of the main drives behind the evolution of design from exclusively focusing on arti-facts and functions to more nuanced considerations of users and use contexts (Ehn, 2008; Sanders & Stappers, 2014). In Feenberg's terms, involving users in design processes reveals the contingency of technical rationality:

> Neither expert nor lay actors have a monopoly on rationality. Rationality is distributed across the lines dividing expert from lay and facts from values. The point is not that these lines are unreal. They are eminently real: without them, no modernity. But they are porous and allow translations in practice.... The expert/lay fragments into which rationality is split meet in a dialogue of multiple rationalities. And when the lines separating lay and expert are crossed, so are the lines separating discourse and material reality. (Feenberg, 2017, p. 134; 172)

Involving the public in design, in other words, appears as a welcome respite from the boundary policing of technical rationality, and in this sense provides a possible antidote to technocracy.

    While Feenberg has little to say about the dynamics of (co)design, the designer's role seems to change in one significant manner: while designers still mediate (between often conflicting demands) and translate (between meaning and function), they have to be much more attentive to the needs and opinions of others by, for instance, developing methods by which to better understand user values and finding ways to empathize with users. The inclusion of such methods (often under the ban-ner of "user-centered design") allows the outcomes of (co)design processes to be more "concrete" (in Gilbert Simondon's terms) by including additional functional-ities or "layers" that represent a wider variety of interpretations of, and expectations from, the technology (Feenberg, 2017, p. 57). But while for Feenberg (co)design is mainly a process of negotiating different forms of expertise and the rationalities to which they are coupled (see for instance Feenberg, 1999, pp. 123–125), design the-ory offers a much richer view of what happens when disparate actors are brought together in and through design. This, of course, also entails a richer view of the designer's role.

    In participatory design as it was developed in Scandinavia during the 1970s (Björgvinsson et al., 2012; Ehn, 1988, 2008),[5] difference in degrees and types of expertise was one of two main rationales for including workers in workplace decision-making and design. Providing space for tacit knowledge to influence the

---

[5] It should be noted that participatory design has more than one origin and "flavour" (see Luck, 2018).

design and introduction of new technologies and work processes could be seen, in Marxian terms, as a way to reverse deskilling and the alienation it causes. The other reason, however, was much more intuitive: "Participatory design started from the simple standpoint that those affected by a design should have a say in the design process" (Ehn, 2008, p. 94). Tasked with resisting and overturning unequal power relations in the workplace, the designer is seen as both a political and a moral agent. But outside the workplace not all those who are affected by a technology could be involved in its design (an issue to which I will return below).

Shifting our view from labour to creativity, we find a different kind of role for the designer in co-design. Here design is seen as a platform for "collective creativity" (Sanders & Stappers, 2012, p. 7), and the designer is viewed as a facilitator or a kind of midwife, creating the conditions for participants' creativity and natural design skills to come to the fore. This is consistent with what Ezio Manzini, one of the lead proponents of co-design, understands as a shift of the discipline from expert problem-solving to collective sense-making:

> Design is a culture and a practice concerning how things ought to be in order to attain desired functions and meanings. It takes place within open-ended co-design processes in which all the involved actors participate in different ways. It is based on a human capability that everyone can cultivate and which for some – the design experts – becomes a profession. *The role of design experts is to trigger and support these open-ended co-design processes, using their design knowledge to conceive and enhance clear-cut, focused design initiatives.* (Manzini, 2015 pp. 53–54; emphasis added)

When "everyone designs", as suggested by the title of Manzini's recent work, designers stop (at least theoretically) designing for others and instead instigate, frame and facilitate collaborative sense-making processes.

Thus articulated, it is no surprise that (co)design has become very popular: "today, in a networked society, *all design processes tend to become co-design processes*" (Manzini, 2015, p. 48; emphasis in origin). But as Rachel Charlotte Smith and Ole Iversen (2018) point out, because participation in design has become so ubiquitous the radical implications of its introduction have been diluted. It turns out that it is quite difficult to create and maintain "a space that permits a heterogeneity of perspectives among actors who engage in attempts to align their conflicting objects of design" (Björgvinsson et al., 2012, p. 102). Not all user-centered design moves beyond commercial optimization,[6] just as not all participatory design is empowering. The question, then, is not necessarily *how much* but *what kind* of participation. Some forms of participation are merely symbolic or tokenistic, while others achieve a lasting redistribution of power. Discerning which is which is far from straightforward, as evident in the proliferation of different frames or 'ladders' that aim to formalize public participation in a more meaningful manner.[7]

---

[6] This is one of the key differences between user-centered design and what Pacey (2001) calls "people-centered" technology.

[7] Sherry Arnstein's (1969) "ladder for citizen participation" was the first and most famous of those, but several others have been developed since (see for instance Cardullo & Kitchin, 2019).

This is perhaps where understanding the (co)designer as a mediator between the contrasting imperatives of operational autonomy and participant interests finds its limits. This is in large part because participant interests themselves are seldom homogenous and may encompass myriad other important dimensions: who gets to participate in (co)design? (or which participant interests are left out?), and how are the needs of vulnerable or marginalized communities addressed in processes that involve more powerful actors? Who represents the needs and perspectives of future generations, a significant issue when considering the long-term social and environmental impact of many of our technologies? These questions reveal the delicate position of the designer, called to equalize *within* the (co)design process uneven power relations that are well entrenched beyond the design process itself. Design becomes a microcosm of society. The picture becomes even more muddled when considering that (co)design increasingly involves nonhumans (robots or algorithms, for instance). How would these perspectives be represented? And lastly, what if a (co)design process, as well-meaning as it may be, is still part of a design culture that is steeped in unequal power relations and othering? This is the premise of Arturo Escobar's (2018) proposal for "autonomous design," "a design praxis with communities that has the goal of contributing to their realization as the kinds of entities they are" (p. 184). Here the designer is called upon not only to respect the community's identity, organizations, relations and practices, but to preserve space for a plurality of ontological and epistemological positions to emerge in and through the (co) design process: "Any redesigned design philosophy must articulate a critique of the rationalistic tradition and reconstruct its own mode of rationality, open to the plurality of modes of consciousness that inhabits the pluriverse" (ibid., p. 211). In Escobar's proposal we find the "lines separating lay and expert" (Feenberg, 2017, p. 172) are not only crossed but entirely redrawn – perhaps even erased – as design is becoming "pluriversal."

5.5   (re)Design

Despite the increasing popularity and purchase of (co)design, technological innovation is still mostly practiced as an exclusionary, exclusively expert undertaking, and democratic rationalizations, if they take place at all, do so after the fact, once the technology is already in the hands of the public. In such cases unanticipated consequences or unmaterialized potentialities may evoke a public response, or, in Feenberg's words, the public provides "a 'reality check' on the work of technical experts" (2017, p. 7) which, if successful, trickles down to the level of the artifact or system itself. Dissatisfaction with a technology may also provoke its hacking by more technically savvy actors. In both cases, what I will call here (re)design refers to a situation in which design codes are called into question and the technology is redesigned to better accommodate public concerns and needs. As Feenberg explains:

Technical artifacts and systems are situated in the lifeworld where they are appropriated or suffered by ordinary people. They become objects of explicit normative judgment when they cause problems or distress. These judgments activate the same or similar rational operations and categories that originally presided over the constitution of the technical functionalities. Everyone, not just experts, is capable of operations such as abstraction. In technosystem struggles rational principles in their original lifeworldly form are reapplied to the technosystem through judgments based on experience, often informed by counterexpertise. *The design process is reactivated through interventions based on the operations as they appear in the lifeworld.* (Feenberg, 2017 p. 169; emphasis added)

There are several important aspects to Feenberg's articulation of (re)design. First, public demands to renegotiate design codes emerge on the background of technology use in everyday life, where the artifact or system may not cohere with users' needs or expectations. In what I have argued elsewhere was evidence of Feenberg's grounding in phenomenology (Bendor, 2013), everyday experience becomes the ontological background for political action. Second, because renegotiating design codes often requires considerable political momentum, (re)design often entails the consolidation of publics around unsatisfactory technological experiences. Participant interests, in this formulation, do not precede but emerge as consequence of technology-in-use. Third, and much like in the case of (co)design, the consequences of (re)design are understood to be more mature or, in Simondon's language, more "concrete" technologies since they are able to accommodate a larger number of interests, meanings and functionalities into a single technical entity. As Feenberg explains, "When successful, the struggle [implied in (re)design] leads to a higher stage of development of the technical artifact or system, higher in the sense that it better fulfills the needs or recognizes the rights of individuals" (Feenberg, 2017, p. 170).

Processes of (re)design clearly hold an important place in the way Feenberg understands democratic rationalizations because it is here that what I call 'sociotechnical malleability' is matched by the public's willingness to shed their natural acquiescence to technology in favour of a more active position. This position would never seem reasonable if technology was not flexible enough to change; and technology would never change if the public did not assume agency in pursuit of their participant interests. It is therefore unsurprising that in Feenberg's work this is the type of democratic rationalization that takes centre stage. But what, according to Feenberg, is the role of the designer here? By and large it remains the same. New demands may not be invited but forced on designers (by pressure groups, new legislation, and so forth), but the latter will still perform the same operations: mediating (between conflicting interests), accommodating (those public interests that force the renegotiation of design codes), translating (between lay and expert terms), and layering (as many meanings and as much functionality as possible in a single artifact or system). Most importantly, designers remain reactive in relation to the entire process. If it was up to them, the technical code would never be renegotiated – at least not until this would be consistent with their own operational autonomy. Critical designers, however, have advocated for a much more active role for designers not only in responding to public demands to (re)design but in anticipating, or even inviting such demands.

Take for example the question of biases in design. Technologists have long been aware that their creations have important consequences for those who use them, or as Terry Winograd and Fernando Flores (1986, p. xi) famously wrote, "In designing tools we are designing ways of being." But instead of waiting for users to (sometimes painfully) discover those biases in everyday life, what started with "value-sensitive design" (Friedman, 1996) and can now be seen part of a more encompassing "design for values" perspective (Van den Hoven et al., 2015), asks designers to consider, anticipate and address such questions pre-emptively:

> Design and designers can be frontloaded with moral and social values and are able to realize these values and can be held accountable for doing so.... The hope is that failure by societal opposition during implementation and adoption can be a phenomenon of the past as value issues are being addressed from the start. (Van den Hoven et al., 2015, p. 3).

While not all possible biases can be detected in time (which, as noted above, is one of the rationales for (co)design), and not every single technology may be able to accommodate all value positions, designers can prepare for controversy by being transparent about, and accountable for, the decisions they make. That said, maintaining transparency has become increasingly difficult with the advent of highly complex, algorithmic technologies in which the technical "black box" becomes so opaque that even designers find it hard to fully understand why their creations behave the way they do (Hirsch et al., 2017). This calls for going beyond transparency to legibility, explainability and contestability: "When an algorithm impacts a person there must be an efficient process to allow that person to challenge the use or output of the algorithm" (Lyons et al., 2021, p. 23).

Other designers have gone even further by suggesting that design should not only accept and prepare for controversy but should actively provoke it, thus materializing a "shift from designing applications to designing implications" (Dunne & Raby, 2013, p. 49), or offering forms of "prefigurative criticism" that already associate a future artifact with its anticipated consequences (Tonkinwise, 2014).[8] Design, in these instances, functions speculatively (Dunne & Raby, 2013) and performs discursively (Tharp & Tharp, 2018, p. 79). Take for instance Jensen and Lenskjold's (2004) suggestion that urban designers encourage instead of attenuate social friction in pursuit of social innovation and change. This is quite consistent with Anthony Dunne & Fiona Raby's (2001) insistence that critical design should generate "complicated pleasures" (a term they borrow from Martin Amis): "A slight strangeness is the key – too weird and they [critical design artifacts] are instantly dismissed, not strange enough and they're absorbed into everyday reality" (p. 63).[9] The preference of dissonance over smoothness in technologically mediated experiences draws

---

[8] Tonkinwise attributes the idea to Anne-Marie Willis and Tony Fry.

[9] "The challenge is to blur the boundaries between the real and the fictional, so that the visionary becomes more real and the real is seen as just one limited possibility, a product of ideology maintained through the uncritical design of a surfeit of consumer goods" (Dunne, 2005 [1999], p. 84).

attention to the mediating role of design itself. This is what Carl DiSalvo (2012) terms "adversarial design", "a kind of cultural production that does the work of agonism through the conceptualization and making of products and services and our experiences with them" (p. 2). Building on Chantal Mouffe's work, adversarial design seeks to foment dissensus by drawing attention to the kind of politics it embodies or performs. By exposing its "formal bias" (in Feenberg's terms) for the world to see (or experience), it anticipates and precipitates (re)design: "For adversarial design, the task is to identify and describe how the qualities of computation are used for political ends and what political issues they bring forth" (ibid. p. 22). An approach such as adversarial design points to the limits of Feenberg's framework for democratic rationalization because the technical artefact does not require public intervention to become more democratic. By exhibiting the discursive nature of technology, it *democratizes by design*. This becomes even more pronounced in the next type of design.

5.6 (un)Design

While the renegotiation of design codes in (re)design processes often requires intermediaries, "hearings, lawsuits, and boycotts" which may "result in changed regulations, designs, and practices" (Feenberg, 2017, pp. 53–54), some design codes can be forced to change through acts of appropriation. This, Feenberg writes, "involves hacking or the reinvention of devices by their users to meet unanticipated demands" (ibid.) In Feenberg's formulation the designer is entirely absent from this type of democratic rationalization so I will refer to it here as (un)design.[10] But as will become clear, this should not necessarily be the case. Just as choosing to eliminate by design or not to design at all can be seen as forms of design, designers can actively (un)design.[11] The latter, in other words, is not "no-design" but signifies the potential for a different kind of design altogether.

On first view (un)design appears to be a more democratic version of (re)design because the public may effectively circumvent the limitations set by technical elites and make use of the technology as they please, and on their own (in line with what Illich (1973) calls "conviviality"). In this sense, (un)design is to (re)design as direct democracy is to representative democracy. This view, however, glosses over a much more complex reality. First, there is a question of skill. While some forms of appropriation are easy enough for almost anyone to conceive and enact, other forms of appropriation aimed at actualizing "ambivalent potentialities" (Feenberg, 1999,

---

[10] While I am aware of previous uses of the term by Pierce (2012) to refer to exnovation; in a symposium held at the University of Applied Arts Vienna in July 2016 to indicate "reducing, simplifying, removing or eliminating design (and complexity)"; and in the title of a recent collection of essays on "critical practices at the intersection of art and design" (Coombs et al., 2018), I use (un) design here as a specific indicator of openness and appropriate-ability.

[11] See for instance Pierce (2012), and Tonkinwise (2014).

p. 105) are complex technical activities that require considerable skills and/or resources. In such cases there is room to question whether "hacking" or "reinvention" are indeed viable options for everyone, and, on the same token, whether we are any closer to what Jones (2000 p. 407) calls "universal despecialisation" just because YouTube is flooded with DIY tutorials. Feenberg, in other words, may be overstating the capacity of laypersons to reshape complex technologies without the help of the technical disciplines.

Second, and as Feenberg himself recognizes in his discussion of the internet (Feenberg, 2017, ch. 4), capitalism has a remarkable capacity to absorb and monetize innovations that initially challenge the prevailing system. The issue goes beyond the myriad forms of commodification and exploitation of immaterial labour that add value to contemporary digital platforms, and gets to the nature of appropriation as a core quality of late capitalist dynamism (Boltanski & Chiapello, 2005). As Rosa et al. (2016 pp. 56–57) explain,

> Capitalist development … always involves the internalization of externals, the occupation of a non-capitalist or non-market (-like) 'other'. What is appropriated and commodified by the capitalist 'machine' is not just new territories, resources and new markets, but also spheres of life and forms of human activity that have as yet not been subject to market exchange, and ever new segments of our personality structures like emotional needs or social relationships.

In Habermasian terms, capitalism is a remarkably capable colonizer of the lifeworld. From this perspective, every hack may potentially turn into a feature and so we may not be able to evaluate the true political extent of a forced renegotiation of the design code without the power of hindsight.

That said, and to the extent that (un)design is indeed a valuable form of democratic rationalization, can designers play an active instead of a passive (or even nonexistent) role in such processes? The question is complicated not only because it implies a relinquishing of operational autonomy by designers, but because it asks designers to let go of much of their professional training as intentional creators of thought and action (Jones, 2002).[12] What does it mean to intentionally give up the intentionality of design? Would this not reduce design to the act of providing materials without form, or, alternatively, providing forms without use? It appears that (un)design poses significant challenges to both designers and to Feenberg's view of the role of designers in democratic rationalizations.

As the essays collected in Eglash et al. (2004) demonstrate, every act of appropriation takes place within complex social and cultural contexts. What these acts have in common is a subject (or user) able to imagine alternative uses for the technical artifact, and a technical artifact whose material properties are flexible enough to support the kind of alternative uses imagined by the user. We can say, then, that appropriation is a phenomenon distributed across users, artifacts, and environments, but which is ultimately an expression of the user's autonomy. Importantly, yet

---

[12] John Chris Jones (2002) defines designing as "thoughts and actions intended to change thoughts and actions".

unaccounted for in Feenberg's framework, designers may support appropriation by creating artifacts that are not only open enough to be appropriated but that actually invite reinterpretation, adaptation or reinvention (Eglash, 2004).[13] Such artifacts may be designed to be ambiguous, thus precluding any singular interpretation, meaning or connotation (Gaver et al., 2003; Sengers & Gaver, 2006). They can manifest "tactical formlessness" (Hunt, 2003) or eschew simple notions of usability, efficiency and optimization by deploying "para-functionality" (Dunne, 2005 [1999]) as a way to provoke critical reflection on design and the social contexts in which it operates. Some artifacts may flaunt their unfinishedness to indicate an ontological incompleteness (Bendor, 2018, pp. 146–148) or to extend their relevance and value to users over time (Tonkinwise, 2004). A useful blueprint for unfinished artifacts can be found in Gilbert Simondon's (2017 [1958]) articulation of the "open machine" as one that "harbors a certain margin of indeterminacy" that allows it to change in relation to its environment (p. 5). "This form of openness," Betti Marenko (2021, p. 219) explains, "whereby the object is worked upon, expanded, amplified and upgraded entails the irruption of the unexpected, the off-grid, the unplanned, the emergent and the accidental in the constitution of machines". More concrete examples can be found in Alejandro Aravena's incremental architecture (Aravena & Iacobeli, 2016) and Luciano Crespi's (2020) proposal for a new interior design made for neo-nomads (see also Bendor, 2021).

Open machines can be seen as signs of technical adaptation, flexibility, sophistication and maturity. But can they be created in "standard" design processes? Tonkinwise (2004) is doubtful, and suggests that creating "unfinished things" requires a fundamental shift in how designers approach their practice. John Chris Jones's (1983) suggestion of "continuous design" seems to get to the heart of the matter by emphasizing process over product, and thus flipping design's means-ends relation on its head. This may allow a more robust, dynamic and adaptable design process, but one that requires designers to be humble and recognize that not only do they not have all the answers, but that they may not even be asking the right questions. Deferring the "closure" of technologies as a means to improve design is also the rationale behind the notion of "metadesign" (Fischer & Giaccardi, 2006; Giaccardi, 2005). Here the extension of the design process allows, on the one hand, to encourage user creativity and autonomy, and on the other, to improve the system itself. In this sense we can locate metadesign on the intersection of (co)design and (un)design, whereby the creation of participatory spaces following a conventional design process would allow the designer to discover unanticipated effects and "mismatches" between users and systems, and then fold those back into the system. "Rather than presenting users with closed systems, meta-design provides them with opportunities, tools, and social reward structures to extend the system to fit their needs", and in this sense, metadesign "has shifted some *control* from designers to users and empowered users to create and contribute their own visions and

---

[13] As Eglash (2004) explains, in reinterpretation the change in the artifact is semantic, in adaptation the semantic change is accompanied by a change in use, and in reinvention the material properties of the artifact are changed to accommodate changes in meaning and use.

objectives" (Fischer & Giaccardi, 2006, p. 430; emphasis in origin). Whether this newfound measure of freedom merely automates (co)design processes or allows for interventions in the context in which the technology will be used, remains unclear. But like continuous design, metadesign asks designers to first acknowledge their limitations, and then deliberately leave some aspects of the design unspecified in order to encourage users to discover and remake the system's meaning and use on their own. In practical terms this means abandoning the pursuit of certainty and welcoming contingency, refraining from planning a complete system and instead "seeding" potentials for emergent interactions. This last element (borrowed from Roy Ascott) is perhaps the most important to our discussion of (un)design for it precipitates a real "cultural shift" as Giaccardi (2005, p. 348) argues. This is because in metadesign, as with Johan Redström's (2008) proposal for "'design' after design", there is a radical folding of future unanticipated use into design activities. The result is a veritable blurring of the boundaries around design – a questioning of designerly intentionality and that which renders design a unique human activity.

Although these initial sketches of (un)design require further clarification and perhaps expansion into something like a theory of sociotechnical malleability,[14] the challenges posed by (un)design should be clear: from Feenberg, (un)design asks that the role of the designer in democratic rationalizations not only be clarified but be expanded to account for the different ways in which designers may seed or precipitate democratic rationalizations. From designers, (un)design asks to let go of some of the most fundamental attributes of their practice, that is, the temporal orchestration of the process and the ability to predict and prescribe use.[15] As hard as it may be, design must either "unlearn" to predict and control as John Chris Jones (1983) suggests, or learn to "unpredict" as Eduardo Viveiros de Castro (2019) puts it. The reward may be transformational for designers and emancipatory for users.

5.7   Conclusion

The aim of this chapter was to suggest a few touchpoints between Andrew Feenberg's critical constructivism and current work in critical design, with the hope of initiating a productive dialogue between the two. Taking Feenberg's notion of the design code as an invitation for such an encounter, I illustrated how designers can take an

---

[14] In general, such a theory should account for the dynamic interplay between an artifacts' capacity to be appropriated and the user's creative strategies of appropriation within multiple contextual factors.

[15] The ability to predict use is part of what Feenberg explains as "anticipation," one of the operations included in his instrumentalization theory: "Design depends on more or less successful prediction of users' behavior. Predictions are necessary to anticipate the conformity of the specified design to its purpose" (Feenberg, 2017, p. 183).

active, meaningful role in facilitating what Feenberg calls "democratic rationalizations" in three different ways. In short, (co)design asks designers to facilitate collaboration and nurture collective creativity while maintaining as much autonomy for participants to make the design process and its outcomes their own. In Feenberg's terms, (co)design can be seen as a space for multiple rationalities to co-produce technologies that better respond to users' everyday life. In (re)design processes designers may be forced to respond to their users' needs post-hoc, but they may also anticipate users' needs and even embrace opportunities to provoke users to reflect on the social, cultural and political implications of the technologies they use. In Feenberg's terms, (re)design is an opportunity for designers to use their "operational autonomy" to help manifest and consolidate their users' "participant interests" by leaving material traces of alternative rationalities or by hinting at their possibility. Lastly, and most radically, (un)design asks designers to let go of (some of) their intentionality and seed potentials for users to appropriate the technologies they design. In Feenberg's terms, this equals an almost complete relinquishing of "operational autonomy" and the valorization of technical rationality that floats it.

Thus articulated, the notion of (un)design offers challenges and correctives to both critical designers and critical constructivism. Engaging with the ideas behind (un)design may shift the self-perception of designers in significant ways – initiating a long needed change in what Feenberg calls "identity" in his instrumentalization theory (Feenberg, 2017, p. 181), away from the rigid models of engineering and closer to the creative, pragmatic openness of bricoleurs (Viveiros de Castro, 2019). Engaging with the practices of (un)design may equally help correct a blindspot in Feenberg's critical constructivism when it comes to the role of designers in democratic rationalizations. More specifically, the design code, a signifier of the political, conflictual nature of design, appears incapable of representing the ways by which designers act as initiators of a more democratic technosphere instead of merely translators of public interests into technical forms. The notion of (un)design, in this sense, offers critical constructivism a new horizon under which designerly pursuits of democratic rationalizations may be recognized and better understood.

Just as critical constructivism stands to benefit from a dialogue with critical design, critical design stands to gain from philosophical programs that articulate design as a field of social struggle. Insofar as a discourse of rationality still pervades the discipline, and I believe this is indeed the case as demonstrated, for instance, by the impact Escobar's recent work has had on the field, critical constructivism offers critical design a potent vocabulary that connects previous challenges to hegemonic deployments of rationality – from Marx and Weber to Marcuse and Foucault – with current struggles to open up and pluralize design. In this sense, critical constructivism offers critical design a particular genealogy with which to anchor a new design culture that respects planetary boundaries and the limits of human agency, and that acknowledges design as a political domain in which futures are continuously remade.

References

Achterhuis, H. (2001). Andrew Feenberg: Farewell to dystopia (R. P. Crease, Trans.). In H. Achterhuis (Ed.), *American philosophy of technology: The empirical turn* (pp. 65–93). Bloomington: Indiana University Press.

Aravena, A., & Iacobeli, A. (2016). *Elemental: Incremental housing and participatory design manual*. Hatje Cantz Verlag.

Arnstein, S. (1969). A ladder of citizen participation. *Journal of the American Institute of Planners, 35*(4), 216–224.

Attfield, J. (2000). *Wild things: The material culture of everyday life*. Berg.

Bardzell, J., & Bardzell, S. (2013). What is "critical" about critical design? In *Proceedings of CHI 2013* (pp. 3297–3306). Paris, France.

Bendor, R. (2013). The role of experience in the critical theory of technology. *Techné: Research in Philosophy and Technology, 17*(1), 47–71.

Bendor, R. (2018). *Interactive media for sustainability*. Palgrave Macmillan.

Bendor, R. (2021). Against Future-Proofing. *Volume 59*, 65–67.

Björgvinsson, E., Ehn, P., & Hillgren, P.-A. (2012). Design things and design thinking: Contemporary participatory design challenges. *Design Issues, 28*(3), 101–116.

Boltanski, L., & Chiapello, E. (2005). *The new spirit of capitalism* (G. Elliott, Trans.). London & New York: Verso.

Cardullo, P., & Kitchin, R. (2019). Being a 'citizen' in the smart city: Up and down the scaffold of smart citizen participation in Dublin, Ireland. *GeoJournal, 84*, 1–13.

Coombs, G., McNamara, A., & Sade, G. (Eds.). (2018). *Undesign: Critical practices at the intersection of art and design*. Routledge.

Crespi, L. (2020). Borders. The design of the unfinished as a new transdisciplinary perspective. In A. Anzani (Ed.), *Mind and places* (pp. 139–157). Springer International Publishing.

Cross, N. (2006). *Designerly ways of knowing*. Springer.

DiSalvo, C. (2012). *Adversarial design*. MIT Press.

Dunne, A. (2005 [1999]). *Hertzian tales: Electronic products, aesthetic experience, and critical design*. MIT Press.

Dunne, A., & Raby, F. (2001). *Design Noir: The secret life of electronic objects*. August/Birkhäuser.

Dunne, A., & Raby, F. (2013). *Speculative everything: Design, fiction, and social dreaming*. MIT Press.

Eglash, R. (2004). Appropriating technology: An introduction. In R. Eglash, J. Crossiant, G. Di Chiro, & R. Fouché (Eds.), *Appropriating technology: Vernacular science and social power* (pp. vii–xxi). University of Minnesota Press.

Eglash, R., Crossiant, J., Di Chiro, G., & Fouché, R. (Eds.). (2004). *Appropriating technology: Vernacular science and social power*. University of Minnesota Press.

Ehn, P. (1988). *Work-oriented design of computer artifacts: Arbetslivscentrum*. Lawrence Erlbaum Associates.

Ehn, P. (2008). Participation in design things. In *Proceedings of participatory design conference (PDC'08)* (pp. 92–101).

Escobar, A. (2018). *Designs for the pluriverse: Radical interdependence, autonomy, and the making of worlds*. Duke University Press.

Feenberg, A. (1999). *Questioning technology*. Routledge.

Feenberg, A. (2017). *Technosystem: The social life of reason*. Harvard University Press.

Feenberg, A. (2020). Critical constructivism: An exposition and defense. *Logos*. Retrieved from http://logosjournal.com/2020/critical-constructivism-an-exposition-and-defense/

Fischer, G., & Giaccardi, E. (2006). Meta-design: A framework for the future of end user development. In H. Lieberman, F. Paternò, & V. Wulf (Eds.), *End user development: Empowering people to flexibly employ advanced information and communication technology* (pp. 427–457). Kluwer Academic Publishers.

Friedman, B. (1996). Value-sensitive design. *Interactions, 3*(6), 17–23.

Gaver, W. W., Beaver, J., & Benford, S. (2003). Ambiguity as a resource for design. In *Proceedings of CHI 2003* (pp. 233–240).

Giaccardi, E. (2005). Metadesign as an emergent design culture. *Leonardo, 38*(4), 342–349.

Haylock, B. (2018). What is critical design? In G. Coombs, A. McNamara, & G. Sade (Eds.), *Undesign: Critical practices at the intersection of art and design* (pp. 9–23). Routledge.

Hirsch, T., Merced, K., Narayanan, S., Imel, Z. E., & Atkins, D. C. (2017). Designing contestability: Interaction design, machine learning, and mental health. In *Proceedings of DIS 2017* (pp. 95–99).

Hunt, J. (2003). Just re-do it: Tactical formlessness and everyday consumption. In A. Blauvelt (Ed.), *Strangely familiar: Design and everyday life* (pp. 56–71). Walker Art Center.

Illich, I. (1973). Tools for conviviality. New York: Harper & Row.

Jensen, R. H., & Lenskjold, T. U. (2004). Designing for social friction: Exploring ubiquitous computing as means of cultural interventions in urban space. In M. A. Eriksen, L. Malmborg, & J. Nielsen (Eds.), *Proceedings of CADE 2004*.

Jones, J. C. (1983). Continuous design and redesign. *Design Studies, 4*(1), 53–60.

Jones, J. C. (2000). *The internet and everyone*. Ellipsis.

Jones, J. C. (2002). *What is desiging?* Retrieved from http://www.publicwriting.net/2.2/digital_diary_02.07.14.html

Latour, B. (2008). *A cautious Prometheus? A few steps toward a philosophy of design (with special attention to Peter Sloterdijk)*. Paper presented at the networks of design meeting of the design history society, Falmouth, Cornwall.

Luck, R. (2018). [Editorial] What is it that makes participation in design participatory design? *Design Studies, 59*, 1–8.

Lyons, H., Velloso, E., & Miller, T. (2021). Conceptualising contestability: Perspectives on contesting algorithmic decisions. In *Proceedings of CSCW, Article 106*. Retrieved from https://arxiv.org/pdf/2103.01774.pdf

Malpass, M. (2017). *Critical design in context: History, theory, and practices*. Bloomsbury.

Manzini, E. (2015). *Design, when everybody designs: An introduction to design for social innovation* (R. Coad, Trans.). Cambridge, MA & London: MIT Press.

Marenko, B. (2021). Future-crafting: The non-humanity of planetary computation, or how to live with digital uncertainty. In S. Witzgall, M. Kesting, M. Muhle, & J. Nachtigall (Eds.), *Hybrid ecologies* (pp. 216–227). University of Chicago Press/Diaphanes.

Margolin, V. (2002). *The politics of the artificial: Essays on design and design studies*. University of Chicago Press.

Pacey, A. (2001). *Meaning in technology*. MIT Press.

Pierce, J. (2012). Undesigning technology: Considering the negation of design by design. In *Proceedings of CHI 2012* (pp. 957–966).

Redström, J. (2008). RE: Definitions of use. *Design Issues, 29*(4), 410–423.

Rosa, H., Dörre, K., & Lessenich, S. (2016). Appropriation, activation and acceleration: The escalatory logics of capitalist modernity and the crises of dynamic stabilization. *Theory, Culture & Society, 34*(1), 53–73.

Sanders, E. B. N., & Stappers, P. J. (2012). *Convivial toolbox: Generative research for the front end of design*. BIS.

Sanders, E. B. N., & Stappers, P. J. (2014). From designing to co-designing to collective dreaming: Three slices in time. *Interactions, 21*(6), 24–33.

Schön, D. A. (1988). Designing: Rules, types and worlds. *Design Studies, 9*(3), 181–190.

Sengers, P., & Gaver, W. W. (2006). Staying open to interpretation: Engaging multiple meanings in design and evaluation. In *Proceedings of DiS 2006* (pp. 99–108).

Simon, H. A. (1969). *The sciences of the artificial*. MIT Press.

Simondon, G. (2017 [1958]). *On the mode of existence of technical objects* (C. Malaspina, Trans.). Minneapolis, MN: University of Minnesota Press/Univocal.

Smith, R. C., & Iversen, O. S. (2018). Participatory design for sustainable social change. *Design Studies, 59*, 9–36.

Tharp, B. M., & Tharp, S. M. (2018). *Discursive design: Critical, speculative, and alternative things*. MIT Press.

Tonkinwise, C. (2004). Is design finished? Dematerialisation and changing things. *Design Philosophy Papers, 2*(3), 177–195.

Tonkinwise, C. (2014). Design away. In S. Yelavich & B. Adams (Eds.), *Design as future-making* (pp. 198–213). Bloomsbury.

Ullman, E. (1997). *Close to the machine: Technophilia and its discontents: A memoir*. City Lights Books.

Van den Hoven, J., Vermaas, P. E., & van de Poel, I. (2015). Design for Values: An introduction. In J. V. D. Hoven, P. E. Vermaas, & I. V. D. Poel (Eds.), *Handbook of ethics, values, and technological design: Sources, theory, values and application domains* (pp. 1–7). Springer.

Verbeek, P.-P. (2015). Beyond interaction: A short introduction to mediation theory. *Interactions, 12*(3), 26–31.

Viveiros de Castro, E. (2019). On models and examples: Engineers and bricoleurs in the anthropocene. *Current Anthropology, 60*(S20), 296–308.

White, D. (2015). Critical design and the critical social sciences. *Critical Design/ Critical Futures Articles, Paper 8*. Retrieved from http://digitalcommons.risd.edu/critical_futures_symposium_articles/8

Winograd, T., & Flores, F. (1986). *Understanding computers and cognition: A new foundation for design*. Ablex Pub. Corp.

Chapter 6
Who Controls the Smart City? From Machines of Loving Grace to a Democratic Transformation from Below

Ryan Mitchell Wittingslow

Abstract In this paper, I argue that smart cities instantiate a form of what Herbert Marcuse called "technological rationality": that is, the process whereby substantive political questions are reduced to ostensibly "neutral" questions of efficiency or cost-effectiveness. Unfortunately, I argue, technological rationality coheres poorly with the necessarily inefficient deliberative and aggregative procedures upon which the legitimacy of democratic systems is premised. Considering that incompatibility, we need to reconceptualise what smart cities are and how they function. These technologies, I argue, need to undergo what Andrew Feenberg calls a "democratic transformation from below"; a transformation whereby citizens can bring smart technologies under collective control, thus preserving the legitimacy of democratic systems. This democratic transformation gives the polis an opportunity to recognise and discuss the affordances that smart technologies offer—and, by extension, an opportunity to collectively and systematically address the philosophical question of what a city *can* and *should* be.

Keywords Smart cities · Democracy · Andrew Feenberg · Technological rationality · Critical theory · Herbert Marcuse · Design · Technical code · Persuasive technologies · Eindhoven

R. M. Wittingslow (✉)
University College Groningen, University of Groningen, Groningen, the Netherlands
e-mail: r.m.wittingslow@rug.nl

6.1   Introduction

In his poem, "All Watched Over by Machines of Loving Grace," Richard Brautigan proposes governance by machines. Instead of liberal democracy, he writes, we will have a "cybernetic ecology" where animals and computers co-exist in "mutually programming harmony." The cost of this service is but a small thing. We need only offer our liberty to Brautigan's titular "machines of loving grace" (Brautigan, 1967). Although it's been a long time since 1967, the utopian impulse that the poem articulates—the notion that we will be fitter, happier, and more productive if we allow ourselves to be administered by intelligent machines—remains compelling. Within the domain of urban design and planning, this impulse has recently expressed itself in a suite of proposals concerning what is generally called the "smart city."

Smart cities are urban areas that use data collection sensors to supply information that is then used to more efficiently manage assets and municipal resources. They employ information and communication technologies to autonomously manage municipal systems with the aid of sensors and actuators managed by artificial intelligences. This information can be either collected from citizens directly, or indirectly from devices such as smartphones. This data is then processed and analysed to manage things like power grids, traffic and transportation systems (trains and buses, or traffic lights, for example), waste management, plumbing networks, information systems, library databases, and law enforcement. Using information and communication technologies, smart cities purport to offer managers and bureaucrats a more efficient means of increasing public safety and managing public assets. Pilot programs are already underway in several countries worldwide.

Smart cities provide a tempting prospect: the optimisation of urban space, seemingly unburdened by ideology or politics. However, at least among philosophers, urbanists, and other similarly minded folk, some of the flaws inherent in this view are well known. Indeed—and speaking from experience—a cheap way to score a chuckle from the right crowd is by quoting former IBM CEO Sam Palmisano: "Building a smarter planet is realistic precisely because it is so refreshingly non-ideological."[1] Regardless of Palmisano's naiveté, smart cities possess politics just like any other artefact (Winner, 1980). But what kind of politics does the smart city possess?

My argument is composed of four parts. First, I will provide an analysis of the "persuasive technologies" that typify smart city projects (Tromp et al., 2011). In response to the Winnerian intuition that artefacts necessarily have politics, several commentators have taken up the task of designing technology with an explicit politics in mind. Unfortunately for citizens, these political artefacts are designed absent sufficient either meta-ethical, collaborative, and/or procedural justification (Engelbert et al., 2019). Even more unfortunately, smart city technologies are often designed with these explicit politics in mind.

---

[1] Sam Palmisano, "Welcome to the Decade of Smart." Remarks presented at Chatham House, London, on January 12, 2010.

Second, I argue that democracy is inherently, and necessarily, both slow and inefficient. This inefficiency is not a flaw. While the collectivist nature of democracy means that decisions are almost always rendered less quickly than in authoritarian forms of rule, this very slowness helps guarantee that new policies and new behaviours are given sufficient deliberative scrutiny. It is upon these profoundly inefficient deliberative and aggregative procedures that the legitimacy of democratic systems is premised (Peter, 2007).

Unfortunately, and as I argue in the third part of this paper, smart city ethics sit poorly with these pluralistic norms. Instead, smart cities are instantiations of "technological rationality," per Herbert Marcuse (2007 [1964]) and Andrew Feenberg (2002). Technological rationality is a discourse wherein substantive political questions are reduced to "neutral" questions of efficiency or cost-effectiveness: a procedural instantiation of modernity's "affirmation of autonomy against every traditional or social authority" (Feenberg, 2002, p. 162). By virtue of purporting to render cities more efficient and manageable, smart city systems are a clear expression of technological rationality in action.

Finally, I will conclude by arguing that the technological rationality that underpins smart city technologies has the non-trivial potential to compromise the necessarily slow, methodical, and pluralistic processes that legitimate structures of democratic governance, unless designed and implemented with adequate deliberative oversight and critical attention. Rather than leaving these technologies in the hands of city managers and bureaucrats, these technologies need to undergo what Feenberg calls a "democratic transformation from below" (Feenberg, 2002, p. 17): a transformation whereby citizens can bring smart technologies under collective control, thus preserving the legitimacy of democratic systems. In doing so, this democratic transformation gives the polis an opportunity to recognise and discuss the affordances (political, economic, existential) that smart technologies offer— and, by extension, an opportunity to collectively and systematically address the philosophical question of what a city *can* and *should* be.

6.2   Choice Architectures

This is an exciting time for researchers in the burgeoning field of "persuasive technology." With the implicit promise of the participatory web and the internet of things finally in a position to bear fruit, there has been an explosive growth in both the literature and the available grant funding allocated to these kinds of initiatives over the past 10 years or so. Uniting these proposals is the notion that designed systems and objects can and should be designed to induce or encourage targeted attitudes or behaviours via "nudges."

Emerging from cybernetics research from the mid-1990s, the idea of "nudging" gained prominence with the publication of a book, by Richard Thaler and Cass Sunstein, appropriately entitled *Nudge*. A nudge, in essence, is a non-coercive means of targeting and influencing given behaviours:

> A nudge, as we will use the term, is any aspect of the choice architecture that alters people's behavior in a predictable way without forbidding any options or significantly changing their economic incentives. To count as a mere nudge, the intervention must be easy and cheap to avoid. Nudges are not mandates. Putting fruit at eye level counts as a nudge. Banning junk food does not. (Thaler & Sunstein, 2008, p. 6)

In targeting those behaviours, planners and choice architects of all kinds hope to passively incentivise desirable behaviours and disincentivise undesirable behaviours. Rather than appealing to crude legislative instruments of control, these choice architects can encourage desirable behaviours without appearing to compromise the liberty of participating subjects. The manipulation of choice architecture is of obvious interest for those interested in designing persuasive smart city technologies. Rather than mistakenly assuming that artefacts are absent political valence, designers can instead take the political character of smart city technologies for granted. In so doing, they can instead design artefacts that encourage or discourage certain behaviours. Consequently, these artefacts are rendered explicitly political, in that they reflect or otherwise instantiate a given notion of the good.

To draw an example from my own country of residence: among the many Dutch streets used as labs to test these persuasive smart city technologies, perhaps the most famous is Stratumseind, in Eindhoven. Stratumseind is one of the busiest nightlife stretches in all the Netherlands. It is also, by Dutch standards at least, quite violent; fistfights and other altercations are, relatively, quite common (Brock et al., 2019, p. 206). It is also one of the smartest streets in all the Netherlands. Lampposts have been fitted with wifi-trackers, cameras, and 64 microphones that can detect aggressive behaviour and alert police officers to altercations.

A previous (failed) experiment changed the light intensity to break up fights, using colour-changing LED bulbs developed and supplied by Philips (Brock et al., 2019, p. 207). A second (aborted) proposal recommended the smell of oranges in the hope that it would disincentivise violence.[2] Contrary to the naive comments made by Mr Palmisano, these are not attempts to render smart cities neutral and non-ideological. Instead, the ethic is baked in: in all cases these are attempts to profile, or target people either exhibiting antisocial behaviour, or nudge people into exhibiting prosocial behaviour. Furthermore, while these strategies may not work (and indeed, in the case of the variable lighting, did not work) what is exciting for urban designers, municipal agents, and other technocrats is the possibility that these strategies will only become more efficient and precise with time.

There exist robust taxonomies for how to categorise these choice architectures. For example: in a much-cited paper, Tromp et al. (2011) attempt to better categorise the nudges by which persuasive technologies function: they provide a robust, and useful, taxonomy of how one might classify these persuasive technologies based on

---

[2] For details of the scheme, refer to "*Trends Uitgaansgeweld: Mei 2016—Augustus 2016*" ("Trends in Nightlight Violence: May 2016—August 2016"), an information sheet published by *Het Centrum voor Criminaliteitspreventie en Veiligheid* (The Centre of Crime Prevention and Security) in 2016 (in Dutch). To access the scholarship motivating the proposal, please refer to Schifferstein et al. (2011).

the intended user experience. In so doing, they give designers the means to better "design for socially responsible behaviour." Tromp et al. then go on to argue that these sorts of technological interventions can be classified into four categories, based upon two dimensional axes: strong vs weak, and hidden vs apparent. These four categories are coercive (strong, apparent), decisive (strong, hidden), persuasive (weak, apparent), and seductive (weak, hidden). Armed with these categories, Tromp et al. then assign "design strategies" to each of these categories: designed to maximise the possibility that a given user will exhibit the desired behaviour, whether by nudging or more explicit means. As part of this programme, they outline a series of means by which user experiences can be constrained and determined, including employing moral shame, the rendering of explicit suggestions, triggering prosocial behaviours by appealing to non-social motivations, activating physiological responses, or eliciting emotional responses (Tromp et al., 2011, pp. 11–17).

It is very easy to apply Tromp's taxonomy to existing smart systems, like those up and running in Eindhoven. Introducing the smell of oranges to make people less aggressive, for example, could either be an example of a decisive strategy or a seductive strategy, depending on the mechanism of action: if it activates a non-binding physiological process to induce a behavioural outcome to induce behaviour, it is seductive; if it activates a binding behavioural response, it is decisive. Conversely, the use of wifi trackers, microphones, and CCTV cameras are either hidden or persuasive, depending on whether they are particularly obvious. In either case, the knowledge that those technologies are in place implicitly forces citizens to police their own behaviour in order to make sure they act in accordance with prevailing norms: they constitute nudges, per the above. Meanwhile, any police action that manifests as a result of these norms being breached is obviously an example of a non-nudging coercive strategy.

That strategic taxonomy, and others like it, are of great interest to choice architects responsible for developing and introducing smart systems into cities. In combining the precise taxonomies offered by Tromp et al. (among others), with the huge amounts of finely grained data available via smart censors and the processing power offered by artificial intelligences, there is a shared hope that these design strategies can be made vastly more efficient within smart cities. In doing so, and in better cultivating socially responsible behaviour, urban spaces will only become safer and more harmonious. It is a desire both utopian and within the realms of possibility. In so doing, smart city schemes like those at work in Eindhoven promises a wonderfully efficient means of policing public space: and, in so doing, enforcing a collective sense of the good. Ah, but there's the rub.

Tromp et al. are for the most part silent on the nature of the good. Indeed, the following sentence is about the extent of their attempts to engage with the deeper ethical questions at the heart of socially responsible design: "A desired social implication, based on collective concerns, defines what behaviour is desired from a social perspective" (Tromp et al., 2011, p. 7). While this statement verges on the truistic, their silence on the good is obviously intentional: their paper is far more interested in assessing the efficacy of given design strategies than in welding those strategies to ethical norms. This attitude is not uncommon amongst researchers interested in choice architecture; if a community has agreed upon a singular standard of the good, it is both reasonable and just to design technologies that encourage observation of that standard of behaviour. However, we should be suspicious of the notion that designers are in fact in possession of the common standards that unite us.

While Tromp et al. assume that designers are in possession of a notion of the good when implementing new systems, it is not at all clear that designers of smart systems (such as those in Eindhoven) actually have a rigorous notion of the good in mind—and certainly not one based upon "collective concerns." Instead, the documentation that accompanies these proposals do something rather more insidious, in that they uncritically conflate the *good* with the *prosocial*. Let us consider, for a moment, the motivations behind installing coloured lighting in Stratumseind, as expressed in "Light the Way for Smart Cities":

> Each weekend, 25,000 visitors head to Stratumseind; on any given Saturday night, there are roughly 850 incidents, 20 of which lead to arrests or detentions. The municipality believes that lighting might be pivotal for de-escalating aggressive behavior and reducing these incidents to increase public safety and the attractiveness of the area. Stratumseind thus offers a unique research and measurement center, where experiments test ways to make the area safer, more vibrant, and more attractive. (Brock et al., 2019, p. 198)

The implicature here suggests that what we expect of public spaces like Stratumseind is that they be both safe and "attractive": attractive for businesses, attractive for tourists, attractive for pleasure-seekers; attractive as a locus of consumption and regulated hedonia. When urban spaces are both safe and attractive, the wheels of commerce and the pursuit of pleasure can function unimpeded. While the paper does not state it baldly, the assumption is clear enough: the prosocial pursuit of safety and attractiveness is isomorphic with the pursuit of the good. As a corollary, the means by which municipalities guarantee both safety and attractiveness is simple: by thwarting, minimising, or disincentivising antisocial behaviour. I am suspicious of these desiderata and their concomitant assumptions.

While most people, in most cases, would agree that what they desire from public spaces is both "safety" and "attractiveness," conflating the prosocial with the good is an obvious category error, in that it fails to take into consideration the moral character intrinsic to certain displays of civil disobedience. Mass protests against unjust policies could hardly be considered prosocial, for example, but certainly function in pursuit of the good. Consider, by means of an example, two events still very fresh in the public psyche: the 2019–20 Hong Kong protests and the George Floyd protests. Although to say so might let slip my not-terribly-well-hidden ideological commitments, it is obvious to me that both sets of protests are in service of the good: the former because the democratic rule of law is valuable and worth preserving; and the latter because systemic racism is a boil that requires lancing. Furthermore, I'm of the view that these protests are in the service of the good in spite of—and maybe even because of!—their antisocial character: faced with dismantling the unjust edifices of state power, gentle reform is simply inadequate to the task.

Unfortunately, smart city systems such as those installed on Stratumseind do not permit these kinds of good-making political expression. This is something of which we should be extremely wary. Not only because the right to protest is hard-won, but also because the right to disagree is baked into the ethic of democracy itself. Unfortunately, expressions of disagreement (particularly when anti-social) are prima facie incompatible with the implicit ethics of most implemented smart city systems. This is because the conception of the good that these systems instantiate is far too monistic and prosocial for such a thing to be possible.

6.3   Democratic Legitimacy

Democracies are inherently, and necessarily, *pluralistic* about the good. Indeed, pluralism about the good is one of the formal features that differentiates democracies from other political systems. Democracy is not a political system with an explicit end in mind but is rather constructed around the notion that there is *more than one good*.

Indeed, there are arguably no goods that are intrinsic to democratic pluralism. So, while certain norms are valued within democratic societies—"tolerance" is one that springs to mind—these should not be considered as goods in and of themselves. Rather, the "goodness" of democratic goods like tolerance are strictly extrinsic. So, for example, in a democratic society it is necessary that people tolerate the existence of other, competing notions of the good, just as they expect their own notion of the good to be tolerated in turn. Tolerance is only virtuous insofar as it helps guarantee that people can continue pursuing their own individual conception of the good without undue interference. The system is premised upon the fundamental assumption that what is good for you may not necessarily be good for me. Indeed, democratic pluralism presumes that no single one of these goods is any "more good" than any other: a sentiment beautifully expressed in E. M. Forster's 1939 *What I Believe*:

> Democracy is not a beloved Republic really, and never will be. But it is less hateful than other contemporary forms of government, and to that extent it deserves our support. It does start from the assumption that the individual is important, and that all types are needed to make a civilization. It does not divide its citizens into the bossers and the bossed—as an efficiency-regime tends to do. The people I admire most are those who are sensitive and want to create something or discover something, and do not see life in terms of power, and such people get more of a chance under a democracy than elsewhere. They found religions, great or small, or they produce literature and art, or they do disinterested scientific research, or they may be what is called "ordinary people", who are creative in their private lives, bring up their children decently, for instance, or help their neighbours. All these people need to express themselves; they cannot do so unless society allows them liberty to do so, and the society which allows them most liberty is a democracy. (Forster, 1939, p. 9)

However, a problem emerges: what happens when the goods valued by different individuals are at odds with one another? What happens when they are, in short, incommensurable? Two values are "incommensurable," as Martijn Boot writes, "if they have different dimensions that cannot be reduced to one dimension so that their amounts cannot be measured and compared on a common cardinal scale of units of value" (Boot, 2017a, p. 315). So, for instance, we could compare the value of a car with the value of a bicycle. There are several relevant evaluative axes upon which they could be compared: top speed (using kilometres per hour), luggage space (using cubic centimetres), resale value (using your medium of exchange of choice), whatever. Consequently, the value of a car and the value of a bicycle are commensurable, in that they can be measured and compared on a common cardinal scale of units of value.

However, not all values are commensurable in this way. By means of example, Boot analyses a legal battle between the British government and a group of people who lived close to Heathrow Airport. In these cases, the residents claimed violation of their right to privacy due to the sharp uptick in night flights from 1993 onwards. The Chamber of the European Court of Human Rights (and then the Grand Chamber, on appeal) were forced to render a verdict on whether the residents' right to privacy outweighed the economic interests of the nocturnal flights. This, Boot argues, is a clear example of a "fourth value relation:" it is not true that the right to privacy (A) outweighs economic interests (B), nor that B outweighs A, nor that A and B have roughly equal weights. Indeed, there is no common standard (like kilometres per hour, or cubic centimetres, or media of exchange) by which A and B can be meaningfully compared. Consequently, we are lacking a means by which we can evaluate and select between these options. A and B are simply and straightforwardly incommensurable; it is "implausible," Boot writes, "that these weights can be determined or that they exist at all" (Boot, 2017b, p. 31).

This is hardly an isolated case. Indeed, many of the most important questions that a polis can ask itself (for example: "Should we increase our refugee intake?"; "To what extent do we have the right to healthcare?"; "What is the right balance between freedom and security?"; "How should we address the threats of anthropogenic climate change?") are questions of this sort, in that opposing sides of the issue in question are also absent clear shared standards by which they can be compared. This is what makes political questions of this sort incommensurable. Because of this incommensurability, for democratic pluralism to function properly there must exist systems, civic structures, and institutions in which competing and potentially incommensurable notions of the good can be discussed, argued, and eventually endorsed. These processes and domains whereby and wherein conversation is facilitated *is* democracy in its purest form. We are familiar with systems of this type (town hall meetings, legal proceedings, parliamentary debates, public protests, letters to the editor, Socrates haranguing passers-by in the agora, and so on); all transparent venues in which different notions of the good can be contested. It is upon these public decision-making processes that the fundamental legitimacy of democratic systems is founded.

Legitimacy is important. Indeed, as Fabienne Peter argues (cf. Peter, 2007, 2008), democratic legitimacy is nothing less than "the first virtue of collective decision-making. [...] [Democratic] decisions have to be legitimate, before anything else" (Peter, 2007, p. 330). Given that our values are potentially incommensurable, we naturally cannot guarantee that collective decisions will satisfy all interested parties. It is for this reason that the polis must decide on a set of norms and procedures—that is to say, deliberative processes—that govern how opposing notions of the good can be contested, by "defining the terms for how the members of a democratic society ought to settle their disagreements about how to organize their life together" (Peter, 2007, p. 330). So, while a particular member of the polis may be unhappy about the outcome of a collective decision-making process, it is imperative that the process *itself*, whatever that might be, be in possession of the trust of the polis.

For Peter, this trust can only be bestowed if those processes meet the criterion of "fairness." Deliberative processes—such as public debates, or protests, or letters to the editor, or whatever—are fair when all citizens can participate in those processes. Citizens, in the appropriate forum, are empowered with the ability to argue for or against whatever value is under debate, after which the outcome is decided democratically. She quotes Gerald Gaus: "In his or her deliberations, each citizen presents what he or she believes is the best public justification; the voting mechanism constitutes a fair way to adjudicate deep disagreements about what is publicly justified" (Gaus, 1997, p. 234, quoted in Peter, 2007, p. 335).

How, then, do we know if the outcomes of these deliberative processes are rationally justified? Peter argues that deliberative processes are rationally justified when the deliberative processes themselves fulfil certain social-epistemological standards. To this end, Peter endorses a strictly procedural approach to social epistemology premised on the idea that there exist normative criteria that apply to knowledge-making practices. This means that she dispenses with the notion that there are procedure-independent criteria for what counts as knowing or not knowing. Instead, it is the procedure itself that guarantees the correctness of the outcome. In this way, correctly designed and implemented deliberative procedures are "knowledge-making," in much the same way that correctly designed and implemented scientific inquiries are knowledge-making without having to be beholden to some external standard of correctness (Peter, 2007, pp. 341–46).

In short, Peter describes legitimate democratic processes as being ideally measured and deliberative, and that these processes permit competing, sometimes incommensurable notions of the good to engage in debate. Furthermore, it is via this deliberative procedure that democracies become able to reconcile their pluralism about the good with the demands of governance. Out of an incommensurable plurality, via deliberation, a smaller number of goods are decided upon and prioritised: an ideally fair deliberative process by which the polis can generate rationally justified collective decisions.

Finally, possessing a robust understanding of "democratic legitimacy" also provides us a principled defence of antisocial behaviour functioning in service of the good. Although expressing antisocial behaviour in response to legitimate democratic processes is clearly not in service of the good, that is not the case when it comes to *illegitimate* democratic processes: procedures that fail to be fair and open (thanks to the coercive powers of economic, political, technological, or other kinds of capital), and that are thus incapable of appropriately resolving disputes (whether incommensurable or otherwise). Antisocial behaviour can then be justified under the condition that it is in service of re-establishing the fairness and legitimacy of democratic procedures. The occasionally-destructive 2019–20 Hong Kong protests and the George Floyd protests are both antisocial expressions of this sort, given that both sets of protests were and are pitched against political and economic systems wherein the few have undue influence over the many: a clear violation of Peter's principle of purely procedural democratic legitimacy.

All of this means that any notion of the "public good" in democracies is in constant flux. It is always up for debate, for revision, for reflection, changing in line with shifting intellectual and moral norms. The good is not a caged creature, but a vital, dynamic entity in which we constantly participate via deliberation, negotiation, and the procedures that constitute our republic. Consequently, there are certain unavoidable administrative costs when it comes to properly implementing fair democratic systems. These costs mean that democracy is inherently, and necessarily, both slow and inefficient. This inefficiency is not a flaw. While the collectivist nature of democracy means that decisions are almost always rendered less quickly than in what Forster calls "efficiency-regimes," the slowness of these procedures helps guarantee that new policies and new behaviours are given sufficient deliberative scrutiny. It is upon these profoundly inefficient deliberative procedures that the legitimacy of democratic systems is premised. "So two cheers for Democracy," as Forster entreats: "one because it admits variety and two because it permits criticism. Two cheers are quite enough: there is no occasion to give three" (Forster, 1939, p. 10).

6.4   Technological Rationality

The monistic ethics of smart city systems sit poorly with these pluralistic norms. This is because smart city systems are structured in such a way as to privilege and encourage an extremely narrow notion of the good: a good that is well-behaved, prosocial, and benignly reconcilable with the political, technological, and economic structures that serve to constitute contemporary market capitalism. In making urban centres more "safe" and "attractive," for instance, smart city systems like those implemented in Stratumseind function in order to guarantee that the rhythms of production and consumption are not endangered.

This should not be a surprise. After all, there exists a long and illustrious body of literature within Marxian critical theory dealing with the graceless intersection of capitalism and democracy. Capitalist systems of consumption and production, claim scholars like Max Horkheimer, are only capable of facilitating bourgeois negative liberty: mere "freedom from" rather than "freedom to." "The limited freedom of the bourgeois individual," Horkheimer writes, "puts on the illusory form of perfect freedom and autonomy" (1972, p. 211). Unfortunately for us, true democratic freedom—that is, the freedom to organize ourselves rationally and consensually, free of the coercive powers of capital—is simply not possible whilst under the aegis of what Herbert Marcuse calls "technological rationality."

Marcuse first addressed technological rationality in 1941s "Some Social Implications of Modern Technology", and further developed the concept in in 1964s *One-Dimensional Man*. As noted in the introduction, by technological rationality Marcuse is referring to the tendency of persons and institutions in developed, liberal economies to reduce explicitly value-laden political questions into "neutral" questions of economics or cost-effectiveness. As Feenberg writes:

> The concept of technological rationality expresses the condensation of social and technical functions implicit in Marx's design critique of technology. It explains how rules and procedures that achieve a certain kind of universality may also represent private interests through the assumptions that form their horizon. These interests are overlooked because they are not expressed through orders or commands, but are technically embodied, for example, in apparently neutral management rules or technical designs. (Feenberg, 2002, p. 66)

A foundational symptom of late capitalism, this tendency emerges from the individualistic rationality of the Enlightenment. Individualistic rationality, Marcuse argues, is a rationality bounded by notions of reason, autonomy (*auto-* meaning "self"; *-nomos* meaning "law"), and individual self-interest. Instead of being subject to the capricious whims of some feudal sovereign, a person with individualistic rationality is afforded the liberty to pursue her own unique ends. These ends are decided not with respect to whatever immediate needs and interests might be in play but are instead the product of non-coerced "autonomous thought and conscience." Importantly, this means that the relationship between the individual rationalist and society is an uneasy one. No longer subject to the iron will of society, the individual rationalist must "break through the whole systems of ideas and values imposed upon them" in order to isolate what is in their rational interest. Thus, the individual rationalist finds herself living "in a state of constant vigilance, apprehension, and criticism," as Marcuse writes, rejecting "everything that was not true, not justified by free reason" (Marcuse, 1982 [1941], p. 140). Applied to the social and economic environment, this new "liberalist" mode of free competition between politically equal agents results in two distinct but overlapping phenomena.

The first effect is both technological and economic. As "free competition" between individually rational agents becomes the byword of economic reform, competitive efficiency "favours those enterprises with the most mechanized and rationalized industrial equipment" (Marcuse, 1982, p. 141). Increased industrialisation and mechanisation in powerful, more efficient firms forces weaker, less efficient, less mechanised participants to either fail or be consumed by their larger brethren. In this way, economic power and technical clout become indelibly linked. Furthermore, although the market remains ostensibly free, the abundance of technological capital that large organisations have at their disposal affords those organisations a powerful institutional inertia: it is enormously hard to freely compete with an economic Goliath when you have the purchasing powers and technological assets of a mere David. In this way, individual rationality tends to concentrate power in the hands of a small number of firms, thereby laying the groundwork for the development of oligopolies and monopolies—at least, absent any legislative or structural restrictions on growth.

The second effect is political and social. Because individualistic rationality is premised upon the setting of "the individual against his society" (Marcuse, 1982 [1941], p. 140), individuals' capacities for collective action become blunted. This becomes particularly apparent when oligopolies and monopolies emerge. Amidst the gargantuan, tectonic struggles of large corporate and government entities, individual rational subjects are rendered largely impotent by their social and intellectual atomisation. In this way—and perversely—individualistic rationality guarantees its own demise: the individually rational "free economic subject" becomes subsumed within the "dominion of the giant enterprises of machine industry" (Marcuse, 1982 [1941], p. 141). Given both the preponderance of monolithic technological systems and the powerlessness of individual subjects, these two phenomena provide the necessary conditions for individual rationality to transmute into "technological" rationality.

By "technology," Marcuse means more than the total aggregate of technical objects. While indeed partially constituted by the totality of "instruments, devices, and contrivances" (Marcuse, 1982 [1941], p. 138), technology (by virtue of subsuming the free economic subject) is also a mode of being that organises behaviour and social relationships. Technological rationality, then, is the form of rationality that develops from, and is enforced by, technology as a mode of organisation.

Imagine, for example, that you are a carpenter, and you have just made a table that you think is particularly fine: perhaps it is unusually elegant, or you have selected a finish that is both hard-wearing and beautiful. Considered in the light of individual rationality, your efforts in producing that table are "independent of recognition and instantiated in the work itself" (Marcuse, 1982 [1941], p. 142); the pride you take in having made the object has nothing to do with what other people think of it. However, that is no longer the case under the aegis of technological rationality. Instead, technological rationalism demands that human efforts can only be "motivated, guided and measured" by the standards of competitive efficiency imposed by industrialisation and mechanisation. Rather than being intrinsic or personal, the worth of any given activity or outcome becomes subject to an economic calculus wherein they are judged according to "the objective requirements of the apparatus" (Marcuse, 1982 [1941], p. 142): that is, judged according to the standards of other people. Money, of course, provides the means by which these external standards are quantified. This means that, as rationality shifts from the individual to the technological, objects and services become assessed only in terms of their commercial value.[3]

This has cascading effects on how we conceptualise human action. Whereas under individual rationality we are empowered to pursue a plurality of ends, technological rationality can only imagine the pursuit of one end: profit, above all else. The technologically rational subject thus is claimed by mass production and mass distribution. We experience, he writes, "an immediate identification of the individual with *his* society and, through it, with the society as a whole" (Marcuse, 2007 [1964], p. 10). In short, technological rationality is the same kind of process as "operationalism" in the physical sciences and "behaviorism" in the social sciences: a kind of "total empiricism" wherein concepts find complete identity with their corresponding set of economic and industrial operations. The technological rationality of total

---

[3] For more on this subject, including the original source of the table metaphor (see Marx, 1992 [1867], sec. 4).

empiricism consequently allows for the elimination of concepts (such as "mind," or "beauty," or "goodness") for which there is insufficient account within technological processes: a procedural instantiation of modernity's "affirmation of autonomy against every traditional or social authority" (Feenberg, 2002, p. 162). Consequently, total empiricism serves to "coordinate ideas and goals with those exacted by the prevailing system, to enclose them in the system, and to repel those which are irreconcilable with the system". As Marcuse writes:

> We are again confronted with one of the most vexing aspects of advanced industrial civilisation: the rational character of its irrationality. Its productivity and efficiency, its capacity to increase and spread comforts, to turn waste into need, and destruction into construction, the extent to which this civilisation transforms the object world into an extension of man's mind and body makes the very notion of alienation questionable. The people recognise themselves in their commodities; they find their soul in their automobile, hi-fi set, split-level home, kitchen equipment. The very mechanism which ties the individual to his society has changed, and social control is anchored in the new needs which it has produced. (Marcuse, 2007 [1964], p. 9)

The net effect of this total empiricism is a loss of political freedom. Politics, after all, is a normative domain rather than a descriptive one. It is a public space—whether physical, conceptual, or organisational—wherein we are able engage in meaningful conversation about the things that we think are important: what is good, what is right, what is desirable; in short, what we *want*. Unfortunately, given technological rationality's emphasis upon competitive efficiency and profit-seeking as the only good worth pursuing, meaningful political decisions are deflated and collapsed into ostensibly neutral questions of cost-effectiveness. And indeed, we see decision-making of this sort all the time. Questions that are fundamentally about contested notions of the good disappear. Instead, they are replaced with questions where the means of evaluation are already, and undemocratically, baked in: questions such as "Does it work as planned?", or "What does it *cost*?"

6.5   Opening the Smart City

Given all that, it should prove absolutely no surprise that smart city systems such as those implemented in Stratumseind are instantiations of *exactly* this kind of technocratic reasoning. Indeed, the whole point of letting a network of sensors, actuators and artificial intelligences take over municipal services is that these slow and inefficient deliberative procedures are no longer required.

Unfortunately for citizens, it is only within the context of these deliberative procedures that the good can be discussed, argued, and decided upon. While it's all very well to design smart cities with a good or an ethic in mind (as in the case of Stratumseind, or the design strategies discussed by Tromp et al.), the pluralistic nature of democracy means that the good is not a set of values but an ongoing process by which the polis can (re)discover and (re)define itself. Although using artificial intelligences to swiftly and opaquely replace deliberative decision-making processes would quite likely produce positive outcomes in certain domains, they are also clearly incompatible with the necessarily slow, methodical, and dull processes that typify democratic governance. Or, as Feenberg writes in *Transforming Technology*:

> What human beings are and will become is decided in the shape of our tools no less than in the action of statesmen and political movements. The design of technology is thus an onto-logical decision fraught with political consequences. The exclusion of the vast majority from participation in this decision is profoundly undemocratic. (Feenberg, 2002, p. 3)

Thankfully, there is a way out of this mess. Contra Horkheimer's pessimism about (liberal) democratic norms being able evade the spectres of capital and modernity, I share Jürgen Habermas' pragmatic conviction that there remains room for hope. For Habermas, critical theory is both diagnostic tool and cure: a means of addressing the ways in which capital distorts democratic processes by fostering the existence of "ideal speech situations" (Habermas, 1970, 1979). These ideal speech situations furnish us with the appropriate conditions for democratic legitimacy—and, consequently, what Horkheimer calls "real democracy" to flourish (1972, p. 250). I am, in short, optimistic about the liberatory powers of democracy—even a democracy stymied by capitalistic systems of production and consumption.

So where to begin? The answer begins with relative (and deceptive) simplicity. Rather than relying on their own unexamined assumptions about what does and does not constitute the good, choice architects could simply design smart city systems in collaboration with the polis, thereby providing some much-needed nuance to these moral assumptions: an example of what Feenberg calls a "democratic transformation from below" (Feenberg, 2002, p. 17). Indeed, this is a recurring recommendation in the literature. Unfortunately, these recommendations rarely influence smart city programmes in any meaningful way. As Engelbert et al. (2019) observe: "many contemporary imaginations of the smart city, as well-intended as they might be, are still cultivating a top-down version of citizen participation and are excluding the interests and perspectives of citizens" (p. 352). While this fact hardly poses a categorical threat to the good-tracking possibilities of the smart city, the need for substantive and robust collective governance structures is certainly a call to action when it comes to implementing future smart city proposals.

So, what might a democratic transformation from below look like? We already know two things. First, it must be achieved via a procedure that is democratically legitimate: a process that is clear, transparent, and sensitive to the needs and preferences of all stakeholders. Second, it is also important that whatever norms are established can be easily and readily revised. After all, it is not enough for this collaboration to only occur *once* if the good is in flux. Instead, the polis needs to be able to intervene in how these systems are designed and managed on a regular basis, in order that the systems in question continue to cohere with both (a) the public good, and (b) a shared notion of what cities can and should be.

This is a fine sentiment but remains unfortunately abstract. We need some better way of thinking concretely about the level on which it is best to get the polis involved in these kinds of decision-making processes, while at the same time resisting the temptation to make serious decisions about what those decisions should be; after all, we don't want to run afoul of democracy's pluralism before we've even started! With that in mind, and whilst outlining this process in full is well beyond the ambit of this paper, for the beginning of an answer as to how we might approach this issue we can turn to Feenberg's "instrumentalization theory."

Feenberg devises instrumentalization theory as an analytical method for understanding and making sense of the rhythms and pressures of technical systems. He begins with what he calls "technical elements;" that is, the elementary technical ideas that are like the "vocabulary of a language; they can be strung together—encoded—to form a variety of 'sentences' with different meanings and intentions" (Feenberg, 2002, p. 78). Understanding what can function as a technical element requires a few things. First, we need to be in possession of certain bodily and conceptual affordances: an upright stance, binocular vision, fine motor articulation of our fingers, a brain capable of making the relevant conceptual discriminations, and so on. We also need the right attitude; "the technical orientation toward reality that Heidegger identified as the technological 'mode of revealing'" (Feenberg, 2002, p. 175). We must begin to think of the world and the things within it in terms of the *possibilities* that it affords: the kind of instrumentalisation that takes place when we begin to think of things in the world as means to our ends. It is a process that "proceeds by decontextualizing objects and simplifying them to highlight those qualities by which they are assigned a function" (Feng & Feenberg, 2008, p. 113). Feenberg calls this "primary instrumentalization."

Feenberg argues that primary instrumentalization is basically neutral when considered in light of different social values. Realising, for instance, that a rock can be made sharper by smashing it into another rock doesn't inherently tell you anything about the panoply of ways in which that sharpened rock could be used, for instance. It is only when we start thinking about the objects of primary instrumentalisation in terms of how we might use them that we begin "reorienting and integrating the simplified objects into a given natural and social environment" (Feng & Feenberg, 2008, p. 113). This process of reorientation—what Feenberg calls "secondary instrumentalization"—is fundamentally what design is all about: the application of the technical elements of primary instrumentalisation into systems of power, practice, and representation. This is the level at which decisions about *how* to use the sharpened rock are made: making arrowheads for hunting, knives for skinning animals, or axes for killing other people. Once integrated into systems of secondary instrumentalisation, the technical elements of primary instrumentalisation are no longer neutral. Instead, they become concrete, socialised, and strongly biased, in that they privilege certain kinds of use.

These two modes of instrumentalization also apply to more sophisticated artefacts, cities included. The raw building blocks of urban spaces (whether smart or dumb) constitute the primary mode of instrumentalisation: a Lego jumble of housing, roads, utility infrastructure, municipal architecture, sensors, actuators, and whatever else. However, understood in isolation, there is little that these objects can tell us. While they're obviously saddled with intended uses, we simply don't know the kinds of political, economic, or phenomenological inertia they exert until they are integrated into a common environment.[4] This process of integration—the secondary mode of instrumentalisation—is the process whereby not only final decisions about how, where, and when urban technologies should be used are made. It is also the process by which the norms implicit to the design and implementation of those technologies—such as the regulated hedonia of the Stratumseind—are enshrined. As Feenberg writes:

> Secondary instrumentalizations lie at the intersection of technical action and the other action systems with which technique is inextricably linked insofar as it is a social enterprise. The dialectics of technology is thus not a mysterious "new concept of reason" but an ordinary aspect of the technical sphere, familiar to all who work with machines if not to all who write about them. (Feenberg, 2002, p. 177)

In short, instrumentalization theory is a useful model from which we can begin to unpack the political and social pressures that technologies exert upon us: a model that commits us to neither the optimism of what he calls "instrumental" theories of technology, nor the pessimism of "substantive" theories of technology.[5] Moreover, instrumentalization theory has undergone a number of revisions and extensions since being first introduced; it is through one of these extensions that we can begin to plan, in a practical and concrete way, how to democratise the smart city.

In their paper "Thinking About Design", Patrick Feng and Feenberg use instrumentalization theory as a lens through which we can ask questions of the ways in which historical choices and shared cultural assumptions about technology "shape the design process" (Feng & Feenberg, 2008, p. 105), as distinct from analyses of design processes that focus on the roles and responsibilities of given stakeholders, such as designers or clients. They do this by contrasting "design space" with "technical codes." By "design space," Feng and Feenberg refer to the sum total of all feasible devices that can be designed with a given set of technical elements. It describes, if you like, the space of possibility. However, when new devices are designed from existing technical elements, it is not simply a free-for-all; instead, these decisions are made in light of "technical heritage": "While in theory there may be hundreds of technically feasible design options for a particular technology, [...] many technically feasible options are non-starters for reasons so obvious that they need no social justification—they are simply dismissed out of hand" (Feng & Feenberg, 2008, p. 115).

---

[4] See the "epistemological problem" of design, per Parsons (2015, pp. 35–39).

[5] In *Transforming Technology*, Feenberg argues that most attitudes towards technology can be characterised in one of two different ways: "instrumental" or "substantive." Instrumental attitudes to technology are grounded in the assumption that technology is a totally neutral instrument of human affairs: objects that extend or expand our capacities without influencing our behaviour. In doing so, these attitudes *overdetermine* the agency of human users. Meanwhile, substantive attitudes (such as Marcuse's model of technological rationality) make the opposite claim: that our technology has an inexorable and unavoidable influence upon our behaviour. That is, these attitudes are united in supporting the idea that our society, its institutions, and its operational norms, are in some sense regulated by the technologies we employ. In so doing, attitudes of this sort *underdetermine* the agency of human users.

These norms, or "technical codes," constrain and shape design processes: they are inherited social, cultural, legal, economic, epistemic, and other standards that influence how and why things are designed the way they are. Some of these codes are explicit, such as building codes and the like: standards that determine certain features regardless of the wishes of stakeholders. Often, though, these codes are implicit. By means of an example, Feng and Feenberg invoke refrigerators, the technical codes of which determine their "size as a function of the social principles governing family size" (Feng & Feenberg, 2008, p. 115). Of course, in most jurisdictions, family size is a combination of preference and capacity; nonetheless, given that families are generally of a certain rough size (itself a consequence of a vast array of social, economic, and design factors), and that refrigerators are designed to be maximally appealing to the largest number of consumers, the standard sizes in which refrigerators can be found are sensitive to those implicit codes.

Although Feng and Feenberg's analysis is primarily descriptive, in it lies the seeds of a normative programme: a way of bringing about the "democratic transformation from below" (Feenberg, 2002, p. 17). While designers and other stakeholders are obviously powerful actors within the design process, more powerful again are the technical codes that constrain and shape the design space. This obviously raises a challenge. If we're interested in making sure that designed things, such as smart cities, are genuinely just, open, and legitimate, we can't simply look at devices in isolation; we also need to understand the technical codes that undergird those devices. The question, as Feng and Feenberg put it, "is not whether to accept or reject technology, but rather how alternative values can be brought into the design process so that the technical codes that determine design are humane and liberating rather than oppressive and controlling" (Feng & Feenberg, 2008, p. 117). This is the challenge that faces us as we start thinking seriously about how to implement smart city systems: not only working to make sure that smart city devices are themselves just and open and legitimate, but also working to make sure that the technical codes shaping these devices are themselves the product of just, open, and legitimate processes.

This analysis of the technical codes that constrain the design and implementation of smart city systems provides the groundwork for a critique of smart cities. This critique is not simply a matter of democratically deciding upon the behaviours that smart cities should or should not facilitate. Instead, democracy plays a foundational role, in that it provides the means by which a given polis can collectively *conceptualise* the smart city on the levels of both primary and secondary instrumentalisation. Furthermore, in conceptualising the smart city—in recognising and discussing the various affordances that smart technologies offer—the polis is also given an opportunity to escape the technical codes that constrain current smart city projects. In this way, democracy's pluralism makes possible the development of hitherto unimagined potentials for smart urban technologies. This, by extension, gives the polis the conceptual and methodological space to systematically address the philosophical question of what a city *can* and *should* be.

These are all fabulously difficult questions: questions that strike at the very heart of what cities are, how they should be used, and under what conditions the polis is constituted. Nonetheless, they are also fabulously important: faced with technologies that promise (or threaten) to radically change the texture of urban landscapes, it is necessary to start thinking concretely about the implicit technical codes that have thus far governed smart city initiatives. These are questions that require careful and sustained community involvement, at all levels of policy and at all levels of research and development. It is only in developing methods that address, analyse, and shift these technical codes that we can ensure that we end up with the smart cities that we deserve.

References

Boot, M. (2017a). Problems of incommensurability. *Social Theory and Practice, 43*(2), 313–342.

Boot, M. (2017b). The right balance. *The Journal of Value Inquiry, 51*, 13–32.

Brautigan, R. (1967). *All watched over by machines of loving grace*. Communication Company.

Brock, K., den Ouden, E., van der Klauw, K., Podoynitsyna, K., & Langerak, F. (2019). Light the way for smart cities: Lessons from Philips Lighting. *Technological Forecasting and Social Change, 142*, 194–209.

Engelbert, J., van Zoonen, L., & Fadi Hirzalla, F. (2019). Excluding citizens from the European smart city: The discourse practices of pursuing and granting smartness. *Technological Forecasting and Social Change, 142*, 347–353.

Feenberg, A. (2002). *Transforming technology: A critical theory revisited*. Oxford University Press.

Feng, P., & Feenberg, A. (2008). Thinking about design: Critical theory of technology and the design process. In P. Kroes, P. E. Vermaas, A. Light, & S. A. Moore (Eds.), *Philosophy and design: From engineering to architecture* (pp. 105–118). Springer.

Forster, E. M. (1939). *What I believe*. Hogarth Press.

Gaus, G. F. (1997). Reason, justification, and consensus: Why democracy can't have it all. In J. Bohman & W. Rehg (Eds.), *Deliberative democracy: Essays on reason and politics* (pp. 205–242). MIT Press.

Habermas, J. (1970). *Toward a rational society*. Beacon Press.

Habermas, J. (1979). *Communication and evolution of society*. Beacon Press.

Horkheimer, M. (1972). Traditional and critical theory. In *Critical theory: Selected essays* (M. J. O'Connell, Trans.). Continuum.

Marcuse, H. (1982 [1941]). *Some social implications of modern technology*. Continuum.

Marcuse, H. (2007 [1964]). *One-dimensional man*. Routledge.

Marx, K. (1992 [1867]). *Capital: A critique of political economy* (B. Fowkes, Trans.). Penguin Group.

Parsons, G. (2015). *The philosophy of design*. Polity Press.

Peter, F. (2007). Democratic legitimacy and proceduralist social epistemology. *Politics, Philosophy & Economics, 6*(3), 329–353.

Peter, F. (2008). *Democratic legitimacy*. Routledge.

Schifferstein, H., Talke, K., & Oudshoorn, D. J. (2011). Can ambient scent enhance the nightlife experience? *Chemosensory Perception, 4*(1–2), 55–64.

Thaler, R. H., & Sunstein, C. (2008). *Nudge: Improving decisions about health, wealth, and happiness*. Yale University Press.

Tromp, N., Hekkert, P., & Verbeek, P. P. (2011). Design for socially responsible behavior: A classification of influence based on intended user experience. *Design Issues, 27*(3), 3–19.

Winner, L. (1980). Do artifacts have politics? *Daedalus, 109*(1), 121–136.

Part II
Trajectories of Contemporary Critique

Chapter 7
Critical (Big) Data Studies

Sally Wyatt

Abstract This chapter critically examines the concept of "big data" and the associated claims of using big data to solve myriad social and scientific problems. Four myths associated with "big data" are discussed. These are: (1) data are given; (2) "big data" is a natural resource; (3) numbers speak for themselves; and (4) everything is already digital. Each of these is critiqued and shown to be wanting. This is done by using examples of big data in different social and scientific application domains, and by drawing on insights from the philosophy of science and technology, science and technology studies (STS), and the emerging field of critical data studies. The conclusion raises a number of questions, the answers to which will enable people to engage critically with big data and to contribute to the development of critical big data studies. The conclusion also reflects on the author's own use of data to support her arguments.

Keywords Big data · Critical theory · STS · Andrew Feenberg · Critical data studies · Technology · Myths · Metaphors · Humanities · Ethnography

7.1 Introduction

The early days of the public internet (from the mid-1990s until the collapse of the first dot.com boom around 2004) were characterised by a great deal of hope and optimism. Politicians, policy makers, journalists and scholars expected that the possibilities of instant and global communication would result in transparency and openness in all spheres of public and private life (Cerf, 1999). Just as with earlier

S. Wyatt (✉)
Maastricht University Science and Technology Studies (MUSTS), Maastricht University,
Maastricht, the Netherlands
e-mail: sally.wyatt@maastrichtuniversity.nl

© The Author(s), under exclusive license to Springer Nature Switzerland AG 2022
D. Cressman (ed.), *The Necessity of Critique*, Philosophy of Engineering
and Technology 41, https://doi.org/10.1007/978-3-031-07877-4_7

"revolutions" in the production and sharing of information, from the printing press onwards (Briggs & Burke, 2009), the reality turned out to be more complex. Totalitarian regimes have also learned how to use digital technologies, and surveillance by private corporations has become routine (Lyon, 2018; Zuboff, 2019). There are fears about individuals becoming overly dependent on small computer devices leading to isolation, anxiety and detachment from reality (Turkle, 2017). And, digital divides remain, between and within countries (Thomas et al., 2018; UNDP, 2020).

Andrew Feenberg (1991, 1999) recognised early that we need critical theory in order to understand the ways in which digital technologies may also reproduce, reinforce or introduce socio-economic inequalities between individuals, social groups, countries and regions. In this chapter, I focus on "big data," one of the recent phenomena in which much hope is being invested, along with artificial intelligence and related developments in digital technoscience. "Big data" (as a concept, thus singular) is made possible by the extraordinary power of computers to collect, produce, store and process data in great volumes and at high speed in order to construct patterns in those data. "Big data," it is claimed, will transform everything from advertising to healthcare as the world becomes increasingly digitised (turning analogue material into digital form) and digitalised (use of digital technologies in government, industry, services and everyday life).

These claims need to be tempered by critique. In this chapter, I discuss four myths of big data: (1) data are given; (2) "big data" is a natural resource; (3) numbers speak for themselves; and (4). everything is already digital. Philosophers and statisticians have already provided a great deal of critique regarding myths 1 and 3. We know that data are never given and never speak for themselves. Nonetheless, the hype around big data suggests that critique needs to be repeated and extended for the "digital age." Myth 2 might be better considered as a metaphor or imaginary, but it is a powerful one that must be challenged. Thinking of data as a natural resource to be exploited for private gain is deeply problematic, as such thinking obscures both the work needed to create data and the destructive un/intended consequences that may arise. It also legitimates exploitation as a worthy goal. Myth 4 is also problematic. Much data are born digital, such as from social media use and data generated from sensors. But much data are hidden, such as those collected by intelligence agencies. Libraries and archives are full of historical documents that have not yet been digitised. And some data are no longer accessible to current ways of knowing.

In the following pages, I first define "big data" and clarify what I understand by critical theory. Each of the four myths is then discussed at greater length, using examples and drawing on insights from the philosophy of science and technology, science and technology studies (STS), and the emerging field of critical data studies. Examples from different domains are used to emphasise the variety of data. Furthermore, data used for scientific research may seem esoteric but those data and the techniques for collecting and analysing such data often find their way into industry and government. To conclude, I raise a number of questions, the answers to which will enable people to engage critically with big data and to contribute to the development of critical big data studies.

7.2   Terms and Debates: Big Data and Critical Theory

"Big data" as a term started to be used in the 1990s (Kitchin, 2014), but its use accelerated after 2007. According to Google Books NGram Viewer (2021), the use of the term grew sixfold from 2007 until 2019. Very simply, "big data" refers to large volumes of data of different types that cannot be collected or analysed in a reasonable length of time by hand or by using conventional data analysis tools and techniques. In business and consultancy literature, the four "Vs" are often mentioned: volume, velocity, variety, and most controversially, veracity. These are summarised in an infographic produced by IBM (n.d.). The volume of data continues to grow. For example, every day, almost 100 million photos are shared on Instagram and more than 300 billion emails are sent (Bulau, 2021). If the Large Hadron Collider (LHC) is running, CERN (*Conseil Européen pour la Recherche Nucléaire*) reads or writes one exabyte (10^{18}) of data in a year (CERN, 2020). To map a single human genome requires many tens of gigabytes of data. Not only numbers, but also words, sounds and images can be born digital or turned into digital formats so that they too can move and be analysed. Veracity is promised, but the uncertainty regarding the provenance and quality of data remains challenging, a point to which I return below. What is important to notice is the dependence on digital technologies to collect and process such volumes of data.

Mary F.E. Ebeling (2016) provides a moving auto-ethnographic account of the ways in which disparate data sources can be combined to draw conclusions of value to advertisers. Her book has the deceptively innocuous title, *Healthcare and Big Data*, and the more evocative subtitle, *Digital Specters and Phantom Objects*. This is a *tour de force* in what could be called "ethnographic noir," a detective story with no simple resolution and very disturbing undertones. Ebeling is an American sociologist who underwent many rounds of invasive fertility treatment, and later received targeted advertising for baby food and all the other paraphernalia that comes with a baby. But the treatments were not successful and she never had a baby. Over the years she continued to receive marketing information for this infant as they became a toddler and started school. The book is a first-person account of how this commodified, marketing baby emerged, as Ebeling's data about credit card payments, and hospital visits circulated between banks, insurance companies, schools, and baby food manufacturers. That this took place in the US is important, as she paid for some of her healthcare with her credit card, and those financial transactions ended up with advertisers who drew unfounded conclusions.

Also starting from her own work experiences in financial services, Cathy O'Neil (2016) identifies many examples of how big data can be used in policing, education, credit rating, and other fields. O'Neil acknowledges that some of these efforts started with good intentions, for example, to reduce crime or improve education. She documents the often undesirable social consequences of using big data to change society. The hopes remain, and are deployed to justify further development and use of big data and the algorithms needed to process them. There are many promising examples, especially in life sciences, where big data might be deployed in the identification of the causes of and treatments for diseases. Climate change researchers and nuclear physicists also use vast quantities of data to test their theoretical claims and make predictions. But, as I demonstrate later, even in these deceptively neutral environments, we cannot accept the promises of big data at face value.

These examples point to the need for critique, but how does one critique "big data" without being dismissive of all data? For the readers of this volume, it will be clear that I do not mean "critique" or "critical" in the everyday meaning of negative. The Socratic sense of examination and reasoning on the basis of evidence and dialogue is closer. But, like other contributors to this volume, I emphasise how theoretical reflection and critique can contribute to challenging and potentially transforming contemporary relations of power and inequality. Radder (this volume, Sect. 1.1, Chap. 1), building on insights from Marx, the Frankfurt School, Habermas (1984), and Feenberg (1999), emphasises the importance of being aware of the potential "gap between [abstract] theoretical claims and the relevant empirical practices and socio-historical developments." In other words, in order to understand the increasingly prominent role of both data and big data rhetoric, we [scholars] need to recognise they are local and global, with a past and a future, and that their meaning and use can be shaped.

When discussing four myths of big data in the following sections, I engage in critique in order to contribute to ongoing debates and actual practices about the use of big data in particular social and scientific contexts. Data are becoming ever more central to how business and public administration are done, how people live their lives, and how science is conducted. Often, as Marx predicted (1976 [1867]) about the extension of capitalist relations, the process of commodification is being extended into the collection and processing of data. Data that people have often voluntarily provided are sold back to them in different forms. Thus, it is important that people learn to engage critically with both data and "big data."

7.3   Myth 1: Data are Given

The original Latin meaning of data is "givens," but data are never given. Data and the means for producing and analysing them are amazing. Human creativity has resulted in the theoretical, experimental and other insights that make it possible to design the means and methods for collecting, categorising, analysing and curating data. Human creativity has also led to supercomputers, satellites, databases and everything else that is required to make data and to make them travel.

Borgman (2015, p. 28) defines data as "representations of observations, objects, or other entities used as evidence of phenomena for the purposes of research or scholarship." Borgman reviews other definitions and provides a number of examples to demonstrate the diversity of data and the interpretative and material work needed to produce data and make sense of them, including the following: Marie Curie's notebook is both scientific and historical data. Astronomical data can only be understood with access to the models used to generate them. Model organisms, such as lab mice, are a source of data, but acquiring useable data and being able to interpret them requires a great deal of theoretical, experimental and practical knowledge (see also Radder, 1996).

Let us compare this to the definition provided by the High Level Expert Group (HLEG) report about open science:

> When the term data is used in this report we refer to digital research objects in a broad sense, including regular research data and also meta-data [data about data], the associated services and workflows, analytics algorithms and all other data-related instruments that modern scientific research uses. (HLEG, 2016, p. 19)

The advantage of this definition is that it acknowledges the work needed to make it possible for data to travel. But there are three problems. First, the definition is circular: data are data. Second, it includes not only data but also the tools and equipment needed to process data. Third, only digital data matter, a point to which I return when discussing the fourth myth.

In common sense understandings of science, the natural world is out there, to be observed and recorded. But the ways in which we explore, categorise and understand the world are neither fixed nor universal. This is self-evident in the social sciences and humanities, where categorisations are not only the product of their time and place but are also performative. For example, notions of "family," "migration," or "work" can and do induce social changes, depending on political, cultural and technical circumstances. This is one of the reasons why historical data from censuses can be hard to interpret. Even in the physical, engineering and life sciences, data are always the end results of observational and experimental processes, conducted by people. Let us take an extreme example of "big science" to see how choices are made. ATLAS is one of four major experiments using the LHC at CERN. It produces 40 million collisions every second, of which only 200 "events" are deemed interesting enough to capture and store permanently. A single event results in 1.6 megabytes of data. What counts as "interesting enough" is decided by a committee of physicists, prior to an LHC experiment, taking into account the prevailing theoretical consensus (CERN, 2011, p. 8).

To say that data are never given is another way of saying that data require cognitive, social and technological work to acquire meaning – in their imagining, their creation, circulation, storage, use and re-use. As Sabina Leonelli (2016) has shown for plant biology, and Pinel et al. (2020) for human genetics, highly skilled people are needed to produce data. Technicians and scientists are the obvious ones, but in the research world there are many others. Libraries, archives, repositories, and publishers are all needed to manage and curate data. They do not look after themselves. Work is required to make data amenable to analysis and research. This is also true for non-scientific uses of data. Recall Ebeling's marketing baby (see above). A great deal of work was required by the fertility clinics, banks and credit card companies, and the producers of baby food and clothing to draw conclusions about the presumed existence of this child. Each and every day, when people are online or are using payment cards, people leave traces that can be turned into data that can be processed to determine what kind of advertising people are sent.

7.4   Myth 2: Data as Resource

"Data is the new oil" has gone beyond metaphor to become almost a cliché over the past decade. Oil is not the only resource to be metaphorically invoked. Puschmann and Burgess (2014) highlight how these nature metaphors (oil here, water later) serve to render technology and data as natural and beyond political control. José Van Dijck (2014) goes further. She points to the "gold rush" metaphor beloved of big data entrepreneurs, and recognises the "peculiar rationale" that presents data as a raw material to be exploited for profit (2014, p. 201), so decidedly not merely "natural." In other words, "data as resource" is not a myth, but a metaphor. The myth is that data and oil are natural, cheap and good for democracy and/or industry.

Both industry and policy makers draw on these resource-based metaphors to emphasise the importance of exploiting the economic potential of data for private or public gain (Rieder, 2018). In a 2012 speech, when she was still responsible for Europe's digital agenda, Neelie Kroes marvelled at the possibilities big data offers to "stimulate [the] market without spending big budgets" (5 March 2012). In the same speech, she heralded the democracy that would be brought by use of social media during the "Arab Spring," the name given to popular uprisings in Tunisia, Libya, Egypt, Syria and Bahrain that began in 2010. Violent repression and state-based misinformation using digital media followed (Howard & Hussain, 2011). Kroes was just as wrong in her choice of metaphor as she was in her expectation that social media use would inevitably lead to greater democracy.

These metaphors could be criticised for not going far enough. Finding and extracting oil and gold are highly knowledge- and capital-intensive activities. Similarly, huge amounts of work and theoretically driven enquiry are needed to make sense of large volumes of data (Leonelli, 2016; Noorman et al., 2018). The negative consequences of oil-based economies have long been visible, such as oil spills and other disasters, and the dependence of the global economy on an unsustainable form of energy. At the very least, this should alert both scholars and policy makers to consider what might go wrong when pursuing data as a resource to be exploited for financial gain. Big data may also have unintended consequences, such as huge financial and environmental costs for data storage, knowledge divides between social groups and between countries, and the datafication of everyday life.

Water-related metaphors are also often used to capture "big data," such as "data flood," "data deluge" and "data flows." These can suggest that the movement of data is fluid and unproblematic. Although flood and deluge can be threatening, they certainly capture the powerful movement of data (and of water). As Leonelli suggests in her discussion of scientific data:

> Not only do data not "flow" toward discovery, but it is the lack of smoothness and predefined direction that makes their travel epistemologically interesting and useful. Furthermore, the idea that data flow seems to suggest that data travel as a cohesive ensemble… In shifting from laboratory to publication, publication to database, and database to new research environment, data can be lost, acquired, misrepresented, transformed and integrated – and the metaphor of a journey seems to better capture those features of mass movement than the notion of flow. (Leonelli, 2016, p. 41)

Leonelli recognises the work required to make scientific data of epistemic value, based on detailed analysis of relevant practices. By proposing "journey" as an alternative metaphor, she contributes to a theoretical and linguistic critique that furthers debate and analysis of the role of big data.

Metaphors describe one thing in terms of another, either because they help to describe something novel or because they have greater rhetorical effect. As speech acts, they are both performative and constative (Sismondo, 1996). Metaphors are not only the preserve of poets but also of scientists, engineers, designers, policy makers, and politicians. And, as I have argued elsewhere (Wyatt, 2004, 2021), metaphors are available to all. It is important for scholars of critical internet and digital media studies to remember McCloskey's warning that "unexamined metaphor is a substitute for thinking – which is a recommendation to examine the metaphors, not to attempt the impossible by banishing them" (1986, p. 81). As Leonelli does by proposing "journey," it is not enough to analyse and critique but also to consider our own words and the work they may do to imagine and create different futures.

7.5 Myth 3: Data Speak for Themselves

The huge volumes of data being generated in the course of scientific experiments, such as those using the LHC or in genetics, have led to claims of a "fourth paradigm" in science, one that is data-intensive and data-driven (Hey et al., 2009). Similarly the large volumes of what are sometimes called "transactional data," the data people generate when using payment cards or leaving messages on social media, have led to claims of the end of traditional social science research methods such as interviews and surveys (Savage & Burrows, 2007). A much-cited essay (more than 2500 times, according to Google Scholar) by Anderson, then editor-in-chief of *Wired* magazine, claimed that big data would lead to a radical transformation of science and the scientific method.

> This is a world where massive amounts of data and applied mathematics replace every other tool that might be brought to bear. Out with every theory of human behavior, from linguistics to sociology. Forget taxonomy, ontology, and psychology. Who knows why people do what they do? The point is they do it, and we can track and measure it with unprecedented fidelity. With enough data, the numbers speak for themselves. (Anderson, 2008)

The title of the article, *The end of theory*, heralds this fourth paradigm in which traditional ways of doing science are challenged by large quantities of data and the digital technologies to process them. Theory development and hypothesis testing will become obsolete and "why" questions will become redundant. Instead, huge volumes of data and some rather basic statistics will provide correlations and insights into everything from the human genome to people's purchasing preferences. But knowing why the genome sometimes mutates and why people buy some things and not others no longer matters according to these big data enthusiasts.

There have been many critiques of this kind of computational colonising of science (Levallois et al., 2013; Berry & Fagerjord, 2017). Boyd and Crawford (2012) offered a critique for science and scholarship based on big data, already in 2012. One example is the use of tweets to gauge public opinion. They rightly point out that much research based on data provided by Twitter highlights the volume of data to justify any conclusions drawn, but such research often neglects to point out that Twitter the company does not make all tweets available for research purposes. Moreover, people who tweet are not necessarily representative of populations, and what people choose to present on Twitter and other social media platforms is "often carefully curated and systematically managed" (Manovich, 2012). The latter is, however, also true of interviews and survey responses. In other words, rather basic statistical research methods about representative sampling, for example, are neglected, meaning that conclusions with such data are, at best, spurious.

In the remainder of this section, I provide examples of three other problems: error, bias and findability. Error is not unusual in science. Reporting possible sources of error and searching for explanations of error (difference between expected and actual results) are important learning mechanisms. When sources of error are known, scientists and those who provide their tools do their best to avoid or correct them. Nonetheless, in 2004, it was realised that Excel (the proprietary spreadsheet software owned by Microsoft) turns gene names into dates. All genomes have an alphanumeric code, such as MARCH1 (Membrane-Associated Ring Finger (C3HC4) 1, E3 Ubiquitin Protein Ligaseand) and SEPT2 (Septin 2). These are automatically rendered as dates in Excel spreadsheets, mar-01 and sep-02 respectively. There was no way to disable this Excel feature, and researchers were left to find their own solutions, with varying levels of success and consistency. Ziemann et al. (2016) found that 20% of articles published in genomics journals between 2005 and 2015 continued to exhibit this rather basic flaw. Microsoft did nothing to fix it, and in August 2020 the Human Genome Nomenclature Committee announced a change in the guidelines for naming genes as a way of avoiding the problem (Bruford et al., 2020). This example illustrates not only how error can be magnified when using digital technologies but also how much work needs to be done by the scientific community to render data (re-)usable. It is also an example of the risks of having the tools for data analysis in private ownership. Many researchers, especially those working in universities and other publicly funded institutions, often use and develop open source software and share it freely via repositories such as Github (though that is also owned by Microsoft). Nonetheless, as the extent of the errors in genetics research indicates, proprietary software continues to be a barrier to realising open science.

Bias, for instance the tendency to treat individuals or groups based on unjustified beliefs or attitudes, often leading to unequal treatment, can also appear in applications using big data. The volume of the data does not automatically lead to correction, as some of the big data enthusiasts claim. Rather it can magnify the bias. The rapidly expanding field of machine learning is full of examples of gender and racial bias (Criado Perez, 2019). If historical data reflecting older gender stereotypes are used to train the machines, then, to give just one example, it is not surprising that

LinkedIn displayed highly paid jobs more frequently to men than to women (Büchel, 2018). More generally, machine learning necessarily represents the past and not the future, and thus cannot be used to support equal opportunities for groups that have been historically discriminated against in labour markets. This example, and many others, support Feenberg's observation that, "[e]xisting science and technology cannot transcend the capitalist world. Rather, they are destined to reproduce it by their very structure. They are inherently conservative" (2014, p. 180).

Some of this gender bias arises because of the very simple international standard that codes the "representation of human sexes," as follows: not known – 0; male – 1; female – 2; not applicable – 9. This is accompanied by qualifying text:

> No significance is to be placed upon the fact that "Male" is coded "1"and "Female" is coded "2". This standard was developed based upon predominant practices of the countries involved and does not convey any meaning of importance, ranking or any other biases that could imply discrimination. (ISO, 2004)

This is an excellent example of how historical "predominant practices" can be built into systems with potentially serious consequences for everyday life and for research. This code is the reason that "male" is the default option on the vast majority of drop-down menus on websites for everything from buying an airplane ticket to booking a concert. This code is also the reason that the long-running Women Writers' Network (n.d.) had enormous difficulties in developing a database of women's literature, dating from the middle ages until 1900. For most of that period, many women writers used male pseudonyms. The researchers involved in the network, from different European countries, wanted a database that captured the complexity of these women's lives, so needed to represent their legal name, their literary name or names, when it became known if they were biologically women. For those building the database, their training and professional standards led them to a simple choice between coding as 1 or 2. The ISO (International Organization for Standardization) standard remains committed to binary gender, even though feminists in the academy and in the political sphere have worked hard over the decades to improve women's rights and demonstrate the fluidity of gender categories.

The third example of how big data are not simply "given" arises from their volume. The vast quantities of data lead to what economists call negative network externalities, unfortunate side effects of more data becoming available. Big data are difficult to manage. The more there are, the harder it can be to find what one is looking for. The FAIR (Findable, Accessible, Interoperable, Re-usable) principles (Wilkinson et al., 2016) set out a technical framework for making data open in order to make it easier for researchers and others to re-use existing data. "Open data" is accompanied by promises of cost-saving, scientific breakthroughs and informed policy making (e.g. one of the foundational documents, Berlin Declaration (2003), and see also Mahrenbach and Mayer (2020) for an analysis of how open and big data are taken up in emerging states). Empirical research by Gregory and her colleagues (2020) explores how researchers actually engage in data search, demonstrating that it is not such a simple practice. Providers of data do not always follow community standards in the production of metadata, making findability and re-usability more difficult. Even when data producers provide high quality metadata and documentation, cross-disciplinary data re-use is not straightforward. Researchers from other fields may not have the experience necessary to understand the methodology used to create the data. The participants in Gregory et al.'s study needed more than a step-by-step guide to how data were collected. For both quantitative and qualitative data, participants needed details about the entire narrative surrounding data creation, i.e. why a method was chosen, and the local conditions of an experimental set-up or study design. This goes far beyond standardized metadata.

In summary, announcements about the "end of theory" were premature. Data are never simply given, nor are they enough by themselves to make any kind of knowledge claim. Theoretical concepts are always built into the classifications and algorithms used to make sense of large volumes of data. Moreover, concepts and theories are necessary for interpreting any kind of big data analysis, especially when researchers are making epistemic, social or normative claims.

7.6   Myth 4: Everything Is Already Digital

One of the biggest fallacies of big data, exemplified in the quote from the HLEG mentioned in Myth 2, is that everything is already digital. Nothing could be further from the truth. Of course, tweets are born digital (though, as boyd and Crawford pointed out it does not mean Twitter will let you have all of them, and it certainly does not mean that tweets are representative of any statistically meaningful population). Sensor data, used to record pollution or traffic, are directly captured into databases. But the location of sensors is not randomly nor evenly distributed, so some pollutants or traffic are more or less likely to be captured, and this can have consequences for health or transport policy.

To support the claim that "everything is already digital," statistics such as the following are presented: in 2020, 1.7 megabytes of data were created every second by every person; 90% of the world's data were created in 2019 and 2020. These numbers come about when, as mentioned earlier, the daily numbers of photographs and videos posted on Instagram and emails sent are counted (Bulau, 2021). By referring to such large numbers, it is indeed tempting to think that data about all of human life and the natural world have been digitally stored to be found when needed. This is not the case.

A perhaps surprisingly low proportion of the collections of national libraries and archives are digitised. In the Dutch National Library, due to copyright restrictions, only 0.8% of the books published since 1960 are digitally available, and only 20% of those published between 1940 and 1959 (L. Wilms, personal communication, 22 January 2021). This has consequences for literary scholars and historians. Humanities scholars are skilled in source criticism, having long dealt with incomplete archives, unreliable witnesses, silent and silenced participants, and in the concluding section, I will return to suggest what contemporary critical data studies can learn from them.

There are other ways that data are not visible and/or not amenable to the digital gaze, defined as the possibility of being captured and analysed by any kind of computational method or digital tool. Below I distinguish between five forms of invisible or shadowy data: disappeared, secret, taboo, unknowable, and machine-generated data and analyses.

First are data that have disappeared or are inaccessible to current ways of knowing. In other words, there are data people can no longer access either because they have disappeared (or been destroyed) or because the necessary equipment, methods, techniques and sources are no longer available. Traces of the data might be present, in publications or lab notebooks for example, but the data themselves can no longer be retrieved. Wylie (2017) elegantly describes how traces of archaeological finds can be stretched or attenuated over time as artefacts and documentation become disconnected. Some of the earliest Dutch census data (the first census was conducted in 1795) only now exist in tabular form, and the "raw data"/completed census forms have disappeared (Ashkpour et al., 2015). Similarly there are classicists and biblical scholars studying the lost works of Aristotle or other texts to which references can be found but not those original, now lost texts.

The second form are secret data that are deliberately kept out of view, such as for security purposes, military intelligence, or competitive advantage. Of course, such data are accessible to those who work with them on a daily basis (cf. Balmer, 2012; Rappert, 2009). Sometimes such data become known at some future point, depending on national rules about public disclosure. Sometimes they are made available to wider publics through the acts of individual and institutional whistle blowers, such as Edward Snowden and WikiLeaks (Coleman, 2014).

Third are data about taboo or very sensitive topics that may be difficult to generate or record, such as data about historic and present child abuse and violence against women. One of the Sustainable Development Goals of the United Nations (UN) is to "achieve gender equality and empower all women and girls." This is laudable, and one of the targets is to eliminate violence against women and girls, as measured by an indicator defined as the "proportion of women and girls aged 15 years and older subjected to sexual violence by persons other than an intimate partner in the previous 12 months, by age and place of occurrence" (United Nations, n.d.). But what about girls younger than 15? Perhaps the United Nations realises that such data would be impossible to gather, but that means violence against young children will remain unrecorded and the UN will not be able to say with any certainty whether progress has been made on achieving this goal.

The fourth type is data that are unknown or unknowable, arising from theoretical lacunae. In 2014, *Capital in the Twenty-First Century* by Thomas Piketty appeared in English to much acclaim, helping to put questions about income and wealth inequality on the political agenda. One of his achievements was to gather together disparate data to support his argument about how such inequalities were widening, especially in the US, UK and France. There were criticisms of some of his data sources, but what these criticisms revealed was the lack of adequate data for economists and policy makers to be able to grasp fully what the rich were doing with their money. Partly this is because the super-rich have the means and motive to obscure their income and wealth from tax authorities. More fundamentally, there is a dearth of data because Piketty depended on the openly available data collected by national authorities, such as those gathered by statistical offices and ministries of finance. These data are based on long-standing but nonetheless faulty ideas held by neoclassical economists about how to theorise and measure wealth. McGoey (2017) persuasively demonstrates that this public data "is incapable of distinguishing between wealth that is earned productively – through making economic contributions – and wealth that is extracted illegitimately through rent seeking" (p. 276). The data simply do not exist because, for more than a century, economists and statisticians have not been taught the theoretical frames that would help them to conceptualise the problem of wealth accumulation and the data needed to measure it.

Finally, there are machine-generated data and analyses. This may sound paradoxical as it would certainly be expected that data generated via computer models and simulations are knowable by digital means. Indeed they might be, but the problem is that they are not knowable by humans if they are produced by black-boxed algorithms, building on patterns within large datasets. Data can be generated by machines, such as models which extrapolate on the basis of past data, or automatic recording of digital events. Examples of the latter include data captured by telephone providers, such as initiating number, called number, length of call, location, etc., and data used to monitor computer and network security, such as log-in data, breaches, etc. Monitoring the quality and provenance of such data can be extremely challenging.

An example of machine-generated analysis is topic modelling, an increasingly popular technique used by computational linguists and others to understand patterns in large digital corpora. It is based on natural language processing in computer science. It is similar in aim to co-word mapping which has its roots in quantitative science studies, and is used to understand semantic relationships between documents in order to map the development of scientific fields. Leydesdorff and Nerghes (2017) compared the results from co-word mapping and topic modelling on a single medium-sized corpus. They obtained quite different results, but the co-word maps performed much better. The major line of defence of topic modelling is that it only works with big data, with very large corpora. Topic models generate a list of words, and humans can always find a narrative to link those words, but the analytic tool itself is opaque and black-boxed.

7.7 Conclusion: What Is to Be Done?

In the preceding pages, I have drawn attention to four myths about big data: that they are given, that they are a natural resource, that they speak for themselves, and that they are universally digital. These myths mean that the promises of big data must be questioned, and attention must be paid to how big data are described, what promises accompany them, and which metaphors are chosen to promote these promises.

Data continue to be gathered by both public and private organisations. People's lives become "datafied," and data shape and reshape business, governments and knowledge production. Data are not flowing freely, neither in the sense of easily nor at zero cost. The latter could perhaps be realised with the current technical means but instead, many data are commodified and sold back to patients, travellers, citizens and consumers in different forms, often indirectly in the form of advertising. Data do not flow easily, for all sorts of reasons, including incompatible standards, proprietary software, and the huge amounts of work needed to make data travel. Whether or not it is helpful to think of these flows of data as "big data" is not the most important question. It is certainly not helpful to think of "big data" as "the new new thing" (Lewis, 1999) since that obscures the political, economic and cultural dynamics that shape the ways in which data are generated and used. The gap between the claims made by enthusiasts about the potential of big data and the specific social and material practices in which data are used remains wide.

Education remains important. Everyone needs to learn what is possible with data, and what remains beyond datafication and the digital. There is an important role for educational institutions and the media, to help people learn what kinds of critical questions to ask when data are presented. For this, we can learn from historians and other humanities scholars who are trained in source criticism. Those techniques can be applied to digital sources and tools (Traub, 2020), and help us all to ask the following kinds of questions: Who created the data? For what purposes? When and how were the data published? Is documentation on collection, curation and provenance available? Who developed the tools for analysing the data, and through which means? Is there any documentation available about the tools used? Are there alternative tools? Do they come up with the same results? How can meaningful search queries be formulated, and the results assessed? What is missing? What has been lost or deleted?

By pursuing these sorts of critical questions, Ebeling was able to understand how the phantom data baby emerged. She had questions for the marketing, financial services and healthcare companies that caused her such pain by creating a phantom that haunted her for years. Drawing on Marxian ideas of commodification, she traced how her intimate health data had become commodified, sold to others, and then used to try to sell her products for her never-born child.

In this chapter, I have presented many different examples of big data, including Ebeling's data baby, and data presented by industry journalists. I have also given examples relating to particle physics, genomics, policing, text corpora, and citations generated by Google NGram Viewer and Google Scholar. I have used these examples to support my critique of the claims made by the many proponents of big data in industry, government and science. I hope that you, the reader, have taken a similar approach, and not simply accepted my claims. Perhaps by following the references I provided, or by considering alternative examples, that may or may not support my claims.

Finally, we need to cherish the heterogeneity of data. This means going beyond "V for variety" in the industry definition of big data. Data can be made to do amazing things that could further knowledge and improve the social world, but that does not happen without human intervention. As the various examples mentioned in the preceding pages have indicated, intervention can take many different forms, from deciding what and how data need to be collected, and then how to process, analyse, store and share them. Paying attention to the decisions made throughout those processes and to the absences, gaps and omissions that are still present is crucial for those engaged in critical big data studies.

Acknowledgements This paper began when I was invited to contribute to the 42nd birthday celebrations of Maastricht University, the *Dies Natalis*, held on 26 January 2018. The title of my lecture was, "Where is the knowledge we have lost in data?", a reference to T.S. Eliot's poem, "The Rock" (1934). A recording of the lecture can be found here: https://www.youtube.com/watch?v=WSOIkHOnzTM. I am grateful to Rianne Letschert, *Rector Magnificus* of Maastricht University, for the invitation.

Thanks are also due to Darryl Cressman for his invitation to turn that lecture into this written text, and for his very constructive and helpful feedback on an early draft. Hans Radder generously provided critical feedback – in the best of the word "critical" – on the text for the lecture and for this chapter.

References

Anderson, C. (2008, 23 June). The end of theory. The data deluge makes the scientific method obsolete. *Wired*. https://www.wired.com/2008/06/pb-theory/. Accessed 6 Apr 2021.

Ashkpour, A., Meroño-Peñuela, A., & Mandemakers, K. (2015). The aggregate Dutch historical censuses: Harmonization and RDF. *Historical Methods: A Journal of Quantitative and Interdisciplinary History, 48*(4), 230–245.

Balmer, B. (2012). *Secrecy and silence: A historical sociology of biological and chemical warfare*. Ashgate.

Berlin Declaration on Open Access to Knowledge in the Sciences and Humanities. (2003). https://openaccess.mpg.de/berlin-declaration. Accessed 3 Jan 2021.

Berry, D., & Fagerjord, A. (2017). *Digital humanities. Knowledge and critique in a digital age*. Polity Press.

Borgman, C. (2015). *Big data, little data, no data. Scholarship in the networked world*. The MIT Press.

boyd, d., & Crawford, K. (2012). Critical questions for big data. *Information, Communication & Society, 15*(5), 662–679. https://doi.org/10.1080/1369118X.2012.678878

Briggs, A., & Burke, P. (2009). *A social history of the media: From Gutenberg to the internet*. Polity Press.

Bruford, E., Braschi, B., Denny, P., Jones, T., Seal, R., & Tweedie, S. (2020). Guidelines for human gene nomenclature. *Nature Genetics, 52*, 754–758. https://doi.org/10.1038/s41588-020-0699-3

Büchel, B. (2018, 1 March). Artificial intelligence could reinforce society's gender equality problems. *The Conversation*. https://theconversation.com/artificial-intelligence-could-reinforce-societys-gender-equality-problems-92361. Accessed 4 Jan 2021.

Bulau, J. (2021, 18 March). How much data is created every day in 2020? *TechJury*. https://techjury.net/blog/how-much-data-is-created-every-day. Accessed 6 Apr 2021.

Cerf, V. (1999). *The Internet is for everyone. ICANN RFC 3271*. https://datatracker.ietf.org/doc/rfc3271/?include_text=1. Accessed 8 Jan 2021.

CERN. (2011). *ATLAS fact sheet*. https://cds.cern.ch/record/1457044/files/ATLAS%20fact%20sheet.pdf. Accessed 7 Jan 2021.

CERN. (2020). *Key facts and figures – CERN Data Centre September 2020_V1*. https://information-technology.web.cern.ch/sites/information-technology.web.cern.ch/files/cerndatacentre_keyinformation_sept2020v1.pdf. Accessed 2 Jan 2021.

Coleman, G. (2014). *Hacker, hoaxer, whistleblower, spy. The many faces of Anonymous*. Verso.

Criado Perez, C. (2019). *Inivsible women. Exposing data bias in a world designed for men*. Chatto & Windus.

Ebeling, M. (2016). *Healthcare and big data. Digital specters and phantom objects*. Palgrave Macmillan.

Feenberg, A. (1991). *Critical theory of technology*. Oxford University Press.

Feenberg, A. (1999). *Questioning technology*. Routledge.

Feenberg, A. (2014). *The philosophy of praxis: Marx, Lukacs and the Frankfurt School*. Verso.

Google Ngram Viewer. (2021, April 3). https://books.google.com/ngrams/graph?content=big+data&year_start=2005&year_end=2019&corpus=26&smoothing=3.

Gregory, K., Groth, P., Scharnhorst, A., & Wyatt, S. (2020). Lost or found? Discovering data needed for research. *Harvard Data Science Review, 2*(2), 1–59. https://doi.org/10.1162/99608f92.e38165eb

Habermas, J. (1984). *The theory of communicative action. Reason and the rationalization of society* (T. McCarthy, Trans.). Polity Press.

Hey, T., Tansley, S., & Tolle, K. (Eds.). (2009). *The fourth paradigm. Data-intensive scientific discovery*. Microsoft Research.

HLEG (High Level Expert Group). (2016). *Realising the European open science cloud*. European Commission.

Howard, P., & Hussain, M. (2011). The upheavals in Egypt and Tunisia: The role of digital media. *Journal of Democracy, 22*(3), 35–48.

IBM. (n.d.). *The four V's of big data. Big data and analytics hub*. https://www.ibmbigdatahub.com/infographic/four-vs-big-data. Accessed 2 Jan 2021.

ISO (International Organization for Standardization). (2004). *ISO/IEC 5218:2004(en) Information Technology – Codes for the representation of human sexes*. ISO.

Kitchin, R. (2014). *The data revolution. Big data, open data, data infrastructures and their consequences*. SAGE.

Kroes, N. (2012, March 5). Digital agenda and open data. From crisis of trust to open governing. *Speech Bratislava*. https://ec.europa.eu/commission/presscorner/detail/en/SPEECH_12_149. Accessed 7 Jan 2021.

Leonelli, S. (2016). *Data-centric biology. A philosophical study*. University of Chicago Press.

Levallois, C., Steinmetz, S., & Wouters, P. (2013). Sloppy data floods or precise data methodologies? Dilemmas in the transition to data-intensive research in sociology and economics. In P. Wouters, A. Beaulieu, A. Scharnhorst, & S. Wyatt (Eds.), *Virtual knowledge. Experimenting in the humanities and the social sciences* (pp. 151–182). The MIT Press.

Lewis, M. (1999). *The new new thing. A Silicon Valley story*. W.W. Norton & Co.

Leydesdorff, L., & Nerghes, A. (2017). Co-word maps and topic modeling: A comparison using small and medium-sized corpora (n < 1000). *Journal of the Association for Information Science and Technology, 68*(4), 1024–1035.

Lyon, D. (2018). *The culture of surveillance. Watching as a way of life*. Polity Press.

Mahrenbach, L., & Mayer, K. (2020). Framing policy visions of big data in emerging states. *Canadian Journal of Communication, 45*(1), 129–141.

Manovich, L. (2012). Trending. The promises and the challenges of big social data. In M. Gold (Ed.), *Debates in the digital humanities* (pp. 460–474). University of Minnesota Press.

Marx, K. (1867/1976). *Capital volume 1* (B. Fowkes, Trans.). Penguin.

McCloskey, D. (1986). *The rhetoric of economics*. University of Wisconsin Press.

McGoey, L. (2017). The elusive rentier rich: Piketty's data battles and the power of absent evidence. *Science, Technology, & Human Values, 42*(2), 257–279.

Noorman, M., Wessels, B., Sveinsdottir, T., & Wyatt, S. (2018). Understanding the "open" in making research data open: Policy rhetoric and research practice. In A. Sætnan, I. Schneider, &
N. Green (Eds.), *The politics of big data* (pp. 292–318). Routledge.

O'Neil, C. (2016). *Weapons of math destruction. How big data increases inequality and threatens
democracy.* Penguin.

Piketty, T. (2014). *Capital in the twenty-first century* (A. Goldhammer, Trans.). Belknap Press.

Pinel, C., Prainsack, B., & McKevitt, C. (2020). Caring for data: Value creation in a data-intensive
research laboratory. *Social Studies of Science, 50*(2), 175–197.

Puschmann, C., & Burgess, J. (2014). Big data, big questions. Metaphors of big data. *International
Journal of Communication, 8*, 1690–1709.

Radder, H. (1996). *In and about the world: Philosophical studies of science and technology.* State
University of New York Press.

Rappert, B. (2009). *Experimental secrets. International security, codes and the future of research.*
University of America Press.

Rieder, G. (2018). Tracing big data imaginaries through public policy: The case of the European
Commission. In A. Sætnan, I. Schneider, & N. Green (Eds.), *The politics of big data*
(pp. 89–109). Routledge.

Savage, M., & Burrows, R. (2007). The coming crisis of empirical sociology. *Sociology, 41*(5),
885–899.

Sismondo, S. (1996). *Science without myth. On constructions, reality, and social knowledge.* State
University of New York Press.

Thomas, J., Barraket, J., Wilson, C., Cook, K., Louie, Y. M., Holcombe-James, I., Ewing, S., &
MacDonald, T. (2018). *Measuring Australia's digital divide: The Australian digital inclusion
index.* RMIT University. https://doi.org/10.25916/5b594e4475a00

Traub, M. (2020). *Measuring tool bias and improving data quality for digital humanities research*
(PhD thesis). University of Amsterdam, Amsterdam, NL.

Turkle, S. (2017). *Alone together. Why we expect more from technology and less from each other.*
Basic Books.

UN (United Nations), Department of Social and Economic Affairs. (n.d.). Sustainable development. *Goal 5.* https://sdgs.un.org/goals/goal5. Accessed 1 Jan 2021.

UNDP (United Nations Development Program). (2020). *Human development reports.* https://hdr.
undp.org/en/indicators/43606. Accessed 8 Jan 2021.

Van Dijck, J. (2014). Datafication, dataism and dataveillance: Big Data between scientific paradigm and ideology. *Surveillance & Society, 12*(2), 197–208.

Wilkinson, M., Dumontier, M., Ijsbrand, J., Appleteon, G., Axton, M., Baak, A., et al. (2016). The
FAIR Guiding Principles for scientific data management and stewardship. *Scientific Data, 3*,
160018. https://doi.org/10.1038/sdata.2016.18

Women Writers in History. (n.d.). *Women writers' networks.* http://www.womenwriters.nl.
Accessed 3 Apr 2021.

Wyatt, S. (2004). Danger! Metaphors at work in economics, geophysiology and the Internet.
Science, Technology & Human Values, 29(2), 242–261.

Wyatt, S. (2021). Metaphors in critical internet and digital media studies. *New Media & Society,
23*(2), 406–416.

Wylie, A. (2017). How archaeological evidence bites back: Strategies for putting old data to work
in new ways. *Science, Technology, & Human Values, 42*(2), 203–225.

Ziemann, M., Eren, Y., & El-Osta, A. (2016). Gene name errors are widespread in the scientific
literature. *Genome Biology, 17*, 177. https://doi.org/10.1186/s13059-016-1044-7

Zuboff, S. (2019). *The age of surveillance capitalism.* Public Affairs.

Chapter 8
The Behavioral Code: Recommender Systems and the Technical Code of Behaviorism

Marit de Jong and Robert Prey

Abstract Our lives are increasingly mediated, regulated and produced by algorithmically-driven software; often invisible to the people whose lives it affects. Online, much of the content that we consume is delivered to us through algorithmic recommender systems ("recommenders"). Although the techniques of such recommenders and the specific algorithms that underlie them differ, they share one basic assumption: that individuals are "users" whose preferences can be predicted through past actions and behaviors. While based on a set of assumptions that may be largely unconscious and even uncontroversial, we draw upon Andrew Feenberg's work to demonstrate that recommenders embody a "formal bias" that has social implications. We argue that this bias stems from the "technical code" of recommenders – which we identify as a form of behaviorism. Studying the assumptions and worldviews that recommenders put forth tells us something about how human beings are understood in a time where algorithmic systems are ubiquitous. Behaviorism, we argue, forms the *episteme* that grounds the development of recommenders. What we refer to as the "behavioral code" of recommenders promotes an impoverished view of what it means to be human. Leaving this technical code unchallenged prevents us from exploring alternative, perhaps more inclusive and expansive, pathways for understanding individuals and their desires. Furthermore, by problematizing formations that have successfully rooted themselves in technical codes, this chapter extends Feenberg's critical theory of technology into a domain that is both ubiquitous and undertheorized.

Keywords Technical code · Behaviorism · Recommender systems · Formal bias · Andrew Feenberg · B.F. Skinner · Algorithms · Data

M. de Jong (✉)
Philosophy Department, University of Groningen, Groningen, Netherlands
e-mail: marit.de.jong@rug.nl

R. Prey
Department of Media Studies and Journalism, University of Groningen,
Groningen, Netherlands
e-mail: r.prey@rug.nl

© The Author(s), under exclusive license to Springer Nature
Switzerland AG 2022
D. Cressman (ed.), *The Necessity of Critique*, Philosophy of Engineering and
Technology 41, https://doi.org/10.1007/978-3-031-07877-4_8

8.1   Introduction

Our lives are increasingly mediated, regulated and produced by algorithmically-driven software that are often invisible to the people whose lives it affects. Online, much of the content that we consume is delivered to us through algorithmic recommender systems (hereafter "recommenders"). Recommenders, as a popular textbook explains, "are software tools and techniques that provide suggestions for items that are most likely of interest to a particular user" (Ricci et al., 2011, p. 1). Although the techniques of such recommenders and the specific algorithms that underlie them differ, they share one basic assumption: that individuals are "users" whose preferences can be predicted through past actions and behaviors. Developers of these systems believe that the collection and analysis of such interactional data provides a representation of individuals that is far less susceptible to the prejudices that plague media and market research predicated on demographic profiling. In forwarding an explicitly "post-demographic" agenda, these systems and their developers promote an anti-essentialist ethos that is dictated by "preferences, not stereotypes" (Riedl & Konstan, 2002, p. 113).

Nevertheless, while based on a set of assumptions that may be largely unconscious and even uncontroversial, in this chapter we demonstrate that recommenders embody a "formal bias" (Feenberg, 2017) that has social implications. We argue that this bias stems from the "technical code" (ibid.) of recommenders – which we identify as a form of behaviorism. In line with behaviorism, recommenders work with a definition of preferences that frames them as behavioral dispositions rather than inner states (such as emotions, meaning or values).

Apps and online platforms have sometimes been accused by critics of employing behaviorist tactics to "hook" users and "nudge" behavior (e.g. Zuboff, 2019). Such critics charge that "surveillance capitalism" manipulates users by adopting the practices of operant conditioning, made famous by the behaviorist B. F. Skinner. Our focus here is somewhat different: We are interested in the *episteme* that grounds the development of recommenders. Studying the assumptions and worldviews that recommenders put forth tells us something about how human beings are understood in a time where algorithmic systems are ubiquitous (Cheney-Lippold, 2011). As we increasingly turn to computational tools to organize much of the information and content we create and consume, we subject our discourse and knowledge to the logics undergirding computation. This means that the presumptions of a small subset of the world's population decides on the logic that significantly shapes our understanding of ourselves and the world around us (Gillespie, 2014, p. 168).

What we refer to as the "behavioral code" of recommenders, we argue, promotes an impoverished view of what it means to be human. Leaving this technical code unchallenged prevents us from exploring alternative – perhaps more inclusive and expansive – pathways for understanding individuals and their desires. By problematizing "formations that have successfully rooted themselves in technical codes" (Feenberg, 2008, p. 52), this chapter extends Feenberg's critical theory of technology into a domain that is both ubiquitous and undertheorized.

8.2   The History and Study of Recommenders

Many early Internet users were affected by a paralysing condition known to psychologists as "overchoice" as a seemingly infinite supply of information, products, and services greeted them in their first forays online. Recommenders soon came to the rescue, bringing a degree of order to the chaos of digital information. The foremost technique used in early recommenders was collaborative filtering. Collaborative filtering is a widely used method to filter information by grouping together users deemed to have similar tastes or preferences. Developed as a research project at Xerox PARC in 1992, "Tapestry" is widely considered to be the first algorithmic recommender to use the term "collaborative filtering" (Goldberg et al., 1992). By the mid-1990s, a team from the University of Minnesota employed the same method for a Usenet news recommender called GroupLens. This team later created MovieLens, which asked users to rate movies on a five-star scale and then recommended movies seen by other users who had provided similar ratings. Soon after, MIT's Media Lab released "Ringo" (later Firefly), which used collaborative filtering to automate music recommendations (Riedl & Konstan, 2002). As e-commerce websites began to proliferate in the 1990s, recommenders fulfilled a need by business to help customers sort through products and make choices. In the March 1997 issue of the *Communications of the ACM*, the guest editors marvelled at how "a flurry of commercial ventures have recently introduced recommender systems for products ranging from Web URLs to music, videos, and books" (Resnick & Varian, 1997, p. 58).

Today, we encounter recommenders seemingly everywhere online: They filter books and other products on Amazon, television shows and films on Netflix, and news and social media posts on Facebook. Indeed, much of the online content that we consume is delivered to us through algorithmic recommenders. While collaborative filtering remains the archetypal recommender, a wide array of different filtering systems, such as content-based and context-based recommenders, have been developed over the years. Most recommenders now use an ensemble or hybrid approach, combining two or more filtering methods. While the information that drives these filtering systems can include explicit signals, such as product ratings, there has been a marked trend in recent years towards favoring implicit feedback, such as clicks or other trackable user interactions (Ekstrand & Willemsen, 2016).

The majority of research on recommenders (cf. Adomavicius & Tuzhilin, 2005; Bobadilla et al., 2013; Lops et al., 2011; Pazzani & Billsus, 2007; Ricci et al., 2011), is focused on explaining or comparing the strengths and weaknesses of different approaches, or offering suggestions on how to improve recommendations (cf. Burke, 2007; Linden et al., 2003; Salter & Antonopoulos, 2006; Tkalčič et al., 2010). More recently, humanities and social science scholars have begun to shine a critical light on recommenders to explore how they produce, reproduce and manage consumer desire (Drott, 2018) and individual subjects (Prey, 2018). Other critical research is concerned with privacy issues that surround recommenders (Perik et al., 2004) and with how such systems exercise influence over the culture we consume

(Beer, 2009, 2013; Morris, 2015; Seaver, 2012). For example, the specific techniques Netflix utilizes to understand its users' tastes and to recommend content could impact the type of television programs and films that get produced (Hallinan & Striphas, 2016). Importantly, scholars have pointed out how algorithms and the recommenders they power are always sociotechnical ensembles that extend and magnify "the all-too-human biases, worldviews, and blind spots of the people who designed, built, and maintained them" (Seaver, 2021). One such bias or worldview, we argue, is the assumption that individual preferences can best be defined as behavioral dispositions and predicted through past action and implicit behavior. Before we develop this argument, we will briefly review Feenberg's concept of "formal bias" and how such a bias emerges out of specific "technical codes."

8.3 Technical Code and Formal Bias

Technology, Feenberg writes, is technically underdetermined (e.g., 1992 p. 305; 2008 p. 51). In designing any technological object, one cannot work solely from the principles of technical logic. This manifests itself first in the availability of a surplus of workable solutions to any problem that a technology is supposed to solve. The social actors involved in the design process make the final choice between (technically) equally viable options. They thus need to motivate their choice by something other than technical criteria. The second manifestation of underdetermination can be found in the many ways in which a problem is chosen and defined. What is perceived as a need or problem arises from the viewpoint of the individual. As Feenberg (2017 p. 6) explains: "people who must commute to work acquire an interest in good roads, while those whose homes are polluted by the cars exhaust acquire an interest in better pollution controls, and so on." Neither the subject of technology nor its approach to this subject can thus be fully determined by technical criteria. Technical development does not follow a definitive path towards advancement since there are multiple branches possible and "the final determination of the 'right' branch is not within the competence of engineering, because it is simply not inscribed in the nature of the technology" (Feenberg, 1992, p. 308). Other choices are always possible in terms of what to make and how to make it.

Since technology is underdetermined by purely rational considerations, developers must turn to social judgment in order to select between alternative feasible technical designs. In making decisions, developers rank "items as ethically permitted or forbidden, or aesthetically better or worse, or more or less socially desirable" (Feenberg, 2008, p. 52). Consequently, Feenberg argues that what guides the selection process is the "political-cultural horizon" (e.g. Feenberg, 1999, p. 87) of our society. This term refers to society's broad assumptions about social values. The realization of such social values in the form of technological specification is what he calls the "technical code" (Feenberg, 1999, p. 87–9). Once these values are materialized in technologies, they work to validate the cultural horizon that they stem from. As such, technologies perform a "formal bias" (e.g., Feenberg, 2017):

Seemingly neutral, they actually offer a material affirmation of – and thus a bias towards – the ruling social values. This does not mean that they lose their claim to being rational, as for Feenberg rationality is relative to social context. He makes this clear in his example of "rational" machine design in the era of child labour. As Feenberg writes in this volume:

> [W]hen the socially accepted definition of the labor force included children, features of the technology such as the placement of controls were designed for small workers. This was technically rational under the given conditions although today we might consider the whole business of child labor a scandal.

Technology can as such be simultaneously technically rational *and* formally biased. In the design of recommenders, as we shall subsequently argue, a certain underlying technical code can be identified; one that may be "rational" yet manifests in these systems' formal bias. In order to analyze recommenders in terms of their technical code and formal bias, we first turn to the workings of these systems.

8.4 Inside Recommenders

Since the algorithms that drive recommenders are largely kept a secret (they are, after all, what determines the success of a digital platform), another approach is needed to study them. We choose to study recommenders from the outside and fill in the gaps by reading texts from within the field of recommender systems. We performed a close reading of educational textbooks (e.g., Aggarwal, 2016; Falk, 2019) and papers from the annual ACM Conference on Recommender Systems (e.g., Ekstrand & Willemsen, 2016; Wan & McAuley, 2018): "the premier international forum for the presentation of new research results, systems and techniques in the broad field of recommenders" (RecSys, n.d.). In what follows we provide an overview of the core techniques and data primarily utilized in the development of contemporary recommender systems.

8.4.1 *Behavioral and Environmental Data*

Demographic markers for identity, such as age and gender, have long been used by media and market research as a proxy for preference. Algorithmic recommenders instigated a break with this method by claiming to circumvent the need for proxies altogether. "Treat customers as individuals, not demographics," two pioneers of collaborative filtering advised their readers: "Let their preferences, not stereotypes, dictate which products and messages you present to them" (Riedl & Konstan, 2002).

Indeed, contemporary recommenders could be described as "post-demographic machines" (Rogers, 2009). As the vice-president of Netflix's Original Series remarked: "We found that demographics are not a good indicator of what people like to watch" (Lynch, 2018). Rather than eliminating proxies for preference altogether however, demographics have been replaced with real-time behavioral data. More specifically, users are typically reduced to (1) *measurable*, (2) *implicit*, (3) *past,* and, increasingly, (4) *contextualized* behavior.

Recommenders only allow for input that can be processed by algorithms. Consequently, they work with a certain type of behavior: the kind that can be digitally observed, or "datafied" (Fisher & Mehozay, 2019 p. 10). In other words, their input is *measurable behavior.* What cannot be directly observed by recommenders, however, are inner states such as thoughts, feelings, motives, and preferences. In other words, recommenders cannot immediately observe the very thing that they are after. One way to get around this difficulty is to directly ask users to communicate their inner states. Indeed, coaxing users to provide explicit feedback, such as ratings and reviews, to express their preferences used to be a common approach.

Over time, however, a different method began to be given primacy: tracking *implicit behavior,* like clicks or other trackable user interactions (Ekstrand & Willemsen, 2016; Seaver, 2019, p. 430). This shift resulted from the discovery that explicit user-data – such as ratings – poses a threat to prediction. It turns out that explicit ratings vary significantly depending on time and setting: a user could give a movie three stars one day and five the next one. In addition, explicit data is relatively scarce as it requires users to take time to express preferences. On the other hand, implicit behavioral data, or interaction data such as clicking and scrolling, is demonstrably good at predicting future user behavior, and is also readily available and thus easier to collect (Ekstrand & Willemsen, 2016). Consequently, explicit ratings have been widely replaced with implicit behavioral data (Seaver, 2019).

Explicit data based on users' subjective interpretation is now often perceived as a hindrance to actually understanding the user. In a recent paper, Nick Seaver (2021, p. 15) describes a conversation he had with "Tom," a product manager for "audience understanding" at a music recommendation company anonymized as "Whisper":

'We don't interview users', he told me. Instead, audience understanding depended on the same aggregated listening data that powered Whisper's recommendations. 'We think we have real science here', Tom said.

Netflix developers likewise explain that their platform tracks activity such as "the time elapsed since viewing, the point of abandonment (mid-program vs. beginning or end), whether different titles have been viewed since, and the devices used" (Gomez-Uribe & Hunt, 2015, p. 4). Listening or viewing logs are considered a more legitimate and reliable form of knowledge that better represents how users "*actually* behaved" (Seaver, 2021, p. 15), rather than what they might claim to have consumed if asked explicitly.

Finally, it follows that recommenders work with *past behavior.* As an influential early book in the field announced: "In order to know what someone wants, what you really need to know is what they've wanted" (Riedl & Konstan, 2002, para. 13). This is typical for algorithmic systems: Existing data is used to predict some future state of affairs. For instance, the music you listened to last week will be used as input by the recommender to make predictions about your future listening behavior.

With collaborative filtering the basic premise is that "people who agreed in their subjective evaluation of past [items] are likely to agree again in the future" (Resnick et al., 1994 p. 176).

However, this premise assumes that taste is static, with many users complaining about being "haunted" by their past preferences. In reaction, the field of recommender systems research has recently taken a "contextual turn" (Pagano et al., 2016). As one paper explains:

> [...] a context-driven recommender system, 'personalizes' to users' context states. In this way, it introduces a disassociation between users and their historical behavior, giving users room to develop beyond their past needs and preferences. Instead, users receive recommendations based on what is going on around them in the moment (situation) and on what they are trying to accomplish (intent). (*ibid.* p. 249)

Developers thus began incorporating contextual factors into recommenders to reflect the recognition that users interact with a system from within a particular context. Here "context" is defined as "a set of conditions under which an activity occurs" (Adomavicius et al., 2011, p. 68). In addition to factors that can be immediately known such as day and time, context-aware recommenders can also infer contextual information from behavioral data gathered from smartphone sensors:

> [I]f a user is listening to music on a smartphone, the system might try to deduce whether the device is moving or not. If it is moving, the person might be exercising or they might be driving or cycling. If the device is stationary, the consumer may be sitting on a sofa at home and the appropriate music might be different. (Falk, 2019, p. 17)

The input that recommenders work with is thus composed of *measurable*, *implicit*, and *past* behavioral data (hereafter "behavioral data"), in combination with data about the *context* (hereafter "contextual data") in which the behavior takes place. In the next section we build from here to identify the underlying technical code of recommender systems.

8.5   The Technical Code of Recommenders

Technologies offer a material affirmation of, and a bias towards, particular values and worldviews. More specifically, Feenberg argues that modern technology is biased by contingent social factors specific to capitalism (Kirkpatrick, 2020). Developers of recommenders, like technologists more generally, do not typically aim at specific social benefits or prejudicial outcomes. Instead, they focus on efficiency gains that are to result from the technology that is developed. Over time, the technologies as well as the systems of thought that underlie them become seemingly uncontroversial. As Bernhard Rieder (2020, p. 253) puts it when describing the history of how observed market behaviour came to stand for consumer preference in economics, "[w]hat users do is what they want and what they want is what they shall receive. How could it be otherwise?" It is precisely the apparent incontestability of this "technical code" that renders recommenders "formally biased."

The highly technical perspective involved in the creation of recommenders makes them vulnerable to the influence of existing systems of thought that are likewise disinterested in values or meanings (Kirkpatrick, 2020). In the case of recommenders, we argue that they work according to the objectivist principles of *behaviorism*. As such they embody what we term a "behavioral code." In this section we therefore provide a brief overview of the core ideas of behaviorism and show how they are reflected in recommenders. In the subsequent section we discuss several existing critiques of behaviorism as a way of proving the formal bias of recommenders while also opening up pathways for imagining alternative directions for recommender systems.

8.5.1   Behaviorism: The Core Principles

Since John B. Watson coined the term in 1913, behaviorism grew into a highly influential school of thought that covers multiple scientific fields. For B. F. Skinner (1904–1990), perhaps the most well-known and influential behaviorist, to know a person means to know "what he does, has done, or will do" in certain contexts (Skinner, 1974, p. 176). According to Skinner, "[a] self or personality is at best a repertoire of behavior imparted by an organized set of contingencies" (1974 p. 149). Contingencies refer to the relationship between three things: events that occur immediately before a behavior (antecedents), behavioral responses, and consequences that take place immediately after the response. Certain behavior can be "reinforced" (e.g., Skinner, 1974 p. 42) when its consequences are positive, or weakened if the consequences are negative. Thus, the self, for behaviorists, is "at best" a set of likely behaviors under certain circumstances. As a result, the ingredients necessary to know someone are their overt behavior and the environment in which this takes place.

The emphasis on overt behavior does not mean that behaviorists deny the existence of inner states such as feelings, thoughts, and preferences. Instead, feelings and thoughts are reduced to bodily states and processes. In other words, inner states are seen as a type of behavior, just as overt actions are.[1] However, behaviorists do object to assigning inner states causal power and, as such, explanatory power. Skinner, for example, argued that inner states are by-products. The following passage from his book *On Behaviorism* (1974) sheds more light on this position:

> When a person has been subjected to mildly punishing consequences in walking on a slippery surface, he may walk in a manner we describe as cautious. It is then easy to say that he walks with caution or that he shows caution. There is no harm in this until we begin to say that he walks carefully because of his caution (p. 161).

What Skinner objects to, then, is the role of inner states as the subject of scientific study. "The objection to inner states," Skinner wrote, "is not that they do not exist,

---

[1] As such, behaviorists deny the Cartesian mind-body dualism.

but that they are not relevant in a functional analysis" (Skinner, 1953, p. 35). Instead, he argues, we should shift our focus from the inside to the outside – to overt behavior and the environment in which people act. For Skinner and his followers, only behavior provides publicly observable data upon which to construct rigorous and scientifically-sound models of how and why people do what they do (Moore, 1999). What is more, behaviorists believe that to understand behavior means to be able to both predict and control behavior. In other words, it is about being able to anticipate what people will do *and* being able to steer this behavior through reinforcement and punishment.

8.5.2   The Behavioral Code

Contemporary recommenders posit the internet user in much the same way as Skinner and other behaviorists posited their test subjects. Behaviorists broadly work with two variables: overt behavior and environmental factors. Regarding the latter, recall the turn towards "context-aware" recommenders. Like behaviorists, such systems emphasize the importance of environmental factors in understanding a person's behavior. If you listen to classical music almost every night before you go to bed, Spotify will very likely recommend playlists of this genre to you around this time.

With regard to the first variable, the primary input of recommenders consists of behavioral data. The idea that past behavior lends itself for predicting the probability of future behavior endorses the behaviorist doctrine. As Skinner wrote: "The probability of behavior depends upon the kind of frequency of reinforcement in similar situations in the past" (1974, p. 69). In addition, by focusing on overt and implicit behavior, recommenders meet the behaviorist "rule" of shifting one's attention from inner states to overt behavior. Recommenders focus their attention on what can be "objectively" and consistently measured. While explicit behavioral data used to be collected by recommenders, as pointed out above, subjective interpretations of inner states are now largely dismissed due to their inconsistency and scarcity. For example, in recommending music, Spotify is not that interested in how users self-identify as music fans, or even in demographic markers that traditionally acted as a proxy for music preferences. Instead, a "taste profile" – a dynamic record of one's musical identity – is constructed for each user. This profile is generated primarily through implicit behavioral feedback that is generated every time you search for an artist, listen to a track, add songs to a playlist, or skip a song.

Combining behavioral data and context, recommenders aim to understand the user by identifying patterns of behavior. In Fisher and Mehozay's (2019, p. 10) formulation of the "algorithmic episteme": "To *know* someone does not mean to analytically and empirically understand the reasons for her behavior, but simply to be able to recognize patterns of behavior." This appears to follow the behaviorist doctrine – that to know someone is to know what someone has done, is doing, and will do in the future.

Nevertheless, recommender systems appear to contradict the principles of radical behaviorists by assigning causal power to inner states. As was mentioned earlier, a popular textbook defines recommenders as "software tools and techniques that provide suggestions for items that are most likely of *interest* to a particular user" (Ricci et al., 2011, p. 1, italics added). Similar explanations of recommenders can be found throughout the literature. Lu, Dong and Smyth (2018, p. 4, italics added) write: "Recommender systems learn to predict the degree to which a user will *like* an item." This could merely be a rhetorical device. Considerable research has been conducted in making recommenders more "persuasive" (e.g., Yoo & Gretzel, 2011) and it appears that we are more comfortable following recommendations from a source that claims to understand our inner preferences than from a system that monitors our behavior. Regardless, while recommenders may use mental concepts to present themselves to end users, this does not take away from the fact that they work according to an objectivist, behaviorist interpretation of such concepts.[2]

8.6 The Formal Bias of Recommenders: Critiques of Behaviorism

Recommender systems, we argue, embody behaviorist assumptions. To make this claim is not to suggest that developers are behaviorists who consciously create recommenders according to Skinnerian principles. To restate what was argued earlier: The technical perspective that focuses on efficiency gains makes recommenders vulnerable to influences by systems of thought that are likewise indifferent to meaning and values.[3] As has already become clear from the brief discussion of behaviorism's core ideas, it is the influence of this particular objectivist system of thought that recommenders are vulnerable to.

The materialization of behaviorist ideas in recommenders has closed off alternative pathways for understanding individuals and their desires. While other directions for the development of recommenders were – and always are – possible, now that the behavioral code is materialized it works to affirm itself and obscure its

---

[2] Skinner preferred to avoid mental concepts, but the underlying idea of (analytical or logical) behaviorism is that a mental state or condition is the idea of a behavioral disposition or family of behavioral tendencies (Graham, 2019). This means that a behaviorist can in principle continue to use mental concepts, but they would refer to a certain behavioral disposition rather than inner states.

[3] Even though developers and behaviorists work with different motivations – developers work according to a commercial incentive while behaviorists are motivated by a certain ideal of "real" science – they eventually both aim for the prediction and control of human behavior. These goals have proven to be greatly compatible; the founder of behaviorism, John B. Watson, joined an advertising agency after he left academia and became highly successful in that field (Baars, 1986; Waldrop, 2001). In addition, Skinner's analysis has been called the psychological equivalent of wage-labor capitalism (Baars, 1986), as the prediction and control of human behavior in order to increase productivity has been a central focus of managerial practices; from "scientific management" to "nudge management" more recently (Ebert & Freibichler, 2017).

contingent nature. As such, recommenders perform a formal bias: Seemingly neutral they offer a material affirmation of the ideas that underlie them. Since recommenders reintroduce a behaviorist understanding of humans, a critical analysis of these systems should draw upon criticism of, and alternatives to, behaviorism. This therefore forms one of our aims for this chapter; to reopen the debate around behaviorism. These critiques not only provide alternative stipulations but also allow us to see how the behavioral code that currently underlies recommenders results in a formal bias with social implications – specifically an impoverished view of what it means to be human.

8.6.1   Existing Critiques of Behaviorism

Between approximately 1920 and the mid-1950s (e.g., Baars, 1986; Chung & Hyland, 2012; Miller, 2003; Reisberg, 2016), the majority of psychologists in the United States were behaviorists. By the mid-1950s, however, the popularity of behaviorism went into fast decline as it was critiqued from several angles. In psychology, behaviorism was largely obliterated by the "cognitive revolution" (e.g., Miller, 2003; Reisberg, 2016; Waldrop, 2001). Psychologists grew convinced that a subject's behavior was guided by how the subject understood or interpreted a situation – not by the objective situation itself. By focusing merely on the objective situation, we misunderstand the motivations people have for their actions and subsequently make mistakes in predicting future behavior. In other words, it became clear that psychologists needed to study mental states after all.

From a philosophical perspective, critical theorists contrasted Skinner's "science of behaviour" with what they viewed as the much richer Marxist concept of "praxis."[4] Praxis, according to one critic of Skinner, "refers to man as an active agent in the world, a world that he constructs and transforms, on which he confers meaning, and to which he responds" (Mishler, 1976, p. 25). In other words, from the perspective of praxis the human subject is an interpretive being engaged in meaningful action. One does not simply run, for example, but rather one runs *because* of a reason – a reason that emerges out of the subjective interpretation of an event. The meaning of the behavior is what defines the behavior; which could be as varied as running from something that scares you or going for a run to clear your head.

---

[4] Apart from the praxis critique, there are roughly three main reasons for the rejection of behaviorism within philosophy (Graham, 2019). First of all, many people were, and still are, sceptical about behaviorism's commitment to the thesis that behavior can be understood without referring to mental processes. A second reason for the dismissal of behaviorism is the existence of "qualia" (e.g., Place, 2000): behaviorism cannot account for the qualitatively distinctive experience underlying overt behavior. Yet another critique came from Noam Chomsky (1967 [1959]). According to Chomsky, behaviorism cannot account for the fact that language does not seem to be learned through explicit teaching. He pointed out that linguistic performance outstripped individual reinforcement histories.

Behaviorists, however, reject a focus on meaning not because they deny subjective or inner states, but because they see them as functionally useless for predicting rates of response. There is an analogous focus on "rating prediction accuracy" in recommender system design. Both can be seen as expressions of what Habermas (1970, p. 105–7) called "technocratic consciousness":

> It is a singular achievement of this (technocratic) ideology to detach society's self-understanding from the frame of reference of communicative action and from the concepts of symbolic interaction and replace it with a scientific mode. . . . This is paralleled subjectively by the disappearance of the difference between purposive-rational action and interaction from the consciousness not only of the sciences of man, but of men themselves. The concealment of this difference proves the ideological power of the technocratic consciousness.

For critical theorists, behaviorism represented the further colonization of the life-world by positivist scientism. As one trenchant critique put it, behaviorism circumvents the necessity of interpretation "by defining a single scalar index as the "behaviour" of interest, and by coding many different types of behaviour in this one category while ignoring other features of the behaviour" (Mishler, 1976, p. 32). It conveniently ignores *why* the human subject gently pushes the lever or smashes it. "Instead of a science constructed so as to be appropriate to its phenomena of study, the phenomena are transformed so as to be appropriate to a particular methodology" (ibid., p. 33).[5]

The principal takeaway here is that the model of human action and motivation becomes defined through the lens that it is perceived through. Like the example earlier of the product manager at a music recommendation company that equated "audience understanding" with aggregated listening data, behaviorism distinguished itself from alternative methods of human understanding by claiming the mantle of "real science." In doing so, it defined the world in its image and allowed for certain questions while ignoring others. Another vision of science – one that sees human beings as meaning constructors and symbols users – would result in an alternative definition of the world.

What made behaviorism especially dangerous was not that it did not work, but rather that it pretended to be the only scientific approach to the study and understanding of humans. As Baars (1986, p. 51–52) put it: "Behaviorism was viewed as the one right way to do psychological science; every alternative was unscientific." As we have shown, however, behaviorists actually worked with a very limited understanding of the meaning and purpose of science and of human-beings. While behaviorists could lay claim to an undoubted objectivity in their observations, they had to pay a very high price for it. They had rejected too many things: "[...] in hot pursuit of scientism, psychology had lost psychology" (Baars, 1986, p. 69). In other words, and to return to Feenberg, even though behaviorism might have been rational, it was also formally biased. As Mishler (1976, p. 29) puts it: "More is at stake than whether information about "inner states" helps to "predict" a discrete and

---

[5] Notice here that behaviorism is a presupposed framework rather than a scientific theory, meaning that it cannot be falsified by any experimental results (Baars, 1986).

meaningless response. Rather, these states are central topics of interest in and of themselves, as are their complex relationships to behaviour and the rules governing the stability and change of these relationships." Behaviorism could have been *a* way of doing experimental psychology to be complemented by other forms of study that focus on meaning and understanding. That way, behaviorists would at least have recognized and respected the formal bias that was integrated into their program. Yet this is exactly what they did not allow for.

Like behaviorism, recommenders get the job done. And like behaviorism, this does not mean that they are not formally biased. Recommenders also embody the same impoverished view of what it means to be human. Interestingly, their developers show a similar attitude toward their method as behaviorists did. Recall "Whisper," the music recommendation company studied by Seaver (2021). This company believed to have overcome the "challenging alterity of their users by appealing to "data," which was taken to provide a putatively objective position beyond individual perspectives" (p. 14). Note the further similarity with behaviorism when the employee says: "We think we have real science here" (ibid.).

While the developers of recommenders, unlike behaviorists, may not explicitly claim that their account of humans is *the* way to view them, the materialization of behaviorist assumptions in these omnipresent recommenders does create a formal bias that reinforces a behaviorist understanding of humans. As such, they might even cause users to see *themselves* through a behaviorist lens. After all, recommenders are said to "personalize" content, which critics have argued "imbues the system with the power to co-constitute users' experience, identity and selfhood in a performative sense (Kant, 2020, p. 12). There is however another concern, namely that "[i]t is the programmers themselves who are more likely to suffer these consequences. It is the objectification of others that is dehumanizing, and this is integral to the behaviourist approach" (Mishler, 1976, p. 34).

To summarize, recommenders embody a behavioral code and are as such biased towards the beliefs and values that underlie behaviorism. This formal bias promotes an impoverished view of what it means to be human – among users as well as developers. As such, the formal bias of recommenders should be of public concern.

8.7 Conclusion

Over a decade ago, Google's former CEO Eric Schmidt pointed out how ubiquitous recommendation was (Jenkins Jr., 2010). Today, on platforms like Netflix, "everything is a recommendation": Not only are the films personalized to fit viewing behavior, but so is the cover art (Mullaney, 2015; Yu, 2019). At the same time, data is drawn from an ever-widening and growing array of interactions. As Nick Seaver (2019, p. 11) writes, "algorithmic recommendation has settled deep into the infrastructure of online cultural life, where it has become practically unavoidable."

If recommenders exert such a ubiquitous and powerful influence on our lives, then – as Feenberg asks of technology in general – "why don't we apply the same democratic standards to it as we apply to other political institutions" (Feenberg, 1999, p. 131). In *Questioning Technology*, Andrew Feenberg outlines three forms of democratic intervention in technology: controversy, innovative dialogue and creative appropriation (Feenberg, 1999, p. 120–9). The potential solutions to the problem of the "behavioral code" in recommender design will rely on both the creative appropriation by users of these technologies, as well as innovative dialogue between users and developers.[6] However, what is first required is that the assumptions inscribed in recommenders be made a controversy; an issue of public concern. That is what this chapter has attempted to do.

As we have demonstrated, recommender systems generally share a basic assumption – that individuals are "users" whose preferences can be understood as behavioral dispositions and whose behavior can therefore be predicted through (past) implicit behavior and contextual cues. Extending Feenberg's critical theory of technology into the domain of recommenders, we call this the "behavioral code" of recommenders – a particular technical code that exerts a "formal bias" with social implications. Other choices are always possible in terms of what to make and how to make it. The way in which recommenders currently work is thus not set in stone. The task that remains is to explore other "branches" of development, which perhaps provide a more expansive way in which to understand individuals and their desires.

---

[6] While the purpose of this chapter is not to explore solutions, there are several interesting proposals and projects underway. For example, academics and developers have called for and experimented with more user-centric recommenders that allow users some degree of control over how they are profiled. One example of user-centric design is *gobo.social*, a social media news aggregator designed by the MIT Media Lab. This tool offers sliders that users control in order to filter information: The user can explore a range of political perspectives on a continuum from left to right, or "the extent of seriousness, rudeness, gender, and other parameters" (Reviglio & Agosti, 2020, p. 6). In another example, Harambam et al. (2018) provide an interesting proposal to grant users greater "voice" in our algorithmically-driven media ecosystem. The authors propose the creation of *algorithmic recommender personae* to "allow people instead to demand from [recommenders] to behave in ways that align with their own specific... interests at each single moment" (ibid. p. 4). It is also possible to involve users in the earliest stages of the design and development of recommender algorithms. The benefits of participatory design are not only in creating more user-friendly technologies, but also in making "explicit the critical, and inevitable, presence of values in the system design process" (Suchman, 1993, p. viii). As Feenberg convincingly argues in *Questioning Technology*, by widening opportunities to intervene, user participation in design serves to limit "the operational autonomy of technical personnel" (Feenberg, 1999, p. 135) who are socialized into the technical codes of the profession (ibid, p. 142).

References

Adomavicius, G., Mobasher, B., Ricci, F., & Tuzhilin, A. (2011). Context-aware recommender systems. *AI Magazine, 32*(3), 67–80. https://doi.org/10.1609/aimag.v32i3.2364

Adomavicius, G., & Tuzhilin, A. (2005). Toward the next generation of recommender systems: A survey of the state-of-the-art and possible extensions. *IEEE Transactions on Knowledge and Data Engineering, 17*(6), 734–749.

Aggarwal, C. C. (2016). *Recommender systems* (Vol. 1). Springer International Publishing.

Baars, B. J. (1986). *The cognitive revolution in psychology*. New York: Guilford Press.

Beer, D. (2009). Power through the algorithm? Participatory web cultures and the technological unconscious. *New Media and Society, 11*(6), 985–1002.

Beer, D. (2013). *Popular culture and new media: The politics of circulation*. Springer.

Bobadilla, J., Ortega, F., Hernando, A., & Gutiérrez, A. (2013). Recommender systems survey. *Knowledge-Based Systems, 46*, 109–132.

Burke, R. (2007). Hybrid web recommender systems. In *The adaptive web* (pp. 377–408). Springer.

Cheney-Lippold, J. (2011). A new algorithmic identity: Soft biopolitics and the modulation of control. *Theory, Culture and Society, 28*(6), 164–181.

Chomsky, N. (1967 [1959]). Review of B. F. Skinner's verbal behavior. In L. A. Jakobovits & M. S. Miron (Eds.), *Readings in the psychology of language* (pp. 142–143). Prentice-Hall.

Chung, M. C., & Hyland, M. (2012). Behaviourism, and the disappearance and reappearance of organism (Person) variables. In M. C. Chung & M. Hyland (Eds.), *History and philosophy of psychology* (pp. 144–169). Wiley-Blackwell.

Drott, E. (2018). Why the next song matters: Streaming, recommendation, scarcity. *Twentieth-Century Music, 15*(3), 325–357.

Ebert, P., & Freibichler, W. (2017). Nudge management: Applying behavioural science to increase knowledge worker productivity. *Journal of Organization Design, 6*(1), 1–6.

Ekstrand, M. D., & Willemsen, M. C. (2016, September). Behaviorism is not enough: Better recommendations through listening to users. In *Proceedings of the 10th ACM conference on recommender systems* (pp. 221–224).

Falk, K. (2019). *Practical recommender systems*. Manning Publications.

Feenberg, A. (1992). Subversive rationalization: Technology, power, and democracy. *Inquiry, 35*(3–4), 301–322.

Feenberg, A. (1999). *Questioning technology*. Routledge.

Feenberg, A. (2008). Critical theory of technology: An overview. In G. J. Leckie & J. E. Buschman (Eds.), *Information technology in librarianship: New critical approaches* (pp. 31–46). Libraries Unlimited.

Feenberg, A. (2017). Critical theory of technology and STS. *Thesis Eleven, 138*(1), 3–12.

Fisher, E., & Mehozay, Y. (2019). How algorithms see their audience: Media epistemes and the changing conception of the individual. *Media, Culture and Society, 41*(8), 1176–1191.

Gillespie, T. (2014). The relevance of algorithms. In T. Gillespie, P. J. Boczkowski, & K. A. Foot (Eds.), *Media technologies: Essays on communication, materiality, and society* (pp. 167–194). The MIT Press.

Goldberg, D., Nichols, D., Oki, B. M., & Terry, D. (1992). Using collaborative filtering to weave an information tapestry. *Communications of the ACM, 35*(12), 61–70.

Gomez-Uribe, C. A., & Hunt, N. (2015). The Netflix recommender system: Algorithms, business value, and innovation. *ACM Transactions on Management Information Systems (TMIS), 6*(4), 1–19.

Graham, G. (2019, Spring). Behaviorism. In E. N. Zalta (Ed.), *The Stanford encyclopedia of philosophy*. https://plato.stanford.edu/archives/fall2019/entries/behaviorism/

Habermas, J. (1970). *Towards a rational society*. Beacon Press.

Hallinan, B., & Striphas, T. (2016). Recommended for you: The Netflix Prize and the production of algorithmic culture. *New Media and Society, 18*(1), 117–137.

Harambam, J., Helberger, N., & van Hoboken, J. (2018). Democratizing algorithmic news rec-ommenders: How to materialize voice in a technologically saturated media ecosystem. *Philosophical Transactions of the Royal Society A: Mathematical, Physical and Engineering Sciences, 376*(2133), 20180088.

Jenkins, H. W., Jr. (2010, August 14). Google and the search for the future. Retrieved from https://www.wsj.com/articles/SB10001424052748704901104575423294099527212

Kant, T. (2020). *Making it personal: Algorithmic personalization, identity, and everyday life.* Oxford University Press.

Kirkpatrick, G. (2020). Technical politics. In G. Kirkpatrick (Ed.), *Technical politics: Andrew Feenberg's critical theory of technology* (pp. 70–95). Manchester University Press.

Linden, G., Smith, B., & York, J. (2003). Amazon.com recommendations: Item-to-item collabora-tive filtering. *IEEE Internet Computing, 7*(1), 76–80.

Lops, P., De Gemmis, M., & Semeraro, G. (2011). Content-based recommender systems: State of the art and trends. In *Recommender systems handbook* (pp. 73–105). Springer US.

Lu, Y., Dong, R., & Smyth, B. (2018, September). Why I like it: multi-task learning for recom-mendation and explanation. In *Proceedings of the 12th ACM Conference on Recommender Systems* (pp. 4–12).

Lynch, J. (2018, July 2018). *Netflix thrives by programming to 'taste communities,' not demo-graphics.* Retrieved 1 Nov 2020, from AdWeek: https://www.adweek.com/tv-video/netflix-thrives-by-programming-to-taste-communities-not-demographics/

Miller, G. A. (2003). The cognitive revolution: A historical perspective. *Trends in Cognitive Sciences, 7*(3), 141–144.

Mishler, E. G. (1976). Skinnerism: Materialism minus the dialectic. *Journal for the Theory of Social Behaviour 6*(1), 21–47.

Moore, J. (1999). The basic principles of behaviorism. In B. Thyer (Ed.), *The philosophical legacy of behaviorism* (pp. 41–68). Springer.

Morris, J. W. (2015). Curation by code: Infomediaries and the data mining of taste. *European Journal of Cultural Studies, 18*(4–5), 446–463.

Mullaney T (2015) Everything is a recommendation. MIT Technology Review, 23 March. Available at: https://www.technologyreview.com/s/535936/everything-is-a-recommendation/

Pagano, R., Cremonesi, P., Larson, M., Hidasi, B., Tikk, D., Karatzoglou, A., & Quadrana, M. (2016, September). The contextual turn: From context-aware to context-driven recommender systems. In *Proceedings of the 10th ACM conference on recommender systems* (pp. 249–252).

Pazzani, M. J., & Billsus, D. (2007). Content-based recommendation systems. In *The adaptive web* (pp. 325–341). Springer Berlin Heidelberg.

Perik, E., De Ruyter, B., Markopoulos, P., & Eggen, B. (2004). The sensitivities of user pro-file information in music recommender systems. In *Proceedings of private, security, trust* (pp. 137–141).

Place, U. T. (2000). The causal potency of qualia: Its nature and its source. *Brain and Mind, 1*(2), 183–192.

Prey, R. (2018). Nothing personal: Algorithmic individuation on music streaming platforms. *Media, Culture and Society, 40*(7), 1086–1100.

RecSys. (n.d.). *15th ACM Conference on Recommender Systems,* from https://recsys.acm.org/recsys21/

Reisberg, D. (2016). The science of mind. In D. Reisberg (Ed.), *Cognition: Exploring the science of mind* (6th ed., pp. 2–27). W. W. Norton & Company.

Resnick, P., Iacovou, N., Suchak, M., Bergstrom, P., & Riedl, J. (1994, October). GroupLens: An open architecture for collaborative filtering of netnews. In *Proceedings of the 1994 ACM con-ference on Computer supported cooperative work* (pp. 175–186).

Resnick, P., & Varian, H. R. (1997). Recommender systems. *Communications of the ACM, 40*(3), 56–58.

Reviglio, U., & Agosti, C. (2020). Thinking outside the black-box: The case for "algorithmic sov-ereignty" in social media. *Social Media + Society, 6*(2), 2056305120915613.

Ricci, F., Rokach, L., & Shapira, B. (2011). Introduction to recommender systems handbook. In *Recommender systems handbook* (pp. 1–35). Springer.

Rieder, B. (2020). *Engines of order: A mechanology of algorithmic techniques*. Amsterdam University Press.

Riedl, J., & Konstan, J. (2002). *Word of mouse: The marketing power of collaborative filtering*. Warner Books.

Rogers, R. (2009). Post-demographic machines. *Walled Garden, 38*(2009), 29–39.

Salter, J., & Antonopoulos, N. (2006). CinemaScreen recommender agent: Combining collaborative and content-based filtering. *IEEE Intelligent Systems, 21*(1), 35–41.

Seaver, N. (2012). Algorithmic recommendations and synaptic functions. *Limn, 1*(2). from https://escholarship.org/uc/item/7g48p7pb

Seaver, N. (2019). Captivating algorithms: Recommender systems as traps. *Journal of Material Culture, 24*(4), 421–436.

Seaver, N. (2021). Seeing like an infrastructure: Avidity and difference in algorithmic recommendation. *Cultural Studies, 35*(4–5), 771–791.

Skinner, B. F. (1953). *Science and human behavior*. Macmillan.

Skinner, B. F. (1974). *About behaviorism*. Knopf.

Suchman, L. (1993). Foreword. In D. Schuler & A. Namioka (Eds.), *Participatory design: Principles and practices*. CRC/Lawrence Erlbaum Associates. vii–x.

Tkalčič, M., Burnik, U., & Košir, A. (2010). Using affective parameters in a content-based recommender system for images. *User Modeling and User-Adapted Interaction, 20*(4), 279–311.

Waldrop, M. M. (2001). *The dream machine: J.C.R. Licklider and the revolution that made computing personal*. Viking.

Wan, M., & McAuley, J. (2018, September). Item recommendation on monotonic behavior chains. In *Proceedings of the 12th ACM conference on recommender systems* (pp. 86–94).

Watson, J. B. (1913). Psychology as the behaviorist views it. *Psychological Review, 20*(2), 158.

Yu, A. (2019). How netflix uses ai, data science, and machine learning — from a product perspective from https://becominghuman.ai/how-netflix-uses-ai-and-machine-learning-a087614630fe

Yoo, K. H., & Gretzel, U. (2011). Creating more credible and persuasive recommender systems: The influence of source characteristics on recommender system evaluations. In *Recommender systems handbook* (pp. 455–477). Springer.

Zuboff, S. (2019). *The age of surveillance capitalism: The fight for the future at the new frontier of power*. Profile Books.

Chapter 9
The Algorithmic Thing, The Real, and Contestation: Tracing the Fringes of Critical Constructivism

Yoni Van Den Eede

Abstract A unique and powerful approach in the philosophy of technology, Critical Constructivism endeavors to show how technologies instantiate and consolidate power relations. This chapter elaborates on an enduring vulnerable spot in the framework, namely, that Critical Constructivism only starts to do its work as soon as some problem, struggle, or controversy is recognized *as such*. Only then, the work of "contestation" can begin. For problems that have not been unearthed yet, but that may be brooding in the background or behind the "screen of attention," the approach is, as it stands, less equipped. This issue is exacerbated nowadays with recent developments around "algorithmic technologies" such as artificial intelligence or the internet of things, whose inner workings are extremely difficult to access from a user's point of view, and even increasingly from an expert one. To start orienting Critical Constructivism toward these issues, this chapter pairs it with an unlikely partner, object-oriented ontology (OOO), in order to help it make sense of the problem of the "discovery of problems." The Heideggerian notion of "breakdown" is crucial in this respect. In fact, it will turn out there is already much "object-oriented potential" in Critical Constructivism, but to start bringing that out fully, we need to take a fresh look at several components of Critical Constructivism from the perspective of OOO, specifically the latter's provocative take on breakdown.

Keywords Critical constructivism · Object-oriented ontology · Philosophy of technology · Andrew Feenberg · Graham Harman · Algorithmic thing · Martin Heidegger · Power · Actor-network theory · Postphenomenology

Y. Van Den Eede (✉)
Centre for Ethics and Humanism, Free University of Brussels (Vrije Universiteit Brussel),
Brussels, Belgium
e-mail: yoni.van.den.eede@vub.be

© The Author(s), under exclusive license to Springer Nature Switzerland AG 2022
D. Cressman (ed.), *The Necessity of Critique*, Philosophy of Engineering
and Technology 41, https://doi.org/10.1007/978-3-031-07877-4_9

9.1   Introduction

Critical Constructivism is unique in the philosophy of technology. Feenberg's work stands relatively isolated as a critical island in a sea of more purportedly politically neutral approaches. Those may be "critical" in the commonsense meaning of the term, to the extent that they seek to unveil ways in which technology works that otherwise stay unnoticed, unreflected, or unopposed. But Feenberg's philosophy is something else. Obviously, it is *critical* in the sense of Critical Theory, given how it develops a critique of rationality building on the Frankfurt School's legacy. But embedding this critique within the insights acquired through STS, Feenberg has singlehandedly devised what could arguably still be called the only truly political-philosophical perspective that has arisen in the slipstream of the empirical turn.

Hence, Critical Constructivism is unique in a positive and negative sense. Its positive uniqueness derives from its distinctive combination of critical and con-structivist perspectives (in this light the new name Feenberg has been giving his approach since *Technosystem* (2017) couldn't be chosen better). Its "negative" uniqueness, however, namely, that it continues to be fairly insular within the philo-sophical study of technology, points to a wider problem that has to do with the posi-tion of critique *überhaupt*. A substantial discussion of this problematic falls far beyond the scope of this essay, but by way of tentative orientation I would like to formulate the following wild hunch: it shouldn't surprise us that critical stances are so scarce, given how *hard* it is to critique technology.

Philosophy of technology has shown throughout its rich history—this includes even the "classic" so-called monolithic approaches over and against which the empirical turn was positioned—how technology in everyday life tends to "disappear from view," whether phenomenologically, discursively, ideologically, politically. That is to say, technology does of course stay "visible" in practical contexts—I reach quite deliberately for my smartphone when I want to check my e-mail—but this is mostly in a merely functional sense: I (know I need to) use device X to accomplish goal Y. The wider impacts and effects of the technology, by contrast, stay under the radar. For instance, my constant smartphone-checking may be mak-ing me stressed and anxious, but since this is happening largely unnoticed, I am not able to point a precise finger at the connection between my smartphone use and its wider psychological and existential effects—or at least not right away. Because of its sheer *use*, as Heidegger's tool analysis in *Being and Time* (1962) already exqui-sitely demonstrated, the "technology" as a clearly delineated, consciously perceived entity sinks away into a constellation of things-to-be-done.

Critical Constructivism is just as much concerned with this condition of invisi-bility. Technologies *in fact* instantiate, in material form, power relations, and they help to solidify and sustain these. But in everyday use this fact remains for the most part cloaked, and we need *problems* to arise in specific use contexts to begin to notice the underlying power structures. At this point, Feenberg's approach starts to pay dividends, analyzing as it does the precise ways in which protests over tech-nologies (can) unfold.

However, what if technology brings with it problems that we do not recognize *as* problems, and which to that extent do not bring about a struggle or controversy? Or if they do so, they do so much too late? For issues less recognizable *as* problems, the theory seems somehow less well-equipped—or at least, this is an aspect heretofore less brought out. What is more, recent technological developments appear to aggravate this conundrum of the recognizability of technology *as* technology. Technology is ever more moving into and becoming "us"—or at least what we thought "we" were. With "algorithmic technology" such as artificial intelligence (AI) or the internet of things (IoT), but also with biotechnology such as neural implants, nanobiotechnology, or genome editing, it is becoming increasingly difficult to tell where "we" end and technology begins. Technology is getting ever more invisible, and harder to trace in everyday life. This complicates the position of Critical Constructivism even further.

That said, Feenberg's work seems to harbor insights that for the time being lie relatively dormant in it, but that with some elaboration may better outfit Critical Constructivism for these complications. What I seek to do in this essay is to start unearthing those elements, by exploring an alternative track running parallel to the critical track, situated, really, on the fringes of Critical Constructivism. I'll survey the critical track, all the while keeping an eye on the parallel track, to eventually—if all goes well—find the two routes merging. The alternative track will be, quite unexpectedly perhaps, object-oriented ontology (OOO), in particular Graham Harman's work.

I will first elaborate on the issue of the discovery of problems. Then I zoom in on algorithmic developments (the emergence of what I call the Algorithmic Thing). Critical Constructivism, I find in a next section, "needs" *breakdown* in the Heideggerian sense, as a gateway into possible *contestation*, or protest; but what if breakdown doesn't happen? OOO, by contrast—and this provides the topic for the following section—argues that breakdown is constantly "there." So what if we try to broaden Critical Constructivism's lens by pulling it toward OOO—a not-so-unlikely move, as mentioned, given that some of Feenberg's analyses and notions line up surprisingly well with the object-oriented perspective? This exploration will take up the next, and largest, section. In closing, I will return to the Algorithmic Thing, and draw some conclusions.

9.2 The Discovery of Problems

As said, Critical Constructivism's unique strength manifests itself in how it points the way to resisting technological designs—that is, specific technological designs. In contrast to "classic" critical theory, it no longer critiques *technology* and *instrumental rationality* in the singular. This is its masterstroke, which it accomplishes by marrying critical theory to the constructivist orientation. Instrumental rationality gets instantiated with every technology—the technology efficiently appears to do nothing else than what it is meant to do—but its apparent "purity" as *only* instrumental-rational is only a construction. In reality, the technology is wrapped up in, or better, incorporates a constellation of values and power relations. Yet these tend to get forgotten over time, and in everyday life this constellation becomes an invisibly working, but still effective background, that can do "its work" precisely because it is able to hide behind merely technical considerations. Thus, Feenberg *multiplies* instrumental rationality: it should no longer be viewed as a kind of Big Fate threatening to swallow all domains of life. True enough, it is the case that nearly all domains of life are shaped nowadays by instrumental rationality. However, everywhere, rationality is interwoven with the social. *Reason* and *experience* are intertwined.

It is exactly at this nexus—because of this nexus, even; *because* there is always experience where there is reason, and vice versa—that resistances against technological designs (in the plural) can arise. Feenberg throughout his career has worked out many cases of such resistance, the study of the French Minitel system perhaps being the most famous (Feenberg, 1995). Those protests, controversies, and struggles over technology illustrate the theory of Critical Constructivism. Conversely, the theory of Critical Constructivism precisely means to show how and where such protests can come about. To this extent its aim is obviously political, emancipatory, even pedagogical.

However, there is a lingering difficulty in the framework. Concisely put: Critical Constructivism seems to start "working" only as soon as people have diagnosed or at least perceived or experienced certain concerns, shortcomings, or obstacles. Some time ago I already remarked upon a similar issue (Van Den Eede, 2013), pointing out how the paradigmatic case studies of Critical Constructivism (back then still called Critical Theory of Technology), such as Minitel, the struggle over experimental treatment for AIDS patients, or the environmentalist fight against polluting technologies, could only be evaluated as successful—and thus as exemplifying Critical Constructivism's central dynamic of experience-mediated reason and reason-mediated experience—in retrospect. It is much harder to assess, that is, from the critical-constructivist perspective, controversies that are still in full swing. And it becomes even harder in the case of developments that may have such an innocent appearance that they are not even recognized (or at least not quickly enough) *as* controversy. My example was a seemingly trivial but possibly impactful move on the part of the Belgian postal services to require that mailboxes meet certain logistic criteria, in order for postal delivery to become more efficient—mail carriers wouldn't have to ring the bell anymore when mail didn't fit the mailbox—but in that way potentially limiting the scope for social contact of elderly or isolated people.

The core issue here is the difficulty of recognizing a controversy before it has become a controversy. More often than not, the examples that Critical Constructivism offers are of circumstances in which the concerned actors have already bumped into certain obstacles, have come to an awareness of concerns, or feel the tensions arising from a particular technological organization. Of course, those circumstances are always contextual: only in specific contexts may people start to develop an awareness of an issue or formulate a concern *an sich*. Feenberg illustrates this in *Technosystem* (2017, p.53) with the example of automobiles: drivers "discover an interest in better roads" and "victims of pollution discover an interest in clean air," and none of this would happen if it were not for the existence as such of the automotive network. "Drivers, sufferers, and cars co-produce a network to which all belong and it is this which makes certain interests salient that might otherwise have remained dormant or had no occasion to exist at all" (*ibid.*). These interests emanate from roles in which individuals within those networks are involved: driver, sufferer. And these interests then "become politically salient where the individuals *have the capacity to recognize them*" (*ibid.*; my emphasis).

But this is of course the crux: what if individuals do not have that capacity, i.e., to recognize their interests? What if they *don't* "discover" their interest in clean air or in better roads? The example at hand is straightforward enough to the extent that evidently, actors in the automotive network *have* discovered those interests. But we can easily imagine many scenarios in which this process hasn't taken place yet and is unlikely to take place on a large-enough scale in the foreseeable future. Or more precisely, and even more thornily, we know all-too well the myriad scenarios in which people have a vague, implicit idea of their interests but undertake no further actions to defend them. On the contrary, the whole of their behavior just keeps consolidating the prevailing course of events—as in fact the automobile example helps to illustrate: many people, if one would ask them, will attest to a clear interest in clean air, but this interest doesn't in any way become a motivation for action. They themselves obviously are also drivers—roles can be mixed up—and, in various degrees of explicitness, prefer fast and easy mobility over air quality, and so the efficiency mandate still has the last word. Likewise, with a previously mentioned example, we may have an inexplicit interest in a less stressful way of life, but if we don't make the connection with our digital media usage (the barrage of text messages is exhausting us but we haven't realized it yet), or for whatever reason we refrain from implementing necessary changes in that usage, no real passageway out of the current situation can open up.

Hence, not only is there the issue of recognizing controversies while they are still "subliminal" (Van Den Eede, 2013), or in Feenberg's terminology, dormant, there is another sense of "recognize" in play here as well: recognizing not just in the sense of "there is a problem," but in the sense of recognizing the problem *as a problem for me* (or *for us*), a problem of which I/we may not see the full contours at the moment, but that may down the line become particularly pressing. Both senses pertain to the same predicament, which, as we will eventually see, has an epistemological character.

9.3 The Algorithmic Thing

Of course, one may say: at that moment, *when* problems become pressing, that is when Critical Constructivism really gets into gear. It is no use anyhow to expect anything else. For if a problem is not seen, or *if* it is seen yet not as a problem that concerns *me* in a sufficiently direct sense, how could it even become something to contest, resist, protest, or struggle over? That seems to make sense. However, seen from another angle, this would lead us to a fairly cynical stance where problems that are not insufficiently perceived *as* problems, are simply unsolvable—at least for the time being.

This counts for the kind of issues I described (mailman example etc.), but technological developments that are currently underway only make this quandary more pressing—or worse, actually: they complicate it, *and* make it eminently significant.

I would like to sum up this evolution under the heading of *the Algorithmic Thing*. It will soon become fully clear why I prefer to speak in terms of "things," but for now this terminology serves to put our analysis right in the middle of everyday life—really also in a good old-fashioned Heideggerian sense—where we find ourselves at any moment *amidst things*. Now, something remarkable has been happening with the things surrounding us, of late. Things are turning into "more-than-objects." They are gradually being made up of elusive networks of data, artifacts, and other components that, however, we don't see directly. To begin with, the typical digital devices that surround us—smartphone, laptop, smart TV, smartwatch, and so on—are, obviously, not "dumb" objects: we use them for a whole array of purposes. But beyond that use, behind the (literal and figurative) screen of immediately apparent interaction, *or* even if they are just sitting or lying there, they aren't merely inanimate chunks of metal, plastic, and silicon. They are "alive" at any time, processing tasks, downloading updates, collecting data, uploading data, et cetera.

Those are of course the digital devices we are familiar with. With developments such as IoT or the already mentioned applications in biotechnology and nanotechnology, we can expect in the not-too-far-off future to become increasingly enveloped by more-than-objects. Things will become more and more double-layered in this respect. On the one hand, a "frontside" will be present to us, still fairly accessible to our perception and practical intervention, and forming a part of our intentional actions (but once again, who or what will "we" be?). On the other hand, behind this screen of relative awareness will be hiding, as is already the case with our digital devices today, a pandemonium of code, algorithms, databases, and information-technological structures. Historically, this is more significant than we might usually realize. We are—almost! —starting to take this double-layeredness for granted, but this is truly a first in human history. Although we might say on a general level that behind things are always hiding other things and that we humans have always tried to *get behind* the things, if only intellectually, to unearth their deeper laws, operation, or meaning (cf. Van Den Eede, 2019, p.133ff.), never before will things have acquired this informational-digital "backside."

The far-reaching consequences of this development in a philosophical sense may not have yet fully seeped through to the collective consciousness, but of course the more evident problems pertaining to the "backside" of digital things are well-known and intensely discussed. Worries are voiced about illicit data collection, racial and gender profiling, loss of autonomy, privacy, moral irresponsibility, loss of creativity, loss of humanness, and so on—all because of the hiddenness of "algorithmic mediation." Questions are asked about how to cope with these problems: how to contest those injustices, inequalities, inaccuracies (of algorithmic measurement and quantification); in general, how to counteract "algorithmic disempowerment"?

Yet, typically these discussions compound two different things: on the one hand, the way in which algorithmic constellations show themselves to us in an everyday use context and, on the other hand, the dynamics of protest and requests for change in a technological environment. Imagine a near-future scenario in which one is surrounded on a day-to-day basis from morning to night with algorithmic "maneuvering": you are woken by your IoT clock that detects a traffic jam on your way to work and therefore decides to wake you up earlier; your shower delivers just the right amount of water and soap based on the cleanliness algorithm that knows how long it has been since your last wash; in the kitchen, the oatmeal machine dispenses the right amount of food, measured to your current weight as well as the activities you've got planned for today; all the while you read a selection of news headlines on your tablet pc, tailored to your interests; the smart radio pours out tunes that are meant to get your spirits up—your bathroom mirror had already performed a micro expression analysis on your face and calculated your mood score, so it's only a matter of correcting from there... and so on.

The person in this scenario might at times be critical toward (or annoyed by) some of the devices that she uses, and what they do. But for the most part, the aggregated "web of influence" constituted by the innumerable algorithm-driven constellations that enfold her will have become largely the "normal" scenery enveloping her daily life. Evgeny Morozov (2013) describes this as "invisible barbed wire." These algorithmic constellations prod, persuade, distract, deter us; they shape us and how we think about the world—for example, inculcating in us precisely an idea of what the "right amount" of this or that should be—in ways that stay opaque to us because of their unobtrusiveness (in fact they are often designed to be unobtrusive, "seamless"). So, starting from our everyday perspective, *how could we even begin to contest* potential injustices, inequalities, or inaccuracies, if these do not penetrate the screen of awareness? In the story, the subject can remain unsuspecting whatsoever. She can be expected to go through her day without much dwelling on the processes that affect or maneuver her—exactly because the algorithmic collaboration-with-things is so neatly locked in with the structures and habits of everyday practice and perception. Autonomy may be lost, privacy may be eroded, responsibility outsourced (or otherwise pinned solely on the individual...), stereotypes confirmed, inequality consolidated, ... but we fail to notice any of that. For it is just "everything in its right place."

9.4 Critical Constructivism Needs Breakdown

Precisely here lies the root predicament, which the "mailman problem" already pointed to. How does one discover problems when everything is "business as usual"? How does one become aware of potential dangers, injustices, inequalities, biases when everything feels and appears just *normal*—however new this normal may furthermore be? The emergence and becoming-mainstream of the Algorithmic Thing exacerbates this. In fact, Algorithmic Things, from a designer's and developer's standpoint, are mostly deliberately devised to do their "background" work unnoticed. Putting it somewhat dramatically: the IT sector seeks to fit out the world with this invisible "double," always with us but largely invisible, from whose reign we receive tidings, glimpses, rumors now and then. But like with the kingdoms of old, the real structures of power hide behind the drawbridge and we are reduced to witless witnesses of the royal spectacle (albeit that of course nowadays we as users rather see *ourselves* as the kings, and all this invisible infrastructure is meant to serve us unquestioningly...).

Admittedly, the analogy may be a bit overblown. Still, "Where is power?" and "Can we get it in view?" remain pertinent questions. Critical Constructivism is exactly about these questions. Like no other approach, it shows how the old power structures today take on technological shape in the first instance. Power is exercised *through* and *in* technology. This is what Critical Constructivism means to show, and what it hopes individuals and groups become aware of. But indeed, the catch is in the word "hope." That this will happen is everything but a guarantee, and this explains perhaps the peculiar mixture of optimism and pessimism that lingers through Feenberg's writing. In *Technosystem*, for instance, he is in some passages confident that "technical politics has become routine" (Feenberg, 2017, p.162), that "[t]echnical citizenship has become a reality" (*ibid.*, p.69), that public involvement in technology is growing (*ibid.*, cf. p.56), and that democratic intervention in technology will increase (*ibid.*, cf. p.110). In other places, however, he exhibits more gloominess, finding the prospects for deep societal change scarce, and indeed pointing to "the invisibility of technical power" (*ibid.*, p.194) as a central difficulty for technical micropolitics. This hybrid of hopefulness and remaining unease (for lack of a better term; "despair" is too strong), is less odd than it appears on paper. It hinges on Feenberg's stance with regard to the old revolutionary schools, where he prefers "not to revive the old-fashioned theory of progress but to appreciate instances of progress where they occur. And they do occur" (p.203).

But, again, what if they don't occur? And they don't occur—in many instances. The complication of the Algorithmic Thing adds insult to injury. We have no idea what is behind the algorithmic curtain, *or if* we do have a vague idea, we are mostly not quickly inclined to go look behind it (whatever this may furthermore mean: hacking, demanding transparency, regulation, et cetera), because of how the Algorithmic Thing is simply part of the unquestioned scenery of everyday life.

What is more, the matter gets even more complex than that. In Feenberg's framework, the lay-expert dialogue is central: to enforce technological change, lay users need to learn to interact with experts, to "speak the language" of the expert to a certain extent, for the experts are "at the steering wheel" of technology (although of course they themselves are just as socially embedded as everyone and everything else). But here is a glitch: algorithmic developments are in a certain sense now dethroning the expert, with AI machine learning systems being developed that "learn" by themselves and start "making decisions" by themselves, to the extent that the programmers or developers do not fully understand anymore what is happening. The expert becomes a kind of lay person herself in the face of AI technology. While one may wishfully expect that this will enhance the lay-expert dialogue, making it more evenhanded perhaps—for the lay person and the expert begin to resemble each other in this regard, and "like attracts like," or something in that style—a more sobering outlook is probably that the possibilities for *discovering problems* become even harder to come by.

Be that as it may, the key question here remains how to pierce through the screen of apparent "normalcy." A situation of "everything in its right place"—no matter if this is truly or actually the case—will rarely become a seedbed for protest or contestation in the critical-constructivist framework. How does "democratic rationalization" (Feenberg, 1999) begin? Critical Constructivism needs *breakdown*, that is, in the Heideggerian sense, as a gateway into democratic rationalization. Famously, in the so-called tool analysis of *Being and Time* (1962) Heidegger analyzes our primordial condition of always already being in a situation, practically engaged in a network of references, calling this structure *equipment*. It is only when equipment breaks down, when tools literally break, or get lost, or cannot be reached—in any case the tool is not available for the work we set out to do—that this implicit relational constellation briefly shows itself. More precisely, it shows itself *as* it disappears. It is because the tool breaks down that we suddenly realize the tool was there all along, at our disposal, but we took its functioning, the functioning of the whole referential structure, in fact, for granted. It is a similar dynamic that can be found in Critical Constructivism at the birth of every act of protest, every call for modification or for regulation: one has to "see" the matter of concern first, and this only happens when a tension, or an obstacle, or a frustration emerges from the otherwise "fluent" "relational totality" (with a Heideggerian term) of our technological environs.

But this means that all unnoticed "barbed wire" (Morozov's term again) hiding within "business as usual" can persist if there is no pressing irritation or a kind of upheaval that suddenly shakes up the reigning balance. By way of contrast, consider the corona crisis. Probably like no other societal evolution in recent times—also given its global dimensions—the corona pandemic has been a worldwide *breakdown*, meant here first and foremost in a Heideggerian sense. Suddenly, for instance, more people seem to be concerned about privacy, specifically in the context of the debate about "corona apps," while a cynical observer may remark that privacy didn't seem to be such a big issue for most internet users in the past ten or fifteen years of Google and Facebook dominance. Indeed, the "normal" is thrown upside down with the pandemic, and this seems to suddenly turn up our antennae with regard to certain issues, albeit that the focus of our alertness may be selective and misguided—which seems to corroborate the point under consideration: it is extremely hard to lift ourselves up from business as usual.

If groups or individuals are disenfranchised in a clearly felt way, *breakdown* takes place, and a given situation becomes "present-at-hand." But what if breakdown doesn't take place, and is not likely to take place anytime soon, but there are still potential disenfranchisements lurking, festering beneath the surface? What about things that grow unnoticed, grow on *you* unnoticed, shape and determine you, unnoticed? What with inherent biases or even subtle nudges toward destructive ways of living that may be incorporated in our use of all sorts of appliances and tools?

9.5   OOO's Ubiquitous Breakdown

We'd need an approach, then, that does not depend so much on the occurrence of breakdown to truly begin its work. To be sure, we'd still have to preserve Critical Constructivism's critical achievements (critical here both in the sense of critical theory and "crucial"). Moreover, Feenberg's "hopefulness" may appear overly optimistic or even misplaced at times, but this doesn't mean that hopefulness is not commendable, ethically and existentially speaking. What is more, theories of societal change, such as Critical Constructivism, *need* some notion of hope. As long as optimism has enough *realism* at its base, it can serve as proper motivation for action. As we will shortly see, Critical Constructivism is at its core a *realist* perspective. Bringing this realism fully out into the light, in fact, really reading it *as* realism *sensu stricto*, will help us in tackling the issues described.

Graham Harman in his 2002 book *Tool-Being: Heidegger and the Metaphysics of Objects* already outlined a good part of his "object-oriented philosophy"—nowadays mostly called object-oriented ontology (OOO). His point of departure is his massive re-interpretation of Heidegger, specifically zooming in on the aforementioned tool analysis. Well-known among contemporary philosophers of technology, the tool analysis, among others by way of the example of the hammer, demonstrates how we are always already engaged in a practical context. For philosophers of technology such as postphenomenologists but also critical constructivists, the analysis is important in diverting our attention away from "technology" as an overarching, reified phenomenon—in contrast to Heidegger's later reflections on technology (Heidegger, 1977) which purportedly give us a monolithic notion of technology as all-encompassing phenomenon. The tool analysis shows us the way into fine-grained analyses of how our tools and technologies relate to us, to our environment, to our understanding and praxis, in fact how all these are primordially (always already) interrelated.

Nevertheless, for Heidegger the analysis does not materialize in *Being and Time* (1962, p.91ff.) in the first instance as a specific exercise on how *tools* work. It appears in the context of the larger analysis of the notion of "being-in-the-world" in its different "items of its structure." "World" is one of those items, and it is in the course of investigating the worldhood of world—what *is* world? —that Heidegger develops the notion of equipment. Equipment is not about tools in a strict sense of the term, notwithstanding the many examples hailing from the workshop or any other "instrumental" context. True enough, Heidegger clearly refers to *practical* activity, distinguishing it from theory (all the while however stressing that theory and praxis cannot be separated). But the leading rationale is to "get behind" our tendency, tainting the whole of our philosophical history, to understand being in terms of presence, as something that can be clearly delineated and "placed before us": substance, natural components, *res extensa*, ... *Before* we start doing this, there is already something, something more primordial: the relational constellation in which we are at any time already wrapped up, and that we invariably "pass over," jumping straight at the abstraction of a concept by way of which we *think* we have grasped the essence of being, though we have merely reduced being to *a* being, thus making the cardinal mistake of confounding the ontological difference. Presence-at-hand is not readiness-to-hand. The latter is more primordial.

At this point, Harman steps in, right in the midst of Heidegger's residual ambivalence with regard to the tool analysis being about practical engagement but at the same time being about *everything*. Harman argues that mostly, interpretations have leaned toward one, namely, the former side of this ambivalence, resulting in a dominant pragmatic-pragmatist orientation in Heidegger studies. However, fundamentally, Heidegger is after a more fundamental insight than merely the primordiality of praxis. Being is deeper than beings. It constantly withdraws. This is what Heidegger tries to make sense of through the notion of readiness-to-hand. Before we start pointing to particular beings, present "before us," being must always already be "there"—yet it can't be there in a present-at-hand way. Whenever we are inclined to analyze being in terms of *this* or *that*, we have already converted it into a present thing. Being must always be absent. And so *tool-being*, as Harman calls it in his first book—but which is simply a synonym for readiness-to-hand, and for *being*—forever withdraws.

Surely, there is the analysis of breakdown, where the famous hammer is broken, cannot be found, or something stands in the way of getting it. At such a moment, the relational structure of equipment shows itself in a glimpse, right as it collapses. You were planning on hanging that picture on the wall to decorate the living room, steadfastly marching to the workbench in your garage, only to find an empty spot where the hammer should have been... Consternation. Suddenly the taken-for-granted equipmental whole reveals itself in its now-demised coherence. As a hands-on analysis of practical tool use, this analysis is pertinent and helpful, and it is in this way mainly that contemporary philosophy of technology has adopted the tool analysis. But Harman goes further, underscoring how the tool analysis is about *being*, not just about (human) praxis. The sudden glimpse of readiness-to-hand is paradoxical: we "see" it *as* it disappears. But we don't *really* see it. Only presence-at-hand can be "seen." Readiness-to-hand (tool-being, being) can *only* withdraw, staying invisible as such. So, any "standard" pragmatist interpretation suggesting that praxis puts us in touch with the ready-to-hand, in contrast to all abstracting activities that convert practical relationality into a concept, a substance, an essence, or a principle, is misguided. Praxis *also* turns readiness-to-hand into presence.

More precisely, whenever we grab a tool, set out to achieve an aim (hanging that picture on the wall), deploy a technology, work out a plan, we are not "in" readiness-to-hand; we only ever deal with derivations of tool-being. Being as such never shows itself, and everything we can notice about things is partial. Phrased otherwise: breakdown is ubiquitous. Breakdown—the turning of tool-being into presence, the turning of readiness-to-hand into presence-at-hand—happens *all the time*, whatever we do: build a cabinet, have a conversation, think out a theory. At any time, we only work with or perceive or grasp glimmers of an eternally withdrawing tool-being—or what Harman will later call the the *real core* or *real object*. For, the real cannot be grasped, perceived, or worked with—except indirectly.

9.6 Critical Constructivism's Object

OOO then goes a step further when it extends this scheme to *all* things, *all* object-object interactions. The human being (Dasein) does not have exclusive access, or actually, rather, *lack* of access to things: *every* thing in its real core escapes other things. This aspect of OOO will concern us less here. We are dealing with the *human* critique of technology, and so our need at this point to surpass anthropocentrism is less pressing (however the all-encompassing scope of OOO finds a sort of pendant in Critical Constructivism, in the fairly wide scope of Feenberg's "technosystem" notion). Here, first and foremost, our priority is the existential "use" of OOO: what can it contribute to our critical clout toward technology?

Put succinctly, we can formulate that contribution as follows: it is the seemingly trivial but seminal insight that our outlook on things, our grasp of things, is *always partial*. We can make this more tangible, in fact, by returning to Feenberg. Already embryonically present in Critical Constructivism are ideas that very much line up with the object-oriented perspective—in an existential, lived-experiential sense, that is. Let's try to bring these to the surface. I will focus on three—albeit interrelated—angles: Feenberg's more or less explicit notion of "reality," his understanding of the (technological) object, and, finally, the question of whether technologies have an essence or not.

First, Critical Constructivism's concern with "reality." Unexpectedly, perhaps, Critical Constructivism is heavily invested in piercing through to, at least a form of, *reality*. By way of M. C. Escher's *Drawing Hands* print, Feenberg discusses the "illusion of technology" (2017, p.11): the idea that through technology we achieve our aims without being affected ourselves—in other words, the instrumentalist idea of technology as merely a means to an end and nothing else. This is still how we often tend to think about technology: we perceive its immediate effects—ideally, that which we aim to realize with it—but stay oblivious to its wider impact. The *Drawing Hands* illustrate how we are always enveloped in "strange loops," with Hofstadter's term for how the hands are drawing each other. Our control over technology is finite, and we ourselves are affected by it in turn. Becoming aware of this, one might say, is the start of philosophy of technology, and it is in particular the purpose of Critical Constructivism to unearth the political-economic-social aspect of technologies' strange loops. This is to puncture the illusion of technology and start coming to grips with how things *really* are. "Only in our fantasies do we transcend the strange loops of technology and experience. In the *real world* there is no escape from the logic of finitude" (*ibid.*, 12; my emphasis).

That doesn't mean Critical Constructivism wants to return to a "classic" metaphysical realism of sorts. It does mean that we need to do some soul-searching when it comes to which presuppositions we nourish, mostly implicitly, about *what is real*. The interrelation of reason and experience forms the guiding thread. According to a certain dominant type of societal discourse (nowadays still strong, yet also under intense attack from populists, sceptics, and discontents of all stripes and colors), science is seen as the ultimate arbiter of reality. If something is "scientific," or "scientifically proven," it is the case; any other stance is hearsay, speculation, or construction. Feenberg's point is that clinging too unilaterally to this position leaves all the power in the hands of those who control technology: experts, technocrats, developers. "The scientific idea of nature involves a systematic negation of experience" (*ibid.*, p.13). Crucially, the idea of what is *real* is directly connected to what is seen to be possible when it comes to societal change. We thus need to rehabilitate everyday experience, as a perspective that is just as important in determining *what is real*. "Science and technology influence our understanding of our experience, but the reverse is also true. […] [Non-technical values such as a livable environment or safe work] correspond to realities science may not yet understand, indeed may never understand, but which *are surely real*" (*ibid.*, 14; my emphasis).

Once again, this is no return to, in Feenberg's words, "preconstructivist realism" (*ibid.*, p.51). It is a realism that is rooted in social critique, but in such a way that it is in fact surprisingly akin to OOO's realism. A crucial observation is this: "Protests formulated in the language of values express *aspects of reality* that have not yet been incorporated into the technical environment" (*ibid.*, 8; my emphasis). Indeed, Critical Constructivism hinges on its capacity, or at least its analysis of the capacity for *making new realities*, or more precisely, actualizing potentialities lying dormant in the current circumstances. This brings us to its conceptualization of objects.

Second, Critical Constructivism's notion of "object." Although it is not Feenberg's intention to develop an account of objects as such, he does work out, by way of among others Foucault and Simondon, an account of the technological object. Against Aristotelian substance, modern science posits how "[t]hings have no essence but are composed of functional units awaiting transformation and recombination" (*ibid.*, 125). This is only one side of the story, as we've just seen. Things acquire their meaning depending on their place and role in social settings. So, while the scientific stance "eliminates purpose and hence also potentiality from the world" (*ibid.*), these "naked" things then *retrieve* potentiality by their being embedded in particular constellations, which are now *thus*, but could just have been otherwise, and can change. Feenberg refers to a passage in Marx about the "negro"[1] only becoming a slave under certain conditions: "A negro is a negro. He only becomes a slave in certain circumstances. A cotton-spinning jenny is a machine for spinning cotton. Only in certain circumstances does it become capital. Torn from these circumstances it is no more capital than gold is money or sugar the price of sugar" (quoted in *ibid.*, 28). A thing (or in the first case, a person) is two-sided, Feenberg goes on: we must distinguish "the thing *qua* thing" from "its meaning in capitalist society" (*ibid.*; original emphasis).

But how does one preserve potentiality without reverting to teleology? Marcuse went down that path, positing over and against "one-dimensional" technological rationality the inherent potentialities of objects (humans, natural systems), whose development rationality stifles, and which we'd need to liberate again in order to

---

[1] The term is obviously used here not in a direct signifying sense but only in quoting from the nineteenth-century writings of Marx.

fully let them unfold. Feenberg does not want to take the teleological path, but finds a suitable "middle road" in Simondon's concept of individuation. This is also an account of potentiality, but the difference is that for Simondon, potentiality is deployed in the relation between an object and its "associated milieu." A technological individual (i.e., object) comes about in an interchange between the internal dynamics of the individual and its environment. This still gives development a direction, but not based on a substance wholly internal to the object. "What remains of Aristotelian teleology is the notion of a progressive direction of change but not the metaphysical idea of essence" (*ibid.*, 75). Simondon, in this way, Feenberg observes, offers a bridge position in between older technology critique and constructivism.

Now, one of the most remarkable but easily overlooked aspects of Harman's object-oriented philosophy, is that it is deliberately substantivist, but surely not in the "classic" Aristotelian sense. The object's real core—i.e., *Zuhandenheit*, substance—withdraws absolutely from relation, but this doesn't mean that it is a substance in the Aristotelian sense. A real core belongs uniquely to each and every object, but in fact this landscape of withdrawing cores on the one hand and relations between objects on the other hand is much more fluid than we would be inclined to think on the basis of our presuppositions with regard to "substance." This is the overlooked aspect: in the OOO outlook, *new objects arise all the time*. In fact, Harman puts it as literally as this: with every new relation, a new object comes about. They may be what OOO calls *composite* or *compound objects*, that is, objects composed of other objects—but those are full-fledged objects nonetheless. And so, "any relation between separate things produces a new composite object" (Harman, 2018, p.167).[2] And with every new object comes a new core. This means, simply: the real core is no otherworldly kernel, in which things in some peculiar way partake; as soon as a new object comes about—and this happens quite easily and frequently—a new core will be there as well. So what is this core? In a way, this shouldn't come as a surprise: right from its beginnings in *Tool-Being*, object-oriented philosophy finds that tool-beings *themselves* are made up of *other* tool-beings—in an infinite regress. "The tool-analysis demands a furious regress of tools within tools within tools" (Harman, 2002, p.279). Harman is happy to embrace this infinite regress; the only alternative would to be to presuppose "a *finite* regress, and hence to posit some natural stopping point in reality" (*ibid.*, p.279; original emphasis), which would amount to the Aristotelian substance he wants to criticize.

Hence, there is no irreducible substance, *but* the real core is still there, withdrawing absolutely. So why posit a real core at all? Harman's answer to that question usually runs along the following lines: the real core is a kind of "energy reservoir," making for change. If objects wouldn't hold anything "in reserve" beneath their graspable, perceivable relations, "nothing would ever change" (Harman, 2009, p.187). So, indeed, Harmanian substances have an Aristotelian "feel" to them to the extent that they harbor the "energy," the "potential" for change, although Harman understands the real core as *actuality* instead of potentiality (for an intricate

---

[2] For more on this, cf. Van Den Eede, 2020.

discussion of this, cf. 2002, p.228ff.). But the "function" stays the same: out of the withdrawn substance, which in itself unfolds into countless other objects, comes the capacity for change. "Change only happens because" a thing "is a reality irreducible to the presence-at-hand of all its relations" (*ibid.*, p.230).

Despite the different vocabularies and the obvious conceptual divergences, there is a great deal of resonance here with Feenberg's Simondon-inspired account of the object in terms of non-teleological potentiality. Feenberg needs to safeguard the idea of potentiality in order to enable a theory of societal change (and he says as much).[3] But this could just as much—and perhaps more straightforwardly!—be conceptualized in terms of OOO. OOO also certainly doesn't run the risk of smuggling in teleology again through the backdoor. Despite its appearances, as mentioned, OOO's outlook is tremendously fluid, given how for it, the tension between "the thing qua thing" and "its meaning" in a relational structure is *constantly* at play among objects. At any moment, things are enveloped in this strife: neither completely "real," neither completely relational. And so these sentences out of *Technosystem* sound remarkably Harmanian: "The object is not "real" in any of the usual senses of the term, but rather it is the correlate of an apprehension or intention. But nor is the object subjective; it is a perspective on experience, a cross-section ordering a segment of reality" (Feenberg, 2017, p.139).

Which brings us to the *third point: do technologies have an essence, or not?* The above leaves us with the thorny question about the extent to which we can say that technologies have an essence. Logically speaking, if there is *some* inherent potentiality, no matter the extent to which this potentiality takes shape in a thing's interaction with its environment, we can define this as an essence. Indeed, Feenberg points out as much, when comparing Latour and Simondon: "Where Latour dismisses the notion of *essence*, Simondon reconstructs something similar by shifting the level of potentiality from the Aristotelian substance to the hybrid of individual and its associated milieu" (*ibid.*, pp.76–77; my emphasis). It is not an essence "of old," certainly, an essence in the "essentialist" sense. It is a *historical* essence, which becomes fully clear when Feenberg goes on to compare to Marcuse's Hegelian notion of essence, which "resembles Simondon's theory of individuation in emphasizing the tensions that arise in the relations of the thing and its circumstances. This is the new meaning of "essence" as a historical rather than a metaphysical category" (*ibid.*, p.77).

We've come full circle, really. A distinction must be made between two kinds of essence: an *apparent* essence and a *real* essence. Critical Constructivism endeavors to show how technology is not just what we think it is, namely, the instantiation of technical values (efficiency first and foremost). In other words, technology appears to have "a pure technical essence" (*ibid.*, p.84)—with the emphasis on "appears": "An artifact's line of development *appears* to reveal the implications of a preexisting essence that unfolds with each improvement in its technical basis" (*ibid.*, p.32; my emphasis). But we know that the real story behind the scenes is different. There,

---

[3] "[T]he notion of potentiality [in its new form that Feenberg puts forward] is still an essential basis of any radical theory of politics" (Feenberg, 2017, p.84).

objects "come to their own" in a process of historical-social construction. This kind of essence is more complex, not "pure" at all, with technologies embedded in "the multiplicity of interests they serve" (*ibid.*). Change can come about when individuals or groups act to unearth potentialities which are cloaked in the existing design: "affordances […] that *lie undetected* until new contexts support them or new actors discover them" (*ibid.*; my emphasis). With this, then, a *new essence* is created.

In OOO terms, then, simply, the result of successful democratic rationalization is… a new object. Only the terminology differs: Feenberg speaks of "potentials awaiting realization" (*ibid.*, p.84), beyond the currently reigning power structures. For OOO, the real is always already there, in infinite variety; and it impinges on the world in unexpected, surprising ways. Hence, there is an essence of technology; it is just much more "liquid," multifaceted, and spread-out than we might have thought.

9.7 The Algorithmic Thing's Essence, and the Epistemological Question

Where does this bring us? We started out pointing to Critical Constructivism's unicity in philosophy of technology. In order to preserve and bolster this unique position, we'd need to boost its capacity for enabling pathways into contestation. The issue of discovering a controversy that is not recognizable *as* a controversy potentially blocks some of those pathways; the Algorithmic Thing complicates the situation even further. But it is surely not Critical Constructivism's conceptual infrastructure in itself that is the problem. Rather, what's at stake is what one might call its epistemological vista—more precisely, what it surmises *people* should be able to *notice*. It is clearly, *as a theory*, removed from Aristotelian substance or the Cartesian subject-object split, and offers an account of objects-never-being-*just*-objects (always acquiring through their integration in a network a particular meaning which surpasses their sheer "objectness"), yet what the theory assumes individuals and groups generally are capable of in terms of grasping these dynamics, is by comparison thin. *Breakdown* needs to occur before the critical gear really kicks in.

What I've tried to do in this essay is broaden the lens with which objects should be looked at by actors, through OOO's theory of ubiquitous breakdown. It is not so much a matter of outfitting or enhancing Critical Constructivism with object-oriented spectacles. Rather, Critical Constructivism, as we saw, already harbors an account of (technological) objects which goes a long way into the direction of the OOO perspective. Truly pushing that account through to an object-oriented perspective that at all times looks at things in the expectation of finding beyond their frontside an invisible backside—in a way cultivating a kind of double vision, where we never constantly take things at face value—should expand the potentials for contestation. As mentioned, critiquing technology is, by itself, hard, because of "the invisibility of technical power" (Feenberg, 2017, p.194). An object-oriented orientation, applied existentially, and so deliberately interpreted anthropocentrically—indeed as a lens with which "we" humans look at the world—makes us observe things with renewed wonder, indeed, *wondering* what's behind them.

The challenge would be not to let this wonder degenerate into continuous paranoia. This is where the idea of immensely rich *multiplicity*, so central to OOO (but also already to Critical Constructivism), comes in. Returning to the case of the Algorithmic Thing: we might say, in light of the analysis above, that there *is* an essence to it. However, Algorithmic Things are not just one "thing"—one autonomous entity. As a rule, they will be composite objects (cf. supra). In fact most things we encounter in the world are composite objects. So it depends on at which "level" we look at things[4]: on which "sub-object" are we zooming in? Or conversely, when zooming out, what is the biggest object the objects under consideration are part of? Here, Timothy Morton's notion of *hyperobject* (2013) becomes tremendously useful: a hyperobject is an object that is so massive that we never perceive it in full. Climate change and capitalism are examples, but the internet may also be thought of as one. We know how Feenberg (2012, 2017, p.89ff.) analyzes the internet as an as yet unfinished technology: we can expect its final potential has not been deployed so far. This lines up with the OOO outlook: an object typically has a "life trajectory" (see for instance Harman's case study of the Dutch East India Company (2016), a gigantic object that grows and evolves over many decades and eventually also "dies").[5] And this bigger object may consist of smaller objects, each with their own particular real core, thus, essence.

And so we wind up with *countless essences*, but again, all depends on where one is looking at any specific moment. The smaller object will have another essence than the smaller object. And thus it is perfectly possible that what defines the smaller object differs substantially from what typifies the bigger one. For example, in smaller subregions the internet may be a beacon of democratic action, while on a larger level—as hyperobject, and like so many other domains and sectors in society—it may be completely in the grip of expectations of efficiency, productivity, commercial profit, and so on. In any case, our vista becomes immensely diversified and manifold—richer than what Critical Constructivism already allows for. *At the same time*, strangely enough, such an analysis of "the biggest object" may in fact lead us *back* to "classic" critical theory, whose all-encompassing notion of technical rationality Feenberg has been endeavoring to "multiply." On the level of the *biggest object*, all-penetrating technical rationality may actually *reign*. Or phrased differently, there may still be a level at which the "classic" analysis makes sense, just like there are levels at which a deterministic analysis is pertinent (technologies do have effects in minute relational/networked ways).

This does leave us, finally, with one thorny obstacle: for our epistemological perspective to broaden in this way, our worldview would have to *become* Harmanian-Feenbergian. In order to see things this way, one would need to fully embrace the "lens," make it one's own (cf. Van Den Eede, 2020, pp.209–10). Feenberg has in

---

[4] There is some convergence here with Feenberg's notion of "layers" (cf. Feenberg, 2017).

[5] It would be interesting to further explore the lines of connection between Critical Constructivism and this first attempt on Harman's part to develop a social theory within the bounds of OOO (Harman 2016), but this exceeds the scope of this chapter.

fact critiqued Latour's political theory along these lines: in order for the theory to work, common sense would have to become Latourian, he observes, which he thinks is unlikely to happen (Feenberg, 2003, p.91). It is probably just as unlikely that common sense will soon become Feenbergian-annex-Harmanian. However, isn't this a hurdle for all pedagogically and/or politically engaged philosophies? The philosophy *an sich* is developed in order to unearth something that was previously concealed (in whatever way). And so one has to incorporate the philosophy at least to a certain extent in order to fully reap its fruits. In a way, one has to go "live" within the object that the philosophy is, become part of the object—and thus, actually, go form a new object, together with it: me+the philosophy. It is precisely this perspective-taking, this perspective-creating, that object-oriented thinking incites us to engage in.

References

Feenberg, A. (1995). *Alternative modernity: The technical turn in philosophy and social theory*. University of California Press.

Feenberg, A. (1999). *Questioning technology*. Routledge.

Feenberg, A. (2003). Modernity theory and technology studies: Reflections on bridging the gap. In T. J. Misa, P. Brey, & A. Feenberg (Eds.), *Modernity and technology* (pp. 73–104). MIT Press.

Feenberg, A. (2012). Introduction: Toward a critical theory of the internet. In A. Feenberg & N. Friesen (Eds.), *(Re)Inventing the internet: Critical case studies* (pp. 3–17). Sense Publishers.

Feenberg, A. (2017). *Technosystem: The social life of reason*. Harvard University Press.

Harman, G. (2002). *Tool-being: Heidegger and the metaphysics of objects*. Open Court.

Harman, G. (2009). *Prince of networks: Bruno Latour and metaphysics*. re.press.

Harman, G. (2016). *Immaterialism: Objects and social theory*. Polity.

Harman, G. (2018). *Object-oriented ontology: A new theory of everything*. Penguin.

Heidegger, M. (1962). *Being and time*. Translated by John Macquarrie and Edward Robinson. Blackwell.

Heidegger, M. (1977). *The question concerning technology and other essays*. Translated by William Lovitt. Harper Perennial.

Morozov, E. (2013). The real privacy problem. *MIT Technology Review*, October 22, 2013. https://www.technologyreview.com/s/520426/the-real-privacy-problem/.

Morton, T. (2013). *Hyperobjects: Philosophy and ecology after the end of the world*. University of Minnesota Press.

Van Den Eede, Y. (2013). The Mailman problem: Complementing critical theory of technology by way of Media Theory. *Techné: Research in Philosophy and Technology, 17*(1), 144–162. https://doi.org/10.5840/techne20131718

Van Den Eede, Y. (2019). *The Beauty of Detours: A Batesonian philosophy of technology*. State University of New York Press.

Van Den Eede, Y. (2020). Imagining things: Unfolding the 'of' in philosophy of technology, through Object-Oriented Ontology. In H. Wiltse (Ed.), *Relating to things: Design, technology and the artificial* (pp. 191–213). Bloomsbury.

Part III
Critical Theories of Technology

Chapter 10
Beyond Efficiency: Comparing Andrew Feenberg's and Byung-Chul Han's Philosophy of Technology

Federica Buongiorno

Abstract In this chapter I will compare Andrew Feenberg's philosophy of technology with that of the Korean-born philosopher Byung-Chul Han to show that their *diagnoses* about the contemporary social and political status of technological development, as well as their *prognoses* about how to deal with it, differ radically. I argue that this radical difference is based on: (i) two different evaluations of modernity and, in particular, two different appraisals of both Martin Heidegger's and Michel Foucault's thought; (ii) an anti-essentialist stance about technology on Feenberg's part, and an essentialist/reductive appraisal of technology on Han's part; (iii) two different accounts of the Internet and of information technology. My aim is to show that Feenberg's reflection represents a more solid and detailed analysis of the phenomenon of technology, from the point of view of its limitations and its potentials, as well as a critical corrective to Han's radicalism, as far as outlining the realistic possibility for future political interventions.

Keywords Andrew Feenberg · Byung-Chul Han · Philosophy of technology · Efficiency · Martin Heidegger · Michel Foucault · Internet · Democratization · Capitalism · Data · Labour

10.1 Introduction

My aim in this chapter is to compare Andrew Feenberg's philosophy of technology with that of the Korean-born philosopher Byung-Chul Han. Feenberg and Han are two of the most prominent theorists of technology writing today and their work represents two poles of the contemporary philosophical critique of technology.

F. Buongiorno (✉)
Dipartimento di Lettere e Filosofia, Università degli Studi di Firenze, Florence, Italy
e-mail: federica.buongiorno@unifi.it

© The Author(s), under exclusive license to Springer Nature Switzerland AG 2022
D. Cressman (ed.), *The Necessity of Critique*, Philosophy of Engineering
and Technology 41, https://doi.org/10.1007/978-3-031-07877-4_10

I wish to show that their *diagnoses* about the contemporary social and political status of technological development, as well as their *prognoses* about how to deal with it, differ radically. By referring to Feenberg's theory of modernity, with special emphasis on *Questioning Technology* (1999) and *Technosystem* (2017), and to Han's books on contemporary society (*Müdigkeitsgesellschaft*, 2010 [*The Burnout Society*, 2015a]) and *Transparenzgesellschaft* (2012 [*The Transparency Society*, 2015b]) I will argue that this radical difference is based on:

   (i) two different evaluations of modernity and, in particular, two different appraisals of both Martin Heidegger's and Michel Foucault's thought. While Feenberg suggests we reassess the notion of rationality on the basis of critical theory, thereby re-interpreting Foucault's (and Marx's) notion of technology in light of current societal developments, Han assumes that Foucault's disciplinarity paradigm is an outdated account of contemporary societies, which are now ruled by psychopolitics;
  (ii) an anti-essentialist stance on technology from Feenberg's part, which results in a kind-of constructivism that does not reduce the essence and function of technology to efficiency, but seeks to bind it to a dimension of meaning; on the contrary, Han employs an essentialist view of technology which—in a fundamentally pessimistic way—considers efficiency and performance/achievement (*Leistung*) to be the essence and goal of technology;
 (iii) two different accounts of the Internet and of information technology. I will analyse this aspect by taking into consideration Feenberg's *Technosystem* (2017, especially its fourth chapter, "The Internet in Question") and Han's book *Im Schwarm. Ansichten des Digitalen* (2013 [*In the Swarm. Digital Prospects* 2017a]), comparing Feenberg's multi-layered view of the Internet with Han's notion of the digital swarm to critically display the different approaches to power and information exchange developed by the two thinkers.

Byung Chul Han's radical "uncompromising" philosophy elicits widely divergent reactions today: enthusiastic endorsement or utter dismissal, with little room for more nuanced forms of appreciation. On the one hand, this reflects Han's ability to see through things and interpret our contemporary moment; on the other, it can lead to a degree of subjective annoyance at the stark tones of his philosophy. This unease only increases when we compare Han's thought with that of Feenberg's, whose reflections are conveyed in an opposite style—a rhetorically more sober yet equally radical one. This is not merely a matter of a different way of evaluating technology and its socio-political effects. Rather, what is at stake is a profoundly different conception of what philosophical thought is required to do in contemporary neoliberal society, and hence this chapter is intended to also contribute to assessing the role of the Humanities in our capitalist system.

10.2   A Different Understanding of Modernity

"Indeed, the prospect of a philosophy of technology had to wait for the dissolution of the widely held belief that philosophy was interested in technology only to condemn it," Feenberg writes in his introduction to the edited collection *The Politics of Knowledge* (1995, p. ix). This claim has yet to be entirely disproven by the development of the philosophy of technology as a discipline, and partly applies to Han's pessimistic view of the essence and function of technology (especially digital technology). In a 2015 interview, Han states:

> I try to describe what is present. It is hard to see through things (...). It is possible that my books hurt, because I show things that people don't want to see. It is not me, not my analysis that is merciless, but the world in which we live; it's merciless, crazy and absurd. (Boeing & Lebert, 2015)

Further in this interview, Han claims that digital technology is largely responsible for the absurdity of the contemporary world: "But life is endangered when everything is automated, when everything is ruled by algorithms. An immortal machine human (...) would no longer be human" (Boeing & Lebert, 2015).[1]

Han's pessimism can be attributed, to some extent, to his deeply humanistic interpretation of modern technology, positing technology as something essentially distinct and opposed to humanity. As Natasha Lushetich (2018) has noted in her review of *Psychopolitics*, "Han's reading of algorithmic logics is too humanistic," in the sense that it is not based on a precise theoretical analysis of digital media - of the sort provided instead by authors such as Benjamin and Adorno with their critical reflections on mass media (the press, radio, films). From the initial exclusion of technology from humanistic discourse—noted by Feenberg—we pass, with Han, to a hyper-humanistic evaluation of it that stretches the two poles of the subject dominated by technology and of the dominant technology, thereby reducing the field of possible transformative interventions. Feenberg's "optimistic" view radically differs in regard to this point: "No doubt modern technology is immensely more destructive than any other. And Heidegger is right to argue that means are not truly neutral, that their substantive content affects society independent of the goals they serve. But this content is not *essentially* destructive; rather, it is a matter of design and social insertion" (Feenberg & Alastar, 1995, p. 16). Significantly—as I will show—Feenberg has developed a profound critique of Heidegger's philosophy of technology, whereas Han has drawn upon it positively.

The most profound difference between Feenberg's thought and Han's revolves precisely around the *essence* of technology. Put differently, the difference in question is between an anti-essentialist view of technology (Feenberg) and an essentialist one (Han). On the one hand, for both authors it is evident that "technology is the medium of daily life in modern societies" (Feenberg, 1999, p. vii): as such, technology is not a neutral phenomenon, but embodies a series of social aspects that must

---

[1] Han voiced similar views in an interview with Vera Tollmann, in 2011, just after the publication of *Müdigkeitsgesellschaft* (see Tollmann, 2011, last retrieved November 18, 2020).

be included in critical-philosophical analysis. On the other hand, although Han acknowledges the social dimension of technology as an explicit component, he traces the technological element back to the paradigm of *performance* (*Leistung*), thereby reducing technology to the function of efficiency. In doing so, Han shares an important assumption with essentialism, namely that technology, despite its many different meanings, can be reduced to efficiency.

Against this, Feenberg (1999) argues that technology cannot simply be reduced to efficiency: "essentialism holds that technology reduces everything to functions and raw materials" (p. viii), maintaining that devices are "essentially functional, and therefore (…) *essentially* oriented towards efficiency. In the pursuit of efficiency, technical disciplines systematically abstract from social aspects of their own activities" (p. ix). In this abstraction lies the limit of the essentialist explanation of technology. According to Feenberg, "the 'essence' of actual technology, as we encounter it in all its complexity, is not simply an orientation towards efficiency" (p. x). Only by detaching the technological object as such from the experience of it is it possible to essentialize the functionality of technological devices. However, according to what Feenberg describes as an experiential perspective (and what I would suggest we describe as a phenomenological perspective),[2] the dimension of meaning, of the *lifeworld*, is inseparable from the technological dimension: both contribute to structuring the *irreducible* complexity of technology.

Feenberg's and Han's divergent interpretations ultimately derive from two different ways of assessing technology, both of which are mediated by Heidegger's philosophy. In Chap. 8 of *Questioning Technology* (1999), Feenberg engages with Heidegger's philosophy of technology, criticising it in view of the development of a socio-critical alternative. His thesis is clear-cut: "To a considerable extent, it is the very authority of Heidegger's answer to the 'Question' [*of Technology*] that has blocked new development" (p. 183).[3] More specifically, the Heideggerian notion of *Bestand* (standing reserve), which is the idea that technology changes the world into "raw materials" and that it is incapable of letting being appear, entails an attitude of

---

[2] In its assumptions, Feenberg's approach offers plenty of phenomenological insights, although these are rarely explicitly formulated. The idea that in order to appreciate the horizon of meaning of technological phenomena in their material objectiveness, these must be traced back to the sphere of the *lifeworld*—which is to say, the moment of the direct experiencing of these phenomena in their social and cultural stratifications—represents a key principle within a phenomenology of technology of Husserlian inspiration. Moreover, in a passage from *Technosystem* discussing Ian Angus's work, Feenberg appears to be drawing upon the notion of formalisation by applying it to the concept of reification theorised by Lukács: formalisation is understood in a Husserlian sense here, as a process that is separate from generalisation and which, unlike the latter, "breaks that connection to the individual and substitutes variables that can refer to any object whatsoever. (…) Formalization and instrumental practice share an alienation from the lifeworld that results in Husserl's 'crisis'" (Feenberg, 2017, p. 128).

[3] Part III of *Questioning Technology* is entirely devoted to the topic of "Technology and Modernity." In particular, Chap. 8 focuses on "Technology and Meaning," providing a detailed critique of Heidegger's views in *The Question Concerning Technology*.

resignation and passivity, not transformation or reform,[4] a perspective that Han shares with Heidegger.

In *Im Schwarm*, Han writes:

> "The hand acts": this is how Heidegger defines the essence of the hand (…). According to Heidegger, the hand is the medium of "Being", which indicates the original ground of meaning and truth (…). The typewriter, which requires us to use only our fingertips, draws us away from Being. (Han, 2013, p. 52)[5]

Han shares this view and radicalises it in his interpretation of the digital era: "No doubt, Heidegger would argue that the digital device further worsens this atrophy of the hand" (Han, 2013, pp. 52–53). In a footnote in *Transparenzgesellschaft*, Han further states: "The virtual world lacks the resistance of *Reality* and the negativity of the *Other*. Against its unfathomable positivity, Heidegger would evoke the 'earth' (*Erde*) again. This describes what is hidden and inaccessible, and closes itself" (Han, 2012, p. 88). According to Han, this concealment, this subtracting of something from the virtual in favour of the real is a necessary form of resistance against the technological positivisation of the world, which translates into practically absolute control over the human form of life.

Here, Han is concerned with what Feenberg describes as "technocracy" and addresses it with reference to Foucault's theory: Feenberg's peculiar interpretation of Foucault allows him, unlike Han, to avoid a totalizing critique of technology (and technocracy). "By 'technocracy' I mean a wide-ranging administrative system that is *legitimated* by reference to scientific expertise rather than tradition, law, or the will of the people. (…) What makes a society more or less 'technocratic' is largely its rhetoric rather than its practice" (Feenberg, 1999, p. 4). This system, which is typical of capitalist society, is an outcome of modernity and of the related notion of rationality, as well as the dilemma it implies: on the one hand, categories such as modernisation, rationalisation, and reification embody the very essence of (post) modern rationality and are difficult to avoid when seeking to interpret post-industrial history; on the other hand, "these are 'totalizing' concepts that seem to lead back to a deterministic view we are supposed to have transcended from our new culturalist

---

[4] Feenberg does not fail to emphasise a certain shift of perspective in Heidegger's later works, such as his *Discourse on Thinking*, where the German philosopher presents a mode of "free relation" with technology: "We can affirm—Heidegger writes—the unavoidable use of technical devices, and also deny them the right to dominate us, and so to warp, confuse, and lay waste our nature" (Heidegger, 1966, p. 54). It is crucial to establish this free relation insofar as technology is the cultural form that makes every element of modernity subjectable to control: this "culture of control" implies "an inflation of the subjectivity of the controller, a narcissistic degeneration of humanity" (Feenberg, 1999, p. 185). However, Heidegger's reflection is still compromised by a number of ambiguities—for example: "He warns us that the essence of technology is nothing technological, that is to say, technology cannot be understood through its usefulness, but only through our specifically technological engagement with the world. But is that engagement merely an attitude or is it embedded in the actual design of modern technological devices?" (*ibid.*, p. 186).

[5] All translations of passages from Han's works are my own. In brackets I am providing a reference to the German original and—when a text is quoted for the first time—to the title and year of the English edition.

perspective" (Feenberg, 1999, p. 14). How can we overcome this dilemma, which is to say the alternative between a universal rationality and specific, situated political values?

Michel Foucault's thought acquires special significance from this perspective: "for Foucault, the essential innovation of modernity is the reliance on forms of knowledge that are simultaneously forms of power" (Feenberg, 2017, p. 19). More specifically, according to Foucault, modernity is characterised by the alliance between knowledge and power, whose union engenders both individual subjectivity and the social order: power/knowledge constitutes a web that takes the concrete form of technologies, architectural structures, and devices, as well as practices, organisations, institutions, and standardised roles. Political, medical, and administrative forms of knowledge, as well as technical sciences such as criminology and psychiatry, deeply shape power relations in modern societies and go hand in hand with the emergence of institutions established for confinement and control purposes. Unlike Weber, however, Foucault believes that rationality is not singular but plural, meaning that it needs to be contextualised: "There is not one rationalization but many, corresponding to the many domains of social life" (Feenberg, 2017, p. 20).

This contradicts an idea of rationality that has been historically modelled on the paradigm of science and mathematics: certainly, the rationality of an institution cannot be compared to that of a scientific discipline, insofar as it is not determined by an intrinsic logic, but by causal and symbolic connections that elude the criterion of exactness. Still, as Foucault noted, rational institutions play the "game of truth (…), an ensemble of procedures which lead to a certain result, which can be considered in function of its principles and its rules of procedures, as valid or not, as winner or loser" (Foucault, 1988, p. 16 [qtd. in Feenberg, 2017, p. 22]). According to Feenberg, Foucault's genealogical approach can be applied to the technosystem, since institutions and artefacts are assemblages of components combined with a functional role in society, and not determined by any intrinsic essence.

Whereas the Foucauldian approach substantiates Feenberg's anti-essentialism and is adopted—albeit in a revised form—by him, in almost all his works, Han confirms the impossibility of applying Foucault's categories to the present. Han devotes a whole chapter of *Müdigkeitsgesellschaft* to the topic, entitled "Beyond Disciplinary Society." At the beginning of this chapter, he writes:

> *Foucault's* disciplinary society, made up of hospitals, madhouses, prisons, barracks, and factories, is no longer the society of today. It has long been replaced by a completely different society, made up of fitness centres, office towers, banks, airports, shopping malls, and genetic laboratories. Twenty-first-century society is no longer a disciplinary society, but rather a performance society. (Han, 2010, p. 17)

In his works from 2010 onwards, Han has engaged with Foucault's disciplinary paradigm. According to the French thinker, from the seventeenth century onwards, power evolved from being the power of death to being the power of life, which no longer threatens to kill, but rather aspires to indefinitely extend and reinforce life through the rigorous governing of bodies and planning of biological life. This change in the form of power is a consequence of the industrialisation process, which requires a new form of control over the body in order to adapt it to suit the requirements of mechanical production. Therefore, human bodies are normatively subjected to a series of rules, obligations, and prohibitions that are functionally productive, while aberrations and anomalies are eliminated, insofar as they hamper the process of production. Disciplinary power thus entails a biopolitical power technique: it is exercised over the population as a "productive and reproductive mass to be governed scrupulously. (…) Reproduction, birth and death rates, health levels, and life expectancy become an object of regulation" (Han, 2013, pp. 97–98).

In *Psychopolitik. Neoliberalismus und die neuen Machttechniken* (2014 [*Psychopolitics. Neoliberalism and the New Technologies of Power*, 2017b]), Han explicitly speaks of "Foucault's dilemma;" Foucault does not make a transition to psychopolitics, for he fails to realise that the biopolitical paradigm is unsuitable for describing neoliberal society and its range of digital power/technologies. According to Han, "Neoliberalism does not deal primarily with the biological, somatic, or bodily; rather, it discovers the psyche as a productive force. This conversion to the psyche, and hence to psychopolitics, also depends on the mode of production of present-day capitalism" (Han, 2014, p. 39). In other words, it is not the body, but the psyche of human individuals that is an object of control by neoliberal power: "the body is dismissed from the process of direct production and becomes an object of aesthetic or technical-sanitary optimisation" (*ibid.*). This occurs because neoliberal performance society increasingly steers clear of the negativity of prohibition typical of disciplinary society and is becoming characterised by a growing degree of deregulation: "prohibition, obligation, and the law are replaced by planning, initiative, and motivation" (Han, 2010, p. 18). Biopower is thus replaced by psychopower, "The call for *motivation, initiative*, and planning exploits more effectively than whips and commands," (*ibid.*) insofar as it coats itself in a veneer of freedom, by invoking an illusory self-planning on the part of individuals. Industrial capitalism—in Han's view—has turned into neoliberalism and financial capitalism, which no longer rules through sovereign power or disciplinary control. In post-capitalistic society it is more important to control the minds of individuals in order to practice a form of *psychopolitics* whose aim is to lead people to interiorize the commitment to work, so that individuals may *exploit themselves*—without even recognizing the exploitation, since now it does not come from "someone else", from the outside but from the subjects themselves. Although individuals think that they are *freely projecting themselves*, they have never been so *sub-jected* to work.

Modernity—which according to Han essentially coincides with the legacy of the Enlightenment—is thus characterised by a three-fold structure, which is to say by three stages of "Enlightenment" that describe the evolution of media technology. Han assumes that the process of Enlightenment is still ongoing, starting from a first phase in which humans tried to separate knowledge from mythology by resorting to statistics, i.e. to computing capacity. At a second stage, the Enlightenment led to the "transparency" imperative, which affirms that "everything must be at the service of data and information." This in turn led to "data fetishism" (*Daten-Fetischismus*), which is at the core of the "third Enlightenment": data replace theory. As a neoliberal form of power, psychopolitics is thus closely connected to IT and digital culture.

Before turning to consider this aspect in the third section of this paper, I would like to focus on a highly problematic theoretical aspect of Han's critique of Foucault. This critique would appear to entail—at least to some degree—the re-establishment of the Cartesian dualism between mind and body. For Han, the point is that the (digital) technologies of neoliberal power imply a dismissal of the body, which becomes only marginally relevant, in favour of subliminal control over individuals' psyche. However, as he himself notes in one passage, "the performance subject remains disciplined. He has passed through the stage of discipline" (*Ibid.*, p. 19). But what this means is that: (a) psychopower *implies* biopower: performance society is not simply a new phase in the evolution of capitalism, which breaks with the previous one, but rather incorporates the stage of disciplinary society; (b) psychopower could never control the psyche if it did not *also* exercise control over the body: for the psyche is embodied and does not exist as a separate entity from the body. As I have shown elsewhere (Buongiorno, 2015), according to Han there is a clear separation between biopolitics and psychopolitics, and the forms and modes of communication also clearly differ from the disciplinary age to the transparency paradigm. However, things would appear to be more complex than Han assumes. I would suggest that the effects of these practices are not merely psychological and subtle, but rather of a *psychosomatic* kind. After all, we cannot really distinguish between mind and body with respect to the effects produced by political practices making use of advanced technologies, without falling back into a form of Cartesian dualism—which, as Han himself assumes, has been overcome by the subject of transparency society.[6]

10.3   Beyond Efficiency

In *Technosystem*, Feenberg observes that the study of technology in the philosophical tradition has been associated with different orientations—Marxism, Pragmatism, Heideggerian philosophy, sociological theories of modernity —that have emphasised the relationship between technology and society by seeking to grasp the specific trait of modernity in relation to science and scientific revolutions. These different orientations have identified the gradual loss of human agency in technological societies as the main pathology of modernity. Feenberg writes: "Their themes are familiar: technocracy, the tyranny of expertise, the substitution of knowledge for wisdom and information for knowledge, a vision of society as a complex

---

[6] If we assume that the very beginning of body and mind control, as a biopolitical practice, dates back to Descartes' dualism of substances, then we should consider dualism itself as an early form of that practice. However, rather than analyzing Cartesian dualism in terms of its biopolitical consequences, Han overturns it in order to show the increasing priority of *sum* over *cogito*, and over all other *activities performed* by the subject in capitalist societies. Han's thesis of the "hypertrophic Ego" is grounded in the reversal of Cartesian dualism and represents a direct consequence of people's life turning into a mere domain of demand and consumption.

of functional systems, the meaninglessness of modern life, the obsolescence of man, and so on" (Feenberg, 2017, p. 39). These are crucial issues for Han, who can easily be included within this interpretive current. We need only recall the afore-mentioned interview Han gave in 2015, in which he states: "It is possible that my books hurt, because I show things that people don't want to see. It is not me, not my analysis that is merciless, but the world in which we live; it's merciless, crazy and absurd" (Boeing & Lebert, 2015). However, technology as such, which is to say technology in its technical aspects, remains broadly excluded from Han's analysis. As I will show in the next section, he does not fully master the particularities of digi-tal technologies and apparently fails to grasp their complexity—which is certainly problematic for a variety of reasons, yet also capable of expressing a potential for transformation. This stems from the fact that whereas Feenberg opts for an anti-essentialist interpretation of the phenomenon of technology, Han reduces the essence of technology to the paradigm of efficiency and performance.

The emergence of biological and social sciences in the late eighteenth and nine-teenth centuries marked a turning point in the debate on technology by anchoring it in the idea of progress, the key concept of modernity; the belief took shape that technological progress would free human civilisation, ensuring happiness and pros-perity. These objectives are natural traits of the biological make-up of human beings which technology is allegedly capable of furthering, yet not altering; hence the neu-tral character assigned to technology, which has led politics to be left out of the debate. We are thus faced with an alternative, as Feenberg notes: "either politics becomes another branch of technology, or technology is recognized as political. The first alternative leads straight to technocracy, public debate will be replaced by tech-nical expertise; research rather than uninformed opinion of the voters will identify the most efficient course of action" (Feenberg, 1999, p. 2). The characterisation of modern rationality as a "pure drive for efficiency, for increasing control and calcu-lability" (*ibid.*, p. 3) has led the essentialist approach to embrace what is a determin-istic view, according to which technological progress has an automatic and unilinear character. To this we should add the assumption that technology is essentially ori-ented towards domination. Against this view, Feenberg assumes that "technological development is not unilinear but branches in many directions, and could reach gen-erally higher levels along several different tracks. And, secondly, social develop-ment is not determined by technological development but depends on both technical and social factors" (*Ibid.*, p. 83).

Against the essentialist perspective, Feenberg endorses a critical-constructivist approach, centred on the combination of STS (mainly social constructivism and actor-network theory) and a revised version of the Frankfurt School's critical theory, with the aim of providing a new interpretation of the technological application of rationality. In particular, Feenberg (1999) draws upon Herbert Marcuse, whose thought he sets in contrast to Heidegger's philosophy: "unlike Adorno and Heidegger, he [Marcuse] thinks human action can change the epochal structure of technological rationality and the designs which flow from it" (p. 153). According to

Marcuse, science and technology can become "vehicles of freedom,"[7] whereas in Heidegger's (and Weber's) works, techno-scientific rationality is presented as non-social, neutral, formal, and as an agent of control. For Feenberg (1999), Marcuse's idea of technical rationality indicates "the most fundamental social imperatives *in the form they are internalized by a technical culture*. This is what, in a constructivist framework, I have called the 'technical code'" (p. 162). Technical codes are embodied by technical systems and reinforce the culture within which they emerge and its constitutive values. It is precisely in this sense that technology is political, and not, as in the essentialist version, neutral. According to Marcuse (and Feenberg), then, technology and normativity coexist in all sectors dominated by science and technology.

Based on these assumptions, efficiency—the distinguishing feature of modern technical rationality—does not exhaust the complexity of technology, but neutralises its social and political dimensions. It is necessary, therefore, to go beyond the paradigm of efficiency in order to offer a more complex view of technology that, in place of the pessimism of absurdness, also enables transformative political intervention.[8]

> The design that eventually prevails in the development of each technology and institution is the framework within which it is rational and efficient. Efficiency is not an absolute standard since it cannot be calculated in the abstract but only relative to the specific contingent demands which bias design. (Feenberg, 2017, p. 59)

Han's interpretation goes in the opposite direction, since he suggests that contemporary performance society and its technological apparatus are essentially characterised by the efficiency with which they exploit the excess of positivity permeating them, i.e. the apparent freedom that the performance subject believes to enjoy but which has actually become the mechanism of exploitation employed in psychopolitics—a mechanism far more effective than the kind of external constrictions typical of disciplinary society. Performance society no longer resorts to prohibitions and prescriptions, but encourages subjects to do and plan things, extolling their capacity to act: "The positivity of the capacity to act is much more effective than the negativity of duty. (…) The performance-subject is much faster and more productive than the obedience-subject" (Han, 2010, p. 19).[9]

---

[7] In Marcuse's own words: "(…) in order to become vehicles of freedom, science and technology would have to be reconstructed in accord with a new sensibility—the demands of the life instincts. Then one could speak of a technology of liberation, product of a scientific imagination free to project and design the forms of a human universe without exploitation and toil" (Marcuse, 1969, p. 19).

[8] "I have proposed the term 'democratic rationalization' to signify user interventions that challenge undemocratic power structures rooted in modern technology. With this concept I intend to emphasize the public implications of user agency" (Feenberg, 1999, p. 108).

[9] This thesis is a recurrent one in Han's writing, from *Müdigkeitsgesellschaft* onwards. Consider the following passages, for example: "Self-exploitation is far more effective than external exploitation, as it is accompanied by a sense of freedom" (Han, 2013, p. 93). "Self-enlightenment is more efficient than the enlightenment that comes from someone else, because it is combined with a sense of freedom" (Han, 2012, p. 80). "Psychopower is more effective than biopower insofar as it disci-

In order to achieve the universalization of a condition based on the exploitation of freedom, performance society alters the nature of democracy by means of a technological and digital apparatus that aims to make society itself as transparent as possible, which entails a profound transformation of the very notion of power. "Transparency and power do not get along well. Power likes to cloak itself in secrecy. The praxis of arcana is one of the techniques of power. Transparency eliminates the arcane spheres of knowledge." (Han, 2012, p. 78). Here Han is referring to Carl Schmitt, according to whom the postulate of openness "finds its specific opponent in the idea that every kind of politics encompasses *arcana* that are in fact just as necessary for absolutism as business and economic secrets are for an economic life based on private property and competition" (Schmitt, 2003, p. 134 [author's translation]). The call for transparency—in politics and in the production and use of information—is a ruse that performance society, which is entirely positivised and empowering, employs to increase its control over individuals, and hence its economic yield. Han engages in a radical interpretive reduction of information to pure positivity: information is essentially "bad" precisely because it is transparent and positive, and free of any traces of negativity and resistance. "The transparency society is an information society. Information *as such* is a phenomenon of transparency, because it lacks all negativity. It is a positivised, operational language" (Han, 2012, p. 66). Han, therefore, generalises his assessment of information *as such*, without justifying this generalisation and thus without specifying what he means by information. It may be inferred that the reference is to digital information, whose mechanisms and complexity, though, are never thematised and discussed in depth. Consequently, Han condemns information processes *as such*, according to an essentialist approach which leaves no room for a more complex description or any possibilities for change.

Moreover, Han does not regard techniques and technology from the concrete perspective of their technical functioning: they are almost invariably understood in the sense of techniques and technologies *of power*, i.e. not as concrete sets of devices — for instance, no reference is made to problems related to their design— but as *practices* for the exercise of power. This approach almost entirely overlooks the material quality of technical and technological devices, reducing the whole argument about techniques to the political-philosophical dimension without considering the strictly technical aspect. An analysis of the technical aspects of the media, which are not neutral but in turn embody power connections and relations, might tone down Han's pessimism with regard to technology and highlight some oversimplistic interpretations on his part, such as those which emerge from his theory about digital communication, and particularly the use of social networks.

---

plines, controls, and influences human beings not from the outside, but *from within*" (Han, 2013, p. 101). "Neoliberalism is a highly effective system for exploiting freedom (…). It is not effective to exploit someone against his will (…). Only the exploitation of freedom produces the highest yield" (Han, 2014, p. 11). "Smart power with a free and benevolent appearance that stimulates and seduces is more effective than that power which orders, threatens and prescribes" (*Ibid.*, p. 27).

10.4   Is the Internet Bad?

Can the Internet contribute to the democratisation of society? Feenberg's affirmative answer and Han's negative one could not be more far apart. The Internet is a particularly important testing ground for both thinkers, whose theoretical assumptions ought to find concrete confirmation in this sphere. Han's radically critical stance has exercised a powerful appeal for people critical of neoliberalism:[10]

> …digital technology is not just a tool. It is an efficient device to rule. It allows a new power to rule without precedents by those who design these technologies or by those who have access to the data produced by them. This power gives rise to the so-called *psychopolitics*, very much in the Californian spirit of Silicon Valley. (…) I approach digital technology as a kind of ruling, although a *smart and soft* ruling, a "soft totalitarianism" in the words of Eric Sadin, or the "capitalism of   like" in Han's words. (Ballesteros, 2020, p. 3)

This kind of appreciation of Han's basic thesis often betrays fear at the prospect that human kind may be completely controlled by technical artefacts, and leads to a rather problematic desire to restore the "superiority of humans over animals and artefacts" (*Ibid.*, p. 2). The idea is that digital tools (particularly online communication and data collection) are replacing human skills everywhere (in the workplace as much as in the sexual field, the medical one, and so on), thereby reducing humans to one sphere of technological application (among others). This interpretation extends to the digital realm, where analyses already developed for automation processes and their progressive expropriation of human labour seeks to confirm the basic mode of functioning of capitalism. However, as Feenberg clearly shows, the thesis in question only holds if we define the activity of online users as "labour" in the Marxist sense of the term—but this is a questionable connection in many respects:

> The "labor" of users is not abstract but totally concrete in the sense that it depends on personality and style. It is not uniform, is not measured by the time expended on the effort, and it cannot be divided into a portion devoted to reproduction and another portion expropriated as surplus value. (…) In a very broad sense we can call these examples "exploitation," but they do not do the harms nor have the political implications of the expropriation of surplus value in capitalist production. (Feenberg, 2017, p. 93)

---

[10] Han expounds his theory about digital technology in *Im Schwarm* (2013). However, in this work the term "digital" is used in a rather generic sense to describe both applications downloaded from the Internet and intended for personal use (e.g. for biometric measurements) and devices such as Google Watch and other self-tracking and Quantified Self tools, as well as the Internet and online communication via social networking. According to Han, the logic of optimisation and performance applies to all these cases. In other words, Han does not provide a theory which distinguishes between the various aspects of digital technology and considers their specific and distinctive traits, which depend on the different technologies used to implement them and their different designs. By contrast, particularly in *Technosystem*, Feenberg specifically focuses on the Internet as a case study. In the following section, therefore, in discussing the digital sphere, I will essentially be referring to the case of the Internet and online digital communication.

Han argues that through digital technologies the neoliberal system can extend work time into leisure time such that online "entertainment" too becomes a form of exploitation. "To the extent that it is used for the regeneration of the workforce, relaxation is nothing but a mode of work: relaxation is not what is different from work, but its *product*" (Han, 2013, pp. 48–49). The way in which the digital "swarm" works makes it suitable for complete psychopolitical exploitation. A swarm is not a crowd or mass in the classical sense; the latter has a unifying ideology, which establishes a "we", and common actions "capable of taking existing relations of domination head on. (…) Digital swarms lack this resolve: they *do not march* (…), they do not develop political energies" (*Ibid.*, p. 23). Even more radically, Han observes:

> The masses could once organise into parties and associations and were animated by an ideology. But now they are breaking up into swarms of *rowdy individuals*, which is to say isolated, digital *hikikomori* who no longer establish a public space or take part in any public discourse (…): the political *We*, which might be capable of action, in an emphatic sense, is falling apart. What kind of politics—what kind of democracy—is still conceivable today, given that the public sphere is vanishing, given the mounting egotistic and narcissistic transformation of man (*Ibid.*, p. 85)?

The notion that digital phenomena do not nourish any kind of political energy is questionable: this view completely blots out the potential of the Internet and of social networks as a means to develop forms of protest and dissent in various spheres of political activism (e.g. environmentalism, the "Me Too" movement, anti-Trumpism). It fails to distinguish between the destructive use of the Internet (for instance, through the creation of so-called "shitstorms," which are one of Han's focuses of interest) and its constructive use. As Feenberg notes, "the Internet has broken the near monopoly of the business- and government-dominated official press and television networks by enabling activists to speak directly to millions of online correspondents" (Feenberg, 2017, p. 108).[11] By contrast, digital communities are an utterly degenerative phenomenon according to Han. Through this perspective one is bound to conclude that the Internet cannot in any way contribute to the democratisation of society and that it is only a cause of human exploitation and degeneration. Feenberg (2017) recognises instead that the Internet is an ambiguous system whose structure results from the "struggles between certain business groups and the public over the design of the system" (p. 87). This approach is based on an acknowledgement of the fact that by now society has a technical character, just as technology has a social character: this reciprocity entails the complexity of the techno-social system and requires an equally complex analysis.

In particular, "a more comprehensive understanding of the Internet" is possible within the framework of a theory that examines "a complex matrix of multiple 'intentionalities' and functions." Among these, we find two fundamental forces:

---

[11] Similarly, Brian Haman (2018) has noted in his recent review of *Psychopolitik* that, "Han overlooks the link between social media, mobile phones, and activism. After all, the myriad ways in which the digitally dependent masses across the world have harnessed these new technologies precisely in order to subvert existing power structures suggest the extent to which digital spaces have become contact zones."

"the business interests such as the major service providers that are attempting to transform the Internet into an entertainment medium [consumption model] and the public actors who employ the Internet to participate in the life of society [community model]" (Feenberg, 2017, p. 100). According to Han, this kind of participation is ruled out by the very structure of the Internet: "digital communication (…) extensively erodes society, the We. It destroys public space and exacerbates the isolation of man: digital communication is dominated by narcissism, not by 'love of one's neighbour'" (Han, 2013, p. 65). Han's condemnation is categorical: digital technology destroys all spirit, action, thought, and truth. In their place, we get operations, calculations, and information. Citizens are no longer such; they are merely consumers: "Citizens are defined by their responsibility for the community, which is something consumers lack. In the digital *agora*—in which polling place and market, *polis* and economy coincide—voters behave like consumers. We can expect the Internet to replace polling places for good" (Han, 2013, p. 90). Ultimately, then, the spread of the Internet amounts to the destruction of democracy by means of turning citizens to consumers and politics into a marketing/shopping enterprise.

Feenberg's analysis goes in the opposite direction. While concluding that "there is no consensus about whether or not the technology can actually support community," i.e. "whether the Internet contributes to community or undermines it," (Feenberg & Barney, 2004, p. 2) over the course of the 2000s the public debate on the environment, health, education, and labour has not only increasingly taken place on the Internet, but has also led to the use of digital and online resources to further these sectors. Therefore, "to the extent that the demands of lay actors gain influence in these domains, the scope of democratic public life expands to include technology. We call this process 'democratic rationalization'" (Feenberg & Barney, 2004, p. 3). Contrary to what Han suggests, the Internet and digital communication not only do not "destroy" democratic spaces, but can even contribute to expanding them: "where groups seek community, they find the means to create it and use the technology to their purposes despite the various obstacles identified by critics of networking" (Feenberg & Barney, 2004, p. 4). The criticism directed against the critical theorist of digital culture Christian Fuchs in *Technosystem* also applies to Han's stance: "Fuchs dismisses the democratic implications of the Internet because of its economic function, but the human significance of online interaction persists despite its place in the capitalist economy" (Feenberg, 2017, p. 94).

Feenberg's markedly more optimistic view is based on the belief that, as an ambivalent system shaped by the struggle between the competing interests of several social actors, the Internet is characterised by a multi-level structure, where each level reproduces the social tensions it encompasses.[12] To take an example also dear

---

[12] Feenberg has identified the following levels: (1) Nonhierarchical structure; (2) Anonymity; (3) Broadcasting; (4) Data storage; (5) Many-to-many communication. All these levels reflect the systemic ambivalence of the Internet, which is to say the struggle between the consumption model and the community model: the two models coexist and the predominance of one over the other in certain contexts does not amount to a permanent achievement, but only to a particular stage in a process of development whose outcomes are still largely unpredictable.

to Han, the collection of data on the Internet serves different purposes. "The consumption model privileges commercial purposes. Data is collected by the owners of social networking sites, analysed, and sold to improve the performance of advertising…stored data is also available to governments for surveillance" (Feenberg, 2017, p. 104). At the same time, "in the community model certain kinds of data access are restricted to protect privacy. This has influenced the use of data by business." The growing attention towards privacy issues—including recent scandals—are forcing corporations to limit their data mining intrusions. At the moment, both models coexist on the Internet and it is still impossible to tell which will prevail. According to Han (2014), the consumption model is already dominating the management of online data: "human beings today are being processed and treated as data packages to be exploited on the economic level. They thus turn into goods. Big Brother and *big deal* gang up: the surveillance state and the market coincide" (p. 89).

The kind of reciprocity which, in the digital sphere, is an indicator of neoliberal exploitation according to Han ("Media such as blogs, Twitter and Facebook demediatise communication. Today's society of opinion and information is based on this demediatised mode of communication: each person produces and spreads information" (Han, 2013, p. 27)), constitutes the defining and positive trait of the community model according to Feenberg. "The essence of the community model is reciprocity. Each participant is both reader or viewer and publisher. To maintain this structure, the community model requires the continued neutrality of the network so that nonprofessionals, unprofitable, and politically controversial communication will not be marginalized" (Feenberg, 2017, p. 106). For Feenberg, then, "the ultimate challenge for democracy on the Internet" is "to preserve the conditions of online community" (2017, p. 110). This possibility of preserving and expanding democratic spaces even in the ambivalent domain of digital media is ensured by the fact that the struggle between the consumption model and the community model is not over and neither model has triumphed yet. Yet, for Han, the neoliberal consumption model has already claimed victory over any possibility of democratic intervention and so we cannot put any hope in digital media, including the Internet.

10.5   Conclusions

In this contribution I have traced a comparison between Andrew Feenberg's and Byung-Chul Han's philosophy of technology in order to highlight the differences that emerge from the essentially opposite theoretical models adopted by the two authors. An analysis of the categories of modernity, efficiency, and power, and of their application to the sphere of digital communication has revealed very different theoretical frameworks, which can now be summed up as follows:

- Feenberg's philosophy is a philosophy of technology in a strict sense: his analysis of technology pays close attention to both the socio-political and the technical aspects of technology, based on an acknowledgement of the fact that these two dimensions entail one another through the concrete ways that various artefacts are designed. Strictly speaking, Han's philosophy is not a philosophy of technology at all because it does not distinguish and investigate the *technical* aspect of technology, but remains on a more general descriptive level. In this sense, at its current stage of development, Han's thought provides a *propaedeutics* to a possible philosophy of technology, to be more widely developed on the basis of Han's assumptions. However, it is likely that a material analysis of the concrete and specific technical aspects of the various technologies mentioned by Han will lead to an at least partial revision of his propaedeutics and of some of its more radical assumptions.

- Feenberg embraces an *anti-essentialist* approach in his assessment of technology, which in his view is neither good nor bad in itself; rather, by incorporating in its design the outcome of social struggles between different interests and actors, technology meets specific requirements while remaining open to possible correctives, precisely insofar as it is based on a structural conflict that cannot be resolved once and for all. On the contrary, according to Han, hyper-modern technology is *essentially* harmful: in all spheres of human life, neoliberal psychopolitics employs a technical apparatus for the purposes of exploitation and domination. Social conflicts have been resolved to the detriment of users and to the sole benefit of producers insofar as the coercion that ensures exploitation has been entirely internalised. However, I have shown why this radical assumption is in many respects a questionable one.

- According to Feenberg, the space for democratic intervention is not simply threatened by technological and especially digital development, but can be directed in such a way as to expand and reinforce democratic participation. Han instead appears to rule out this possibility. In his view, political representation is not merely in crisis (precisely on account of the spread of Big Data), but may be regarded as an experience of the past which is no longer an option. What Han fails to provide is an indication as to "what is to be done," given these assumptions. Indeed, an answer to this question falls beyond the scope of philosophy as he understands it.

Han's thought has acquired considerable and increasing popularity in Europe and Latin America. He possesses an unquestionable ability to identify certain highly problematic mechanisms in the current neoliberal regime, which he has succeeded in presenting to a public that transcends the narrow confines of academia. However, his critique of technology is based on an overly humanistic and rhetorical approach focused on a substantive notion of technology that overlooks the concrete, material functioning of technolog*ies* as well as the issues (power relations) specifically embedded in their design. Feenberg's reflections, on the other hand, focuses on the technical and material aspects of technologies in the first place and derive from them a theoretical critique. Therefore, it represents a more solid and detailed analysis of the phenomenon of technology from the point of view of its limits as much as of its potentials, as well as a critical corrective to Han's radicalism as far as the outlining of realistic political interventions are concerned.

References

Ballesteros, A. (2020). Digitocracy: Ruling and being ruled. *Philosophies, 5*, 9. Online at: https://www.mdpi.com/2409-9287/5/2/9

Boeing, N., & Lebert, A. (2015). Byung-Chul Han: "I'm sorry, but those are facts'". *Zeit Wissen, 5*. Online at: https://skorpionuk.wordpress.com/2015/11/03/byung-chul-han-im-sorry-but-those-are-facts/

Buongiorno, F. (2015). Communication in the digital age. Byung-Chul Han's theory of power and information exchange. *Azimuth. Philosophical Coordinates in Modern and Contemporary Age, III*(5), 119–138.

Feenberg, A. (1999). *Questioning technology*. Routledge.

Feenberg, A. (2017). *Technosystem. The social life of reason*. Harvard University Press.

Feenberg, A., & Alastar, H. (1995). *The politics of knowledge*. Indiana University Press.

Feenberg, A., & Barney, D. (2004). *Community in the digital age: Philosophy and practice*. Rowman & Littlefield.

Foucault, M. (1988). The ethic of care for the self as a practice of freedom: An interview with Michel Foucault. In J. Bernauer, & D. Rasmussen (Eds.), *The final Foucault* (J. D. Gauthier, Trans.). MIT Press.

Haman, B. (2018, May 16). B. Haman reviews Byung-Chul Han's *Psychopolitics* and explores a regime of nomination that has discovered the force of the psyche. In *Hong Kong review of books*. Online at: https://hkrbooks.com/2018/05/16/psychopolitics/

Han, B.-C. (2010). *Müdigkeitsgesellschaft*. Matthes & Seitz. English edition: Han, B.-C. (2015a). *The Burnout Society* (E. Butler, Trans.). Stanford University Press.

Han, B.-C. (2012). *Transparenzgesellschaft*. Matthes & Seitz. English edition: Han, B.-C. (2015b). *The Transparency Society* (E. Butler, Trans.). Stanford University Press.

Han, B.-C. (2013). *Im Schwarm. Ansichten des Digitalen*. Matthes & Seitz. English edition: Han, B.-C. (2017a). *In the swarm. Digital prospects*. MIT Press.

Han, B.-C. (2014). *Psychopolitik. Neoliberalismus und die neuen Machttechniken*. Matthes & Seitz. English edition: Han, B.-C. (2017b). *Psychopolitics. Neoliberalism and new technologies of power* (E. Butler, Trans.). Verso.

Heidegger, M. (1966). *Discourse on thinking* (J. M. Anderson, & E. H. Freund, Trans.). Harper & Row.

Lushetich, N. (2018, October 18). Natasha Luschetich on *Psychopolitics: Neoliberalism and new technologies of power*, by Byung-Chul Han. *Media Theory*. Online at: http://mediatheoryjournal.org/review-natasha-lushetich-on-psychopolitics-neoliberalism-and-new-technologies-of-power-by-byung-chul-han/

Marcuse, H. (1969). *An essay on liberation*. Beacon Press.

Schmitt, C. (2003). *The Nomos of the Earth in the International Law of the Jus Publicum Europaeum*. New York: Telos Press.

Tollmann, V. (2011). The terror of positivity. An interview with the philosopher and media theorist Byung-Chul Han. *Springerin*, (4) Online at: https://www.springerin.at/en/2011/4/der-terror-der-positivitat/

Chapter 11
The Varieties of Praxis: Marx, Lukács, Feenberg, and Czechoslovak Marxism

Ivan Landa and Jiří Růžička

Abstract The chapter addresses the question as to whether East-Central European Marxism was a distinctive intellectual phenomenon by focusing on the case of Czechoslovak Marxism. It gives an affirmative answer, as it claims that the philosophy of praxis makes up its conceptual core. In a first step, Czechoslovak Marxism is situated in a broader plane of East-Central European Marxism and contrasted with Western Marxism. Then, two important intellectual sources of Czechoslovak Marxism are discussed: Marx's early project of practical Materialism and Lukács's theory of revolutionary spontaneity. In a third step the claim is substantiated that until the appearance of Marxist humanism in the late 1950s, one cannot *sensu stricto* speak of any comprehensive philosophy of praxis within Marxism. Thereafter, it is argued that it was specifically Czechoslovak Marxist humanism that came up with a comprehensive philosophy of praxis, including a theory of subjectivity. After that, Marxist philosophy of praxis is confronted with existential phenomenology. In a final step, a critical eye is cast on Andrew Feenberg's efforts to accomplish a synthesis of the Marxist philosophy of praxis and Heidegger's philosophy.

Keywords Czechoslovak Marxism · Marxism · Georg Lukács · Western Marxism · Praxis · Onto-creativity · Karel Kosík · Soviet Marxism · Martin Heidegger · Andrew Feenberg · *Theses on Feuerbach*

I. Landa (✉) · J. Růžička (✉)
Institute of Philosophy, Czech Academy of Sciences, Department for the Study of Modern Czech Philosophy, Prague, Czech Republic
e-mail: landa@flu.cas.cz; ruzicka@flu.cas.cz

© The Author(s), under exclusive license to Springer Nature Switzerland AG 2022
D. Cressman (ed.), *The Necessity of Critique*, Philosophy of Engineering and Technology 41, https://doi.org/10.1007/978-3-031-07877-4_11

11.1   Introduction

It is commonplace to speak of "Western Marxism," construed broadly, to cover all sorts of thinkers and currents such as critical theory, phenomenological Marxism, psychoanalytical Marxism, existential Marxism, and even post-Marxism. At the same time there are no quarrels with attaching the privileged attribute "theory" to conceptions of Western Marxism. With East-Central European, and perhaps even much more with Soviet Marxism, the situation is different. Both are still approached with a certain reservation with regard to content and overall theoretical potential. Both are approached either as examples of the ideological armature of Soviet-type societies or as eclectic intellectual endeavours. While the first perspective contains obvious residues of the Sovietologist idiom, the second implies an Orientalist narrative that goes as follows: even if some of the conceptions of East-Central European or Soviet Marxism are indeed interesting, this is because of the apparent fact that those interesting elements are indebted to influences stemming directly from Western Marxism, or from phenomenology, existentialism, psychoanalysis, and logical positivism.

Most likely none of above-mentioned perspectives does justice to the intellectual phenomenon itself, and thus does not provide us with a high-resolution roadmap of the intellectual territory of East-Central European and Soviet Marxism. This deficiency raises the question we shall address in the following text and attempt, at least partially, to answer: Is it possible to construe an East-Central European Marxism as a distinctive intellectual phenomenon that is not reducible to mere borrowings from Frankfurt School-style critical theory, existentialism, or phenomenology? We offer an affirmative answer, and take this as our initial default hypothesis that requires more extensive substantiation than we intend to provide here, since we focus only on the single case of Czechoslovak Marxism.

In Czechoslovak Marxism, various philosophies of praxis flourished in late 1950s and 1960s. At that time the concept of praxis was frequently used and became one of the fundamental concepts within East-Central European Marxism. Marxist humanists, especially, placed a great deal of emphasis on the concept of praxis. We shall therefore outline the core idea and the systematic motivations that stand behind the varieties of philosophies of praxis. First, we situate Czechoslovak Marxism on a broader plane of East-Central European Marxism, and characterise it by contrasting it with Western Marxism. Second, we discuss two important intellectual sources of Czechoslovak Marxism: the project of practical Materialism outlined in Marx's *Theses on Feuerbach* (1844) and the theory of (revolutionary) spontaneity voiced by Lukács in *History and Class Consciousness* (1923). Then we turn to Czechoslovak Marxism, showing that until the appearance of Marxist humanism in the late 1950s one can hardly speak of a fully formed philosophy of praxis in a strict sense within Marxism. This might sound like a somewhat strange and unsubstantiated exaggeration. The founders of Western Marxism, Georg Lukács, Karl Korsch, and Antonio Gramsci, emphatically stressed the importance of praxis in their polemics with theoreticians of the Second International and in their theoretical attempts. Besides that, the notion of "praxis" has not been wholly neglected by various Marxist currents, Stalinist orthodoxy included. In spite of this, we shall argue that none of those attempts ultimately succeeded in elaborating a comprehensive philosophy of praxis, since, taking the case of Lukács and Korsch into account, praxis was narrowed down to the revolutionary praxis manifested by masses or classes, or – in the alternative case of Stalinist orthodoxy – praxis was reduced to instrumental practice, exemplified by industrial labour performed by masses or machines, hence Stalinist orthodoxy did not require any elaborated conception of the subject. We argue that it was Czechoslovak Marxist humanism that came up with a comprehensive philosophy of praxis, including a theory of subjectivity. In the following steps, we confront this Marxist philosophy of praxis with existential phenomenology, and conclude by casting a critical eye on Andrew Feenberg's efforts to accomplish a synthesis of the Marxist philosophy of praxis and Heidegger's philosophy.

11.2 Marxism Between West and East

In his book *Considerations on Western Marxism* (1976), British historian Perry Anderson famously argued that it is possible to discern at least three defining features of Western Marxism: methodologism; an exclusive interest in superstructure phenomena, specifically art and artistic practice; and a deep pessimism regarding the prospects of revolutionary politics (Anderson, 1976, pp.52–55).[1]

Concerning methodologism, in the early 1920s Marx was recognised as an important methodologist and social epistemologist. In *History and Class Consciousness*, Lukács, following Marx's lead, proposed a methodological understanding of Marxism, arguing that Marxism is not a doctrine or set of theorems, but rather a method. Accordingly, it is possible to imagine that all the theorems formulated not only by Marx, but also by Engels, Lenin, Stalin, or Trotsky can be falsified and dismissed with reference to discoveries in the social sciences. Nevertheless, it will still be possible to proclaim oneself to be a Marxist on the condition that one employs the dialectical method coined by Marx (Lukács, 1972, pp.1–2). In Western Marxism, many attempts, inspired by Lukács, have been made at extracting or distilling the basic elements of the dialectical method from the body of Marx's writings on political economy. Dialectics has been approached as a method, or better as a methodology of both philosophical thought and scientific discovery.

These methodological principles were not derived for their own sake. They were employed within analyses of cultural phenomena, and more specifically art and artistic practice. As an effect, methodologism and the emphasis on art led to losing touch with reality, with politics and economic and social struggles, so that from the late 1940s onwards, revolutionary politics was no longer the main focus of Western

---

[1] What Anderson has especially in mind are the forms that Western Marxism took in the 1950s and 1960s.

Marxism, or better to say, the idea of social revolution was rearticulated in terms of the cultural revolution.

Henceforth, genuine political action was voiced largely in cultural terms, as intellectual interventions that operate on the level of consciousness and its forms. In this respect, politics and art were intimately linked together like communicating vessels, since writing a novel, composing a symphony, or construing a philosophical theory were now understood as genuine political actions, capable of revolutionising forms of consciousness, our concepts or worldviews, and thereby changing reality. Hence the difference between political and mere artistic practice and intervention collapses upon a merging of politics and culture together. This strategic move creates the mere illusion that the gulf between subject and object, theory and practice, consciousness and reality has been *aufgehoben*, so that it is seemingly enough to compose a symphony, write a novel, or construe a philosophical theory in order to change the world, slightly shifting the meaning of the eleventh *Thesis on Feuerbach*: by interpreting or representing the world it is possible to change it. This is the revolutionising of the world through the revolutionising of worldviews.[2] Social revolution has been conceived as an epiphenomenon of cultural revolution, and not the other way around. This ultimately led to pessimism with regard to the idea of political or social revolution.[3]

Analogically, Eastern-Central European Marxism can be also characterised by methodologism, inspired by early Lukács and further by Lenin's Philosophical Notebooks. The main focus was on dialectical method, understood not solely as an abstract mode of thinking, but more robustly as real or concrete dialectics, which operates on the level of individuals and collective subjects who always possess some degree of self-consciousness and who are capable of some type of self-understanding. The articulation and conceptualisation of such a real or concrete dialectics was at the top of the agenda in East-Central European Marxism in the 1950s and 1960s, so that a philosophy of praxis also occupied a prominent place in realising such a programme, as it conceptualised concrete dialectics of human practices. These Marxist philosophers came up with various conceptions of praxis, focusing on labour and the productive forces, political and artistic practice, the phenomenon of play and creativity, and more generally on human spontaneity, productive imagination, or onto-creativity. Basically, the different conceptions of praxis were modelled either according to the model of objectification, or according to the model of intentional, instrumental activity, through which human beings – with the help of tools and materials – realise their aims and goals in space and time, as well as in social "space."

Generally speaking, this philosophy of praxis reacted critically to the traditional and Stalinist versions of historical materialism, which placed too strong an emphasis on technological progress, economic structure, and social laws or revolutions,

---

[2] It is perhaps no coincidence that Marx's *Theses on Feuerbach*, as Anderson notes, did not play an important role in Western Marxism in the 1950s and 1960s as it did in East-Central European and Soviet Marxism.

[3] See also Boltanski (2002, pp.1–22).

forgetting about concrete human beings living their life here and now in the medium of everydayness. In this respect, the philosophy of praxis was part of a much broader anthropological turn that was taking place in East-Central European and Soviet Marxism from the mid 1950s, which stimulated a growing interest in existential questions, such as the meaning of human life with respect to its finitude.

This anthropological turn brought about an interest in superstructure phenomena, such as culture, law, religion – but above all politics. The primacy of politics in East-Central European Marxism is no surprise, given the politicisation of all spheres of society in Eastern Europe during the 1950s and 1960s and the spreading of the discourse of revolution. In fact, East-Central European Marxism can be distinguished by a revolutionary optimism, in which revolution was not limited to the building up of democratic institutions and securing political freedoms and rights. Besides the hopes invested in a scientific-technological revolution, much of the emphasis was placed on both individual and collective subjectivities, on their self-management and especially on their capability to carry out transformations within the forces of production, social structure, and forms of consciousness. At the same time, Marxist philosophers understood themselves as politically engaged intellectuals, whose aim was to contribute – theoretically and practically – to the success of socialist revolution.[4]

The philosophy of praxis can be considered one of the philosophical contributions of East-Central European Marxism. "Praxis" was topical for the members of the Yugoslav group *Praxis*, for the Polish Marxist philosopher Jan Szewczyk, for the members of the *Budapest School*, such as Ágnes Heller and György Márkus, for the Soviet-Marxists Genrich Batischew and Evald Ilyenkov, and for Karel Kosík, Robert Kalivoda, Vítězslav Gardavský and Radovan Richta in Czechoslovakia.[5] In most of these cases, praxis was conceptualised as onto-creativity, an activity that manifests itself in the production of artefacts and institutions, material objects and normative statuses. The revolutionary praxis that was the sole interest of Lukács or Korsch was now subsumed under a broadly construed conception of praxis, which dealt with the question of how social reality is produced or constituted, and not only of how it can be revolutionised.

---

[4] The majority of them belonged to the ranks of party intellectuals, who were actively engaged in politics, e.g. during the Prague Spring of 1968.

[5] Some of the philosophers who can be classified as belonging to Western Marxism proposed a philosophy of praxis of their own, such as Guido D. Neri in Italy, Adolfo Sánchez Vásquez in Mexico, Michel Henry in France, or the early Herbert Marcuse in Germany, some of them being inspired by developments in Eastern European Marxism. The later Lukács, who witnessed the flourishing of the philosophy of praxis in the 1960s, which he occasionally described as a "renaissance" of Marxism, came up with the ontology of labour of his own in *Ontology of Social Being*, understanding labour as a paradigm of teleological positing. See Lukács (1973, pp. 162–174).

11.3   Making Sense of Spontaneity

As we suggested earlier, the philosophy of praxis of East-Central European Marxism in some sense shares its origins with Western Marxism, as it drew inspiration from the philosophical debates that took place in the Marxism of the 1920s, especially those found in Lukács's *History and Class Consciousness*. However, in contrast with Western Marxism, Marx's *Theses on Feuerbach* played an important role in the formation of both East-Central European and Soviet Marxism. In the *Theses*, Marx outlined the philosophical project of practical Materialism and laid the foundations for a materialist ontology and epistemology, a semantic conception of truth, a philosophical anthropology, a theory of revolution, and finally for the conception of subjectivity (Marx, 1978, pp.143–145). We shall limit ourselves to the core of such a project with respect to the later adoption and elaboration of practical Materialism in the form of the philosophy of praxis in Czechoslovak Marxism.[6]

The basic argument of the *Theses on Feuerbach* aims at completing and reconciling two rival philosophical traditions: the idealist tradition that culminated in Kant's transcendentalism and Hegel's absolute idealism on the one hand, and on the other, the materialist tradition that culminated in Feuerbach's materialism of sensuousness. The polemic with traditional materialism makes a case for an alternative materialistic approach, which emphasises human spontaneity. The hitherto existing materialism is described by Marx as merely "sensuous" or "contemplative," since it takes for granted that reality is objective, existing in space and time independently of our consciousness and activity (Marx, 1978, p.143). Hence our primary epistemic access to reality is through the physiological senses and our primary practical contact with reality is via coping with ready-made objects existing in space and time.

Hence for Marx, the obvious failure of traditional materialism consists in promoting only a one-sided ontology, where "to exist" means to be a material object in space and time. If we apply such ontology to social reality, we end up with the statement that there are certain social institutions and normative statuses given out there in social space, which we can compile into a list of what exists, including among others wage labourers, courts, schools, factories, money, etc. By the same token, we would also be able to state that, for example, this individual or that material object is an instance of a particular institution or normative status. And last but not least, we would be also able to cope with money, go to the school or work etc.

This brings us directly to the second objection raised by Marx against traditional materialism, its one-sided epistemology. Accordingly, traditional materialism prioritises physiological senses. If we apply such epistemology to our knowledge of social reality, we end up with the claim that we receive most of our information about reality in a receptive way through social "senses", through which social institutions and normative statuses appear as if they are detached from us and from an "active side" of human beings, from spontaneity of subject.

---

[6] For the elaboration of the project of practical Materialism in context of Karel Kosík's philosophy of praxis see Landa (2022).

Thus for Marx, traditional materialism lacks sufficient conceptual tools to theorise praxis adequately. As a consequence, it is also not capable of conceptualising reality as a product of the human subject, or as a manifestation of her or his spontaneity.[7] Although practical Materialism concedes that human beings are both sensuous and material objects among others in space and time and endowed with both physiological and social senses, it further adds that human beings are ontologically special objects with a special way of acting and knowing reality that is full of her or his products: artefacts, institutions and normative statuses. Hence, practical Materialism is designed by Marx to propose a theory that would explain the sources of social properties of various objects and subjects by reference to human spontaneity, in this way destroying the myth of the subject's passivity that is allegedly uncritically accepted by traditional Materialism. In other words, practical Materialism provides a theory of human praxis.

Here again, Marx points to traditional Materialism and criticizes its understanding of praxis as mere manipulation of objects that are ready at hand. He objects that this is praxis "in its dirty, wheeler-dealing manifestation," or merely instrumental action, driven by aims and goals set up beforehand by a human subject, and in some cases even motivated by the *quid pro quo* principle (Marx, 1978, p.143). As a matter of fact, instrumental action really pertains to praxis. However, at the same time, human subjects cope with objects that are mostly artefacts and as such are products of praxis and thus manifestations of onto-creativity. For Marx, instrumental action can easily be performed by machines, while praxis cannot, although machines themselves are products of onto-creativity. Human spontaneity thus constitutes both a technological difference, distinguishing human activity from the running of a machine, and an anthropological difference, distinguishing intentional and instinctive behaviour. Accordingly, neither animals nor machines are engaged in onto-creative activity, as they "produce" either instinctively, or monotonously.

In the *Theses on Feuerbach*, Marx articulates his understanding of spontaneity especially in terms of "practical-critical activity." Thus he assumes that spontaneity manifests itself eminently in revolutionary praxis, which transforms pre-given and allegedly fixed social institutions and normative statuses. Of course, he also drops some hints in the direction of the ontology of labour understood as *objectification*, but this was elaborated by him in more detail in the *Economic and Philosophical Manuscripts of 1844*, we shall leave it aside here.

In the 1920s Georg Lukács, Karl Korsch, and also Antonio Gramsci pushed the project of practical Materialism further, following Marx's lead, and proposed their theories of spontaneity. Nevertheless, they theorised praxis either too narrowly or too broadly. Lukács and Korsch concentrated almost exclusively on revolutionary praxis and wholly ignored the ontology of labour, whereas Gramsci used the term

---

[7] See the first *Thesis* (Marx 1978, p.143). In this sense, Sebastian Rödl has recently argued that the *Theses on Feuerbach* can be interpreted as offering a materialist alternative to contemporary versions of naturalism and eliminative Materialism, since they highlight precisely spontaneous knowledge: We know ourselves as material objects differently – namely spontaneously, subjectively – than we know other material objects in space and time (Rödl, 2007, pp.121–123).

"philosophy of praxis" in his *Prison Notebooks* to refer to historical materialism or even Marxism *tout court*. The narrowing of the theory of spontaneity to revolutionary praxis, failing to consider labour as objectification, was not only due to the reception of the *Theses on Feuerbach* and to the lack of any knowledge of the *Economic and Philosophical Manuscripts of 1844*, but also due to the methodological understanding of Marxism explicitly voiced by Lukács. If Marxism consists in method and not in any propositional content, and if one understands method as dialectics, operating directly within social reality because self-conscious human subjects are an integral part of it, as Lukács claimed, then it makes good sense to develop a theory that accounts for dialectics in terms of revolutionary spontaneity. For Lukács, spontaneity consists precisely in "practical-critical activity," or in revolution. So he is also at pains to provide an answer to the following question concerning how it is possible to radically transform social institutions and existing normative statuses, instead of asking how it is possible for social institutions and normative statuses to come into existence, to be produced and reproduced by human praxis. For this reason, Lukács focuses mainly on social struggles, on revolts, upheavals, and on social, political and cultural revolutions.

At the same time, he assumed that the philosophy of (revolutionary) praxis informs revolutionary praxis as such, since theories do not hover within a social or political void, but play causative roles, contributing to the transformation of existing institutions and normative statuses. Theories become involved in dialectical interactions between subjects and objects, informing the subject's self-understanding. For Lukács, subjects are spontaneous knowers and doers, who always develop some minimal understanding of social reality and of themselves, which then informs both their thought and action. Since such understandings are influenced by ideologies and infused with prejudices and biases, the theory can play both roles: correcting self-understanding and playing the role of a motivational factor, thus making social reality transparent, contributing to the subjects finally becoming aware of their real place within social structure. Such self-consciousness and self-knowledge is a precondition for revolutionary praxis.

11.4 Czechoslovak Marxism: The Quest for a Philosophy of Praxis

Lukács's *History and Class Consciousness* exerted a strong influence over Czechoslovak Marxism throughout the 1950s and 1960s, as it attempted to provide Marxism with new conceptual foundations and to propose a conception of Marxism not as a doctrine but as a method, which was in both cases combined with a far-reaching diagnosis of the present moment. Such a combination strongly influenced many Marxist thinkers in Czechoslovakia in the era of de-Stalinisation, when socialism as both an idea and reality had been shaken to its foundations. Nevertheless, the discourse of de-Stalinisation remained revolutionary, as Stalinism was sometimes understood as an interruption or betrayal of the revolutionary vision, as a period that should be overcome. This vision as such was in any case retained, albeit in different form.

Czechoslovak Marxist thinkers assumed that Lukács's philosophy of (revolutionary) praxis did not explain the most basic forms of praxis, especially the (re) production of social reality as such. Lukács neglected such a variety of praxis in his philosophical theory, although he was well aware of it, since his understanding of revolutionary praxis did not deliberately focus only on its destructive side, but also thematised its constructive component. Social revolutions usually bring down the old institutions, normative statuses and social relations, yet they also give rise to new social institutions and normative statuses. Lukács's aim was to offer a comprehensive account of the rationale that stood behind revolutionary changes, and to execute an anatomy of revolutionary praxis, intentionally leaving aside the productive or constitutive moment of praxis, such as its "existential", everyday dimension.[8] Here, previously vilified Western non-Marxist philosophy such as existentialism and phenomenology could provide some sort of inspiration, because these had elaborated conceptions of human activity and subjectivity, albeit in idealistic terms, which could serve as a form of contrasting platform for Marxists' own undertakings in this theoretical area.

Czechoslovak Marxism concentrated its theoretical efforts precisely on these aforementioned neglected aspects, developing a more thorough philosophy of praxis as onto-creativity, which would include a conception of its existential dimension, confronted with both existentialist and phenomenological theories of praxis. In what follows, we shall outline with broad strokes the contours of the attempts to elaborate a philosophy of praxis. We will also shed some light on its origins in the mid 1950s, because the Czechoslovak philosophy of praxis did not appear as a mere offshoot within the development of Marxist theory, it was a reaction against Stalinist orthodoxy, and existentialist and phenomenological philosophies.[9]

The philosophy of praxis was a product of the socio-political situation in the former Czechoslovakia and in the Eastern bloc. During the second half of the 1950s, Czechoslovak Marxism entered the stage when an economic, political, social and cultural crisis hit nearly all of the countries in the Eastern bloc. At that time, it was becoming clearer that the Stalinist project of socialist modernity had run out of fuel and needed to be replaced by a different paradigm. This pertained also to the philosophical sphere. The Stalinism (or Marxism-Leninism) that until that time had dominated the intellectual and ideological scene of the Eastern bloc was placed under severe scrutiny. Although in their beginnings, the critical activities and objections raised against the Stalinist orthodoxy were significantly circumscribed by the party establishment, a new form of Marxist thought began to emerge and assert itself. The

---

[8] Lukács focuses on the "existential" moment of praxis, emphasising "everydayness", in his writings from the 1960s, especially Lukács (1963).

[9] Due to lack of space, Polish, Hungarian, and Yugoslav attempts to formulate a philosophy of praxis must be put aside in this chapter, although they addressed the same issues and came up with similar conceptions.

ubiquitous critique of dogmatism, bureaucratism, and the cult of personality, promulgated even by the party itself, was just the first step. The subsequent reconceptualizations of key Marxist concepts became crucial for the post-Stalinist Marxist movement.

Among the most notable Marxist currents which made up the post-Stalinist "epistemic field" in Czechoslovakia, were the following: (1) Marxist humanism, represented by thinkers such as Karel Kosík, Robert Kalivoda, Ivan Sviták, and Vítězslav Gardavský, who mainly focused on an elaboration of philosophical anthropology or ontology of human being; (2) dialectical determinism, whose main representatives Josef Cibulka, Václav Černík, and Jindřich Zelený were equally at pains to reinterpret the concept of dialectical law; and (3) techno-optimism, promoted by Radovan Richta, Radoslav Selucký, Jiří Zeman, and Juraj Bober, based on a reinterpretation of the conception of the forces of production, which besides technology, includes science and knowledge as key forces of production.[10] Although Marxist humanists were not the only ones endeavouring to develop the concept of praxis, they were probably the only ones who raised the question of praxis with the utmost urgency, and consequently were able to build an entire philosophy around it. Therefore, we shall limit our exposition of the post-Stalinist critique of the Stalinist orthodoxy to this intellectual current.

To be sure, the Stalinist orthodoxy did not wholly lack a conception of praxis. However, the problem explicitly voiced by Marxist humanists was that the way in which praxis was characterised and the function ascribed to it evinced systematic flaws and distortions of the most central Marxian insights. First, praxis was reduced to productivism in a sense of quantitative growth of the means of production.[11] Such an understanding promoted by the Stalinist orthodoxy was probably most aptly denounced in the words of Ivan Sviták:

> Finally, the most important aspect, philosophically and ideologically, is that only the tragic distortion of Marxist views has given rise to the idea that the meaning of human life lies in labour, and that Marxist humanism adopts the Protestant view of work as the path to salvation. This stance triumphed in the petty-bourgeois morality of the victorious industrial revolution, in economic utilitarianism and the predominance of object-oriented art (Sviták, 1963–1968, p.73).

The adoration of labour, performance, and adherence to the plan was an integral part of the socialism-building ideology in the era of Stalinism, although traces of this conception go back to Engels, who explicitly defined praxis as productive activity, or even as an industry and as an experiment, by which he refuted the Kantian notion of the "thing-in-itself" (Engels, 1969). Both Lukács and the Czechoslovak Marxist humanists broke with such a conception of praxis. However, whereas Lukács

---

[10] The case of Radovan Richta is complicated. He represents a typical case of how one philosopher can move easily between (seemingly contrary) intellectual currents. While his work on the STR (Scientific-technological revolution) promotes science and cybernetics, his other smaller and more philosophy oriented papers are more than compatible with the philosophy of praxis. We shall refer more to his later texts, in which he offers a philosophy of praxis of his own.

[11] Such productivism was critiqued by Arendt (1998, pp.126–135).

interpreted and criticised this conception for depriving praxis of its active aspect, reducing it to a mere contemplative stance towards reality, Marxist humanists turned against the technicist image of the human "machine," which only fulfils the given plan and thus sustains the narrative of salvation. At the same time, some Marxist humanists like Kosík, defined human being as praxis, whose paradigm was precisely labour. Nevertheless, labour was understood rather ontologically as a spontaneity objectifying itself within space and time. For these philosophers, pure spontaneity and its objectification constitute the defining features of human essence.

Another flaw typical of the Stalinist orthodoxy consists in conflating praxis with the application of a theory to reality or its verification by reality, which also goes back to Engels and was later promoted by Lenin in his *Materialism and Empirio-criticism* (1909), one of the most authoritative texts for Stalinism, stating that praxis is a central ontological and epistemological criterion.[12] Here again, in his critique of Engels, Lukács stressed the following point: the idea that praxis as industry proves or disproves scientific discoveries and theories simply by producing physical, chemical, or biological substances which serve our own purposes, overlooks the fact that industry and science in themselves unconsciously serve the purposes of capitalism. Marxist humanists placed strong emphasis on the fact that praxis as industry renders not only theory, but also people and nature, instruments of certain political goals: every theory and every human being or natural entity can be instrumentalized and manipulated.[13] Against this, Czechoslovak Marxist humanists aimed at reasserting – according to their perspective – the proper philosophical meaning of praxis as spontaneity and its objectification, namely onto-creativity.

In order to accomplish such a task, they brought back the philosophical authority of Hegel, whom the Stalinist orthodoxy had rejected as an obsolete and outright conservative thinker. The main lesson they learned from Hegel was the idea that the

---

[12] The relationship between theory and praxis in Marxism-Leninism is obviously more complicated than is presented above. First of all, it could be argued that Lenin developed the concept of praxis closely following the second Thesis on Feuerbach, and therefore following from practical Materialism. But this does not mean he endorsed the primacy of praxis over theory. It seems that quite the opposite is the case. Treating praxis as the sole criterion of something else is not a commitment to the primacy of it. Even in his later writings, Lenin never progressed beyond this perspective. The same applies to Stalinist theoreticians, who verbally committed themselves to the primacy of praxis, but then in turn endorsed the determinist character of historical laws, which can theoretically be mastered and consequently used as a tool in reshaping reality. However, this is a classical contemplative (theoretical) stance towards reality in general and praxis in particular. On the other hand, something different is represented by the disregard or very loose relationship Stalinism very often showed towards (some) scientific theories and knowledge. In this case, theory was sometimes forcibly adapted to praxis, or better stated to some practical goals. This was by no means pure pragmatism, because that perspective was already defended by the so-called "teleologists" against the "geneticists" in the late 1920s in their battle for the future form of economic planning. To sum up, we believe that Stalinist theoreticism on one hand and Stalinist voluntarism on the other are not mutually exclusive "qualities", but rather form a contradiction of Stalinism itself. For the debate between the teleologists and geneticists see Collier (2011, pp.55–61) and Spulber (1965).

[13] This has been portrayed by Koestler (1940).

essence of human reality lies in the human being's tendency to objectify him/herself in both social and natural reality, and furthermore the idea that the result of this activity is a concrete material object in the broadest sense of the term. By doing so, they wanted to point out the orthodoxy's neglect of the very process of objectification, or as they called it, the genetic aspect of praxis, simply put – creation. The Stalinist orthodoxy, on the other hand, placed excessive stress on the role of objectively existing impersonal laws of social development, where the genetic, creative aspect was suppressed. In their view, the result of this could be nothing other than a reified and therefore possibly alienated conception of society (as well as nature).

However, unlike Lukács, Marxist humanists did not perceive reification primarily or exclusively as a product of the division of the social totality into separated areas, but as a result of certain oblivion that omits the process of creation of an object (tools, artefacts, social relationships). For Lukács and the first generation of Western Marxists, the philosophy of praxis referred to the theory of revolutionary action, which aimed at the resolution of the existing social contradictions through radical social transformation, while traditional philosophy sought to resolve them only in thought.[14] That was not enough for the Czechoslovak Marxist humanists. They sought to formulate their own philosophy of praxis, not as a rival to Lukács's or the Frankfurt School's undertakings, but rather as a complement that would provide Marxism with its own materialistic version of the "existential analysis of praxis."

11.5   Marxist Humanism Meets Existential Phenomenology

From this point of view, the confrontation with existentialism seemed to be inevitable. The reason was obvious. Existential phenomenology, through two of its leading figures, Martin Heidegger and Jean-Paul Sartre, introduced a very similar set of philosophical problems that Marxist humanists also wanted to solve, namely – to speak in the broadest terms – the intricacies of the relationship that humans engage in with the world. The situation even took on a different twist when Sartre himself stated in *Critique of Dialectical Reason* that the existentialist philosophy of human being begins where the Marxist analysis of society ends, and that Marxism needs an existentialist ontology of man for the grounding of its knowledge (Sartre, 1963, pp.179–181). Karel Kosík quickly responded that it was precisely Marxism, not existentialism, that created the "final, i.e. historically insurmountable" ontology of man (Kosík, 1965, p.47). Yet this response did not merely represent a traditional and frequently unsubstantiated defence of Marxism against a critique aimed from a non-Marxist philosophical camp. It represented a forward step and also a call for a (re)

---

[14]Of course, this did not mean cancelling any action whatsoever. Rather, traditional philosophy took the resolution of contradictions as a theoretical problem. Action then occurred as an application of the result of the said resolution. According to Marxism, this is precisely the essence of reformism and a contemplative attitude towards action. See Feenberg (2014, pp.16–19).

construction of Marxist philosophy. Nevertheless, it should be noted that despite significant differences between the philosophical currents in question, for the Marxist humanists, existentialism served as an undeniable source of inspiration. We will now dig deeper into the central issue of the dispute between these two currents.

First, Marxist humanists considered the two types of activities that existential phenomenologists regarded as essential to *Dasein* or *realité humaine*, namely care (Heidegger) and projecting (Sartre), as a distorted or even reified picture of more substantial human activity – praxis. In the case of Heidegger, Marxist humanists interpreted the notion of care as manipulation, coping with or handling of ready-made objects. Accordingly, the concept of care could not suitably describe the origi-nal attitude that human beings "entertained" with reality, because, despite Heidegger's explicit efforts, it still harbours elements of a "contemplative" attitude. Thus, according to Kosík, in Heidegger's concept of care, the individual human being engages with the world in terms of "dealing" with ready-made things, or in terms of "procuring" them. This leads to two observations: (1) dealing/procuring indicates mere manipulation of objects and (2) objects themselves appear to the subject as meaningful only as use values. If there is any production at all, then the "product" of human activity is not a new object but a (new, if at all) meaning, which is, in the process of engagement, given to *already existing* objects. In short: such an engagement is not objectification in a strict sense (Kosík, 1976,p.40; Bodnár, 1965, p.563).

But why is it necessary to talk about objectification? Why do the Marxist human-ists prioritise praxis as production or creation, and not care led by implicit knowl-edge (know-how) or understanding? The reason is that the activity of care and of meaning-giving does not allow us to take a critical stance towards existing reality, since objects themselves appear to us from this standpoint as something already given. We do not recognise them as something that would be the product of our doing and making. From the point of view of care, the reality out there cannot be taken as a product of praxis qua creation that can in principle be undone or remade, but only as a background that we must accept and presuppose in our dealing with objects. Briefly stated, if the essential aspect of care is procuring, dealing with or using tools and materials, then our encounter with reality is simply a matter of *use*, and the result of this encounter is not a new object or a completely novel constella-tion of objects, but a pre-given meaning which corresponds with the functional properties of objects. To put it in more "objectivistic" terms, we recognise the place that objects occupy within reality and the web of their functions, but we do not know how they appeared in this place, why they have certain functional properties, and to what extent these properties reflect some general features of the social form as such.

The Heideggerian emphasis on terms such as "revealing" instead of creating or producing is quite telling within this context. Heidegger aims to avoid any possible reference to the subject's imposing of any meaning or objectifying of her or his will onto the objects out there, since meaning is not something we randomly confer upon the objects around us. Instead, meaning is the result of our "interaction" with social reality, a process Heidegger calls "revealing" (Feenberg, 2005: passim). However, for Marxist humanists, such "revealing" indicates a reified and solely contemplative stance towards reality and in the last instance a reactionary explication of the concept of praxis. Terms such as "care," "procuring," and "revealing" barely indicate anything other than activity aimed at ready-made objects – which is a cardinal sin if, according to Marxist humanists, we wish to truly understand praxis in its essence and variety. Even if we take into account the interesting, lucid, and perhaps even charitable interpretation of Heidegger presented by Andrew Feenberg, who places not revealing, but the activity of producing, at the centre of Heidegger's philosophy, we still do not progress beyond the albeit slightly expanded ancient Greek conception of *poiesis*, which Heidegger reformulates in terms of *"her-vor-bringen"*, as bringing forth something that is already there, although still in concealment and in potentiality. To put it crudely, Heidegger does not describe human being as onto-creative being – it is not praxis. But why should Heidegger, according to Marxist humanists, propose a philosophy of praxis instead of fundamental ontology or – in his case – thought of being?

The answer brings us to the Hegelian roots of Marxist humanism. According to Hegel, we cannot think and talk about human/social reality unless we specify who or what is a human being, as opposed to an animal. Thus, to speak of a second nature (sociality), we should expound the difference between the first and second. The whole sphere of the latter, for which Marxist humanists chose the term "society" or "civilisation", including artefacts, institutions, normative statuses, rules and laws, but also works of art, in the broadest sense of the term created by human beings. As such, those creations necessarily, sooner or later, take on a certain form of independence from their creator – they are both actually and potentially separable, capable of living a life of their own. It is only upon this condition that a subject can occupy a critical distance from its own situation and class position, as well as from its own creations, such as artefacts, works of art and social institutions, and proceed to place them under scrutiny and make them objects of criticism. Radovan Richta made this point clear:

> A work must enter into human life as detached and independent from human beings, external to them, simultaneously being commensurate with them, at least to the extent that it can be measured. There must be even a civilisation which is transformed by human beings, where human forces are objectified and socially developed to the extent that it [civilisation] transcends human beings, dominates them and at the end of the day it calls into question the existence of man himself. (Richta, 1968, p.644)

This perspective also sheds light on the objection Marxist humanists addressed towards the existentialist conception of the reification of nature and the alienation of human beings – the decisive philosophical, social, and civilizational issues of the day, which were diagnosed by both philosophical currents alike. According to the Marxist humanists, the existentialists were satisfied with a mere description of the given state of affairs, but did not provide any explanation of the possibility of such phenomena. As both Kosík and Richta suggest, the reason for this was that their diagnosis was still too empirical – they perceived human activity in an immediate, contemplative way and not in its creative and mediated form (Richta, 1968, p. 648). How then are we to describe this creative and mediated form that makes up a defining feature of praxis? What are the constitutive moments of praxis? In the next section we aim to combine all of the observations together and form (in reconstructive mode) a more or less coherent concept of praxis, as was the intention, we believe, of the Marxist humanists.

11.6 Praxis as Onto-Creativity

Marxist humanists were at pains to develop a philosophy of praxis to counter the Stalinist orthodoxy's emphasis on impersonal development and objective conditions as the driving force of human action. This was by no means a purely theoretical issue. According to Marxist humanists, the Marxist-Leninist insistence on the determining role of the objective aspect of reality was responsible for two preventable outcomes that weakened Marxist theory as well as the Communist movement: social-democratic reformism with a more or less implicit acceptance of capitalism on the one hand, and a state-propelled socialism with its coercive and centralised apparatus on the other. To avoid such consequences, Humanists sought the solution in the re-evaluation and even strengthening of the subjective element in Marxist theory. As we have already outlined above, they found support for this undertaking in Marx's texts themselves: in the *Economic and Philosophical Manuscripts of 1844*, in certain passages of the first volume of *Capital*, but especially in the *Theses on Feuerbach*, presumably because of its cursory yet assertive character, which left plenty of room for imaginative elaborations. It was the attentive reading of the first and also the third thesis that provided them with the guidelines for a new direction. So what does "subject" and "subjective" mean in the Marxist humanist philosophy of praxis?

First of all, Marxist humanists concede to some extent the importance of conscious intention in order to speak of praxis as a human activity. Marx's *Economic and Philosophical Manuscripts of 1844*, with its emphasis on the self-conscious aspect of human existence, and the famous passage from *Capital I*, distinguishing between bees and an architect by means of the fact that unlike a bee or a spider, humans are able to construct a building already inside their own head, seemed to provide a solid cornerstone for this concept. However, to stop here would have meant to submit to contemplative subjectivism, even if it had been presented as a strict correlation of the intending subject to the intended object. Conversely, emphasising the aspect of the object would have meant a retreat into an uncritical "objectivism." For Marxist humanists, the resolution of the problem consisted neither in subjectivism nor in objectivism or, to use the terminology of that time, idealism on the one hand or vulgar materialism on the other, because both made the same mistake – they considered the subject, object and even the relationship between them to be something simply given and ready-made.[15] Hence the philosophy of praxis

---

[15] This form of critique applies also to philosophical perspectives which aim to avoid both subjectivism and objectivism and formulate some sort of middle ground, as is the case with both Husserlian (at least the standard version) and Heideggerian phenomenology.

should not focus on either of the poles of this relationship, but on the relationship itself. But this theoretical gesture alone would make little difference compared either to Husserlian or Heideggerian phenomenology.

As we already know from above, contrary to (existential) phenomenology, the relationship between subject and object is not a relationship of intention, clearing, revealing, care or use, but one of creation or genesis. What makes it specific, to speak in Hegelian terms, is its "mediated" nature. Creation is not just a simple reali-sation of a plan or intention, nor a mere revelation of hidden potentialities. Creation involves, and one might even say embraces, negativity – resistance and challenges, contingencies and "accidents" put forth by the matter itself.[16] But these negativities are by no means just obstacles to be overcome, as Sartre's position in *Being and Nothingness* suggests. Andrew Feenberg's own words about Marcuse's interpreta-tion of Hegel are more than appropriate here in order to characterise what Marxist humanists had in mind:

> A challenge is not simply an obstacle although it may at first appear such. Unlike the object of work, the challenge is not overpowered and transformed, it is 'met.' Nor is meeting a challenge the annihilation of the challenge as such. To meet a challenge is to encounter the apparent obstacle as an opportunity for growth, self-development. The challenge must therefore be respected in its right to challenge and not evaded simply to achieve a goal. Indeed, the true goal can only be reached by maintaining the challenge in its challenging nature." (Feenberg, 2005, p.65)

These are inner conditions of creation itself. In this case, even the much abused term "openness" might be justified: creation is in its essence an open form of activity, because it always encounters possible detours, impasses, and digressions that divert and even dissociate creative activity from its original intention. This is exactly what Kosík had in mind when he wrote: "Though it is a specific human reality that is formed in the process of praxis, a reality that is independent of man also exists within it *in a certain way*. In praxis, man's *openness* toward reality, in general, is formed" (Kosík, 1976, p.139).

In this view, "matter" (the object) cannot be grasped as something that is simply there, prepared to take any form, nor is it a mere source ready for extraction. Matter demands respect from the creative process, though not so much in the sense of "careful" revealing of the already present potentialities as in the sense of finding new ways of resolving (in the Hegelian sense of *Aufhebung*) the said negativities. In this process of creation, negativity must be understood as "the potentiality to pro-duce new qualities" (Kosík, 1976, p.14). The philosophy of praxis seeks to accu-rately understand the process of how something new is produced or constituted: be it an artefact or a social institution (Kosík, 1976, p.14).

The second aspect concerns the difference between Husserlian and Heideggerian phenomenology on the one hand, and Marxist humanism on the other, namely the supposed existence of an "independent" object or reality. Why "independent"? What does this concept even mean within the Marxist humanist framework? Let us

---

[16] Lubomír Nový points out this peculiar space between intention and its realisation, which gener-ates a "field of indeterminate potentialities, becoming without guarantees" (Nový, 1968, p.129).

start with phenomenology. The constitutive moment of phenomenology, both in its Husserlian and Heideggerian form, is a substantial and seemingly indissoluble relationship between subject and object (Husserl's intentionality), or Dasein and the meaningful world (Heidegger's care). Although neither Husserl nor Heidegger denied independent existence outside the realm of this relationship, from the phenomenological point of view there is little to be said about this realm as a "thing in itself". In other words, if there are any objects, they exist first and foremost "for us" as meaningful entities. This is the original human condition according to phenomenology, and the essence of the so-called correlationism.[17]

At first glance, it seems that Marxist humanists state something very similar. For them, praxis is also a form of fundamental relationship between subject and object (or Dasein and world to use Heidegger's terminology), which on top of that possesses one essential quality we do not encounter in Heidegger (or in phenomenology in general): its transformative character, which pertains to both poles of the relationship – subject and object alike. As Kosík points out: "Something essential happens in man's praxis, something that contains its own truth *in itself* rather than merely pointing elsewhere, something that is also of ontological importance" (Kosík, 1976, p.136).[18] But this is just one aspect of the concept of praxis, and by no means the end of story. Another aspect that we highlighted in the previous section is that of the object produced within the process of creation, which enjoys a certain independence. Radovan Richta voices this point quite clearly: "Through dissociating itself from man and immediate nature, human work, the technical world created by man, acquires an independent practical dimension that differs not only from the original laws of nature, but also from the original human intention" (Richta, 1968: 643). Praxis is then marked by two moments: there is a relationship of creation that produces both object and subject; and furthermore, there is a process we call dissociation or decoupling, which dissolves this bond between subject and object. The object as a product of human praxis obtains a life of "its own," and is open for a re-appropriation by potentially many other subjects. This life is nothing other than social reality and history; and it is precisely this dissociation that makes both society and history possible. However, for Marxist humanists such a dissociation can by no means be understood as total or absolute, since otherwise we would quickly descend into the Kantian agnosticism of the "thing-in-itself," which would make revolutionary politics impossible. Accordingly, Marxist humanists try to avoid both the Scylla of radical constructivism and the Charybdis of metaphysical realism.

Although Marxist humanism is often presented as an intellectual current that promotes human subjectivity and individuality, in truth it is equally interested in objectivity and generality. For most humanists (Karel Kosík, Ivan Sviták, Lubomír Nový), the central concept of praxis captures a more complex ontological constellation, in which both subject and object undergo substantial changes. In this respect, praxis cannot be properly conceptualised as a one-sided transformation of the

---

[17] For a definition of the correlationism and its critique see Meillassoux (2010).

[18] (italics ours)

properties of some given object, since it is a creation of both individual and collective subject and object (artefacts, social institutions). Subjects are thus an integral part of the objectivity they create and recreate. To describe a subject as praxis implies such an integration that is in no way to be conceived as factuality or Heideggerian "thrownness". By the same token, an object as a product of praxis is not simply "ready-at-hand." Praxis as an ontological process, as characterised by Kosík, concerns both subject and object, so that praxis is neither doing nor coping performed by a subject (an expression of so-called spontaneous subjectivity), nor an anonymous process performed by a mysterious "das Man". Praxis is definitely a relation between the subject and object, but a relation that is not external but constitutive to both of its terms.

11.7   The Pitfalls of Heideggerian Marxism

Andrew Feenberg has shown in his reconstruction of the Lukácsian tradition of Marxism that it is plausible to reread the Marxist conception of praxis through the lens of Heidegger's account of the Greek *techné*. His rereading makes Heidegger appear much closer to the concerns of Marxist humanists than our presentation suggests. Feenberg explicitly writes that Heidegger,

> ...developed a consistent analysis of production (techné) in which nature is grasped as the essential object of the producing subject. In that scheme the object is not indifferent to its transformation by craft but is appropriate to its finished form in which its own potential is realized. Form is not a fact in the world on the same terms as a thing, nor is it simply the realization of the maker's will. A reciprocal interaction and exchange takes place joining maker and materials in a unity in diversity, a totality. (Feenberg, 2005, p.130)

Although we admire the lucidity and originality of his attempt to Heideggerise Marx (or is it rather a Marxisation of Heidegger?!), we are still not entirely convinced by the feasibility of his project. This does not mean that there are no possible crossovers between the two philosophies, but we consider these similarities to be still more of an analogical or even superficial nature. The simple fact that both of these philosophies tackle the problem of human activity as one of their central issues does not necessarily imply congruency between them. Or to put it differently, the Marxist understanding of praxis is not (necessarily) the same as the Heideggerian one.

As we attempted to show above, the essential feature of the Marxist humanist concept of praxis consists in the explanation of praxis as productive spontaneity, or as onto-creativity: the creation of something new. This is not the case with Heidegger, for whom the activity of "producing" is more about the meaningfulness of the world, and not about new things. Does this mean that Heideggerian "production" produces nothing at all? Certainly not. But the outcome of production in the form of real existing objects has never been the central focus of Heidegger's philosophy, which was always about "caring" and "revealing" as some form of fundamental relation between human existence and world.[19] The Czechoslovak Marxist humanists did not stop there, and they incorporated into their conception of praxis another conceptual element, which we refer to as dissociation or "decorrelation" of the relationship, as its necessary moment.[20] By doing so, they introduced into the concept of productive activity (praxis) an element which allows us to occupy a critical distance towards existing (social) reality and consequently make room for its change – a moment, we believe, which is impossible within the Heideggerian framework. We would now like to show at least one concrete example which will indicate a difference between the Heideggerian (in Feenberg's interpretation) and Marxist (in the humanist interpretation) concept of productive activity.

Let us suppose then for the moment that Heidegger's philosophy is about the production of objects, as Feenberg wishes it to be. Classical Heideggerian examples of production in the sense of *techné* are usually taken from premodern and pretechnological handicrafts like shoemaking. However, if we take these examples just as they are, it is hardly possible from a humanist point of view to assess whether anything *new* is really produced. Of course, there is a new pair of shoes as a product of shoemaking, but is anything new really created? This depends upon the social context and the level of technology. If a thing is just a product of an already established working process, "production" is an activity in which the elements of praxis are reduced to a minimum. Nothing *new* is really happening here – production is mere re-production. But if the producing activity opens up new horizons for the human being, then we can talk about production in the strong sense of the term. Praxis finds its truest expression in an activity which frees a human from the present limitations of everyday life. Praxis in the Marxist humanist sense does not disclose the meaningfulness of things within the horizons of reality, it transcends these horizons and creates a new reality.

---

[19] Despite all the talk of the famous "*Kehre*" in Heidegger's philosophy, there seems to be some profound unity in his philosophical project. This might best be displayed in the *Letter on Humanism* (1949) after he already announced his turn. Heidegger writes in the passage on the concept of being in *Being and Time*: "In *Being and Time* (p. 212) we purposely and cautiously say, il y a l'Être: "there is / it gives" ["es gibt"] being. *Il y a* translates "it gives" imprecisely. For the "it" that here "gives" is being itself. The "gives" makes the essence of being that is giving, granting its truth. The self-giving into the open, along with the open region itself, is being itself." (Heidegger, 1998, pp.254–255). Bearing in mind what Heidegger writes earlier and later in the text, it becomes more apparent that all Heidegger's "concepts" such as "clearing", "care", "being", "Lichtung", "ek-sistence", "openeness" point to the one and the same thing. For the unity of Heidegger's philosophy and the foundational role of human existence in his concept of being see Sheehan (2015).

[20] We need to stress here that we must not confuse the concept of decorrelation with that of alienation in a fashion similar to that by which Lukács allegedly confused alienation with objectification. Decorrelation is a necessary component of any praxis, and does not denote anything negative or wrong. Only through decorrelation, as we understand it, is society and even critical thought possible.

11.8   Conclusion

In our view, Czechoslovak Marxist humanism elaborated a philosophy of praxis that built on the Lukácsian turn in Marxist philosophy, but developed it in its own distinct direction. It shifted the focus from the original goal of Lukács's philosophy from the resolution of social contradictions through revolutionary action, to the notion of praxis itself. The concept took on not only a novel, diversified structure, but also a certain independence it had not enjoyed before. No longer would praxis be subsumed under other concepts such as labour, historical law, or totality, or under other forms of activities like application, verification, or resolution. From now on, the concept stood on its own and even formed a centrepiece of the new materialist philosophy. What does its structure look like, what were its constitutive elements and dynamics? First, praxis is an objectifying activity. The result of activity is some form of a real object, existing independently of its creator. In this sense we can speak about its onto-creative character, because through it a whole new layer of reality is established. Secondly, praxis is *at first* a relationship which links the creator to his/her future work – subjective intention and (new) object, which are mediated through matter itself. Praxis is therefore not an application or an imposition of an idea or form upon inert matter, but a process of creation, in which the subject meets the "challenges" of the matter as a site of negativity (contingency, resistance, possibility). Thirdly, this process only amounts to one of the aspects of praxis, while the second is of equal importance but is usually overlooked – the dissociation of the created work from its creator (the cessation of the relationship). The object now becomes independent and begins to "live its own life" – it is open to re-appropriation and reinterpretation by others throughout history. Only through such a dissociation or decorrelation of the original bond and independent character of the created work can a subject even adopt a critical stance toward existing reality and one's position within it. Finally, praxis is not only onto-creative, but also freedom-creative activity. It transcends the horizons of the established world and creates new perspectives for human beings.[21]

References

Anderson, P. (1976). *Considerations on Western Marxism*. Verso.
Arendt, H. (1998). *The human condition*. London University Press.
Bodnár, J. (1965). Fundamentálna ontológia Martina Heideggera. *Otázky marxistickej filozófie* [Fundamental Ontology of Martin Heidegger. Questions of Marxist Philosophy], *20*(6), 557–569.
Boltanski, L. (2002). The left after 1968 and the longing for total revolution. *Thesis Eleven, 69*(1).

---

[21] The enumeration of the elements should not suggest a ranking of their importance. Together they are tightly interconnected and form the totality of the concept. They are its moments.

Collier, S. J. (2011). *Post-soviet social. Neoliberalism, social modernity, biopolitics.* Princeton University Press.

Engels, F. (1969). *Ludwig Feuerbach and the end of classical German philosophy.* Progress Publishers.

Feenberg, A. (2005). *Heidegger and Marcuse.* Routledge.

Feenberg, A. (2014). *Philosophy of Praxis. Marx, Lukács and the Frankfurt School.* Verso.

Heidegger, M. (1998). Letter on "humanism". In M. Heidegger (Ed.), *Pathmarks* (pp. 239–276). Cambridge University Press.

Koestler, A. (1940). *Darkness at noon.* J. Cape.

Kosík, K. (1965). Človeka filozofia. In J. Bodnár et al., *Člověk, kto si?* [Philosophy of a Man. In Man, Who Are You?] (pp. 40–50). Bratislava: Obzor.

Kosík, K. (1976). *Dialectics of the concrete. A study on problems of man and world* (trans: Karel Kovanda, K. and Schmidt, J.). Dordrecht: Reidel.

Landa, I. (2022). Labour and time. Karel Kosík and temporal materialism. In J. G. Feinberg, I. Landa, & J. Mervart (Eds.), *Karel Kosík and the dialectics of the concrete* (pp. 75–106). Leiden.

Lukács, G. (1963). *Die Eigenart des Ästhetischen.* Luchterhand.

Lukács, G. (1973). The dialectics of labor: Beyond causality and teleology. *Telos, 3*(6), 162–174.

Lukács, G. (1972). *History and class consciousness.* MIT Press.

Marx, K. (1978). Theses on Feuerbach. In R. Tucker (Ed.), *The Marx-Engels Reader* (Vol. 1978, pp. 143–145). W. W. Norton & Company.

Meillassoux, Q. (2010). *After finitude. An essay on the necessity of contingency.* Bloomsbury.

Nový, L. (1968). Marxova filosofie dějin v otevřených dějinách. In Urbánek, E. (ed.), *Marx a dnešek* [Marx's Philosophy of History in the Era of Open-ended History. In Marx and our Present] (pp. 117–143). Praha: Svoboda.

Richta, R. (1968). Technika a situace člověka. *Filosofický časopis* [Technique and a Situation of a Man. Philosophical Journal], *16*(5), 641–657.

Rödl, S. (2007). *Self-consciousness.* Harvard University Press.

Sartre, J.-P. (1963). *Search for a method.* Alfred A. Knopf.

Sheehan, T. (2015). *Making sense of Heidegger. A paradigm shift.* Rowman & Littlefield International Ltd..

Spulber, N. (Ed.). (1965). *Foundations of soviet strategy for economic growth. Selected soviet essays, 1924–1930.* Indiana University Press.

Sviták, I. (1963–1968). Antropologické předpoklady moderní kultury. In I. Sviták, *Vlna pravdy ve filmu.* Praha 1963–1968 (nedatovaný rukopis) [Anthropological Presuppositions of Modern Culture, In *Wave of Truth in Film*, undated manuscript].

Chapter 12
What Place for Nature Within the Critique of Technology?

Adeline Barbin

Abstract Andrew Feenberg's work regularly makes references to the Marcusian notion of potentiality. It offers a way of normatively evaluating technologies by referring to nature. But how can nature and its potentialities be a useful concept in a normative evaluation of technologies? Does it not condemn us to factual statements instead of allowing us normative assessments and choices? To answer this difficulty, I draw on the French tradition of philosophy of technology, which granted a large place to the idea of technology as a biological activity of the livings and apply itself to understand how nature and culture are intertwined. Through the distinction between trend and fact, the work of Leroi-Gourhan helps us to locate how technologies are constrainted by natural laws as well as how they are submitted to contingency. Then, the notions of genesis and concretization of Simondon enable us to distance ourselves from any residual determinism and to specify what can be a potentiality for a technological object. Yet we are still left with the difficulty of the choice of the potentiality to develop. Through the idea of vital normativity developed by Canguilhem, it is possible to sketch an answer: technologies can serve or destroy the vital normativity of the living, which offers us a way of assessing it from a point of view which is both natural and cultural.

Keywords Andrew Feenberg · Technology · French philosophy of technology · André Leroi-Gourhan · Gilbert Simondon · Georges Canguilhem · Concretization · Progress · History · Herbert Marcuse

A. Barbin (✉)
Lycée Renaudeau, Cholet, France
e-mail: abarbin@inventati.org

12.1 Introduction

Large parts of Andrew Feenberg's work are concerned with the question of how to critically assess technologies in a normative way. He is dedicated to bringing a normative dimension to assessing the contingent values materialized in technical designs – a normative dimension that he wants to ground in rationality. Feenberg (2017) finds elements to build a normative critique through Marcuse's notion of potentiality. This notion is linked to a specific conceptualization of nature: Marcuse considers nature to hold "potentialities," which means that we can develop different relationships with it depending on the technologies that we use and that we are not doomed to a relation defined by control and domination. Human reason, Marcuse argues, has preferences for relationships that develop potentialities characterized by beauty, harmony, or peace (Marcuse, 1964).

How can nature help us in our critical evaluation? Nature is frequently used as a static and conservative concept, non-sensitive to cultural variations. What is natural is universal and unavoidable. To use nature to critique technology is, consequently, to examine how nature and culture are intertwined and to locate the contingency in technological creation and change. The goal of this paper is not mainly to underline contingency – STS has done it – but to see how a reference to nature and the understanding of technology as a biological process could yet help us to choose between different paths instead of leading us to an ineluctable step of some improved human being.

To better explore the relation between nature and technology, I will draw on the French tradition of philosophy of technology, a tradition that is particularly rooted in the idea that technology is an activity of the human being as a living being. In other words, technology is a biological process, which is why nature, human, and technology are deeply intertwined. By examining the relationship between technology and the living, the French tradition offers ways of locating the contingency of our technologies and to assess them. For this consideration, I draw on the work of André Leroi-Gourhan and clarify what parts of technology are due to nature and what parts are due to culture, pointing to what can be called "possibilism," which explains which parts of our technologies are constrained by physical laws and prescribed goals and which parts are cultural and contingent.

However, Leroi-Gourhan's thought is marked by a technological determinism linked to an understanding of technologies as a gradual externalisation of human abilities. Consequently, I analyse how Gilbert Simondon conceives technological objects in a way that enables us to overcome the limits of Leroi-Gourhan's thought and to think transformations of technical objects through the notion of concretization and the possibility of invention, notions which give us space to think about contingency. This leave us with a difficulty: how can we decide which potentiality has to be develop? By referring to the nature and the activity of the living, are we not doomed to record facts instead of discussing normativity? Addressing this question, I investigate the work of Georges Canguilhem and his idea of "vital normativity" for it provides us a way of specifying our assessment's criteria.

12.2   Evolutionism as a Technological Tool

The prehistorian, paleoanthropologist, and technologist André Leroi-Gourhan bases his analysis of technology on the same methods used in biology. He offers a theory of technology that he calls technical determinism.[1] He uses the concept of *trend* to organize what we know about technologies and to highlight what we do not know. Technology, in the strict sense of the study of technological artifacts, must do what zoology and botany did between the seventeenth and nineteenth centuries, namely, classify discoveries. *Trends* make this classification possible because a given environment offers to any kind of physical matter a limited number of possibilities of how to exist and in which form to exist. Physical and chemical conditions delimit the range of what is possible. Swimming animals are fusiform, for example, be they fish, reptile, or mammal, their body is spindle-shaped.

Leroi-Gourhan claims that what is true for living beings is also true for tools. From the blade or the stone that is attached to a handle, we can conclude what the shape of the handle was. In a similar vein, the anatomist Georges Cuvier was able to infer the shape of a skeleton from a jawbone he discovered (Leroi-Gourhan (1971 [1943], pp. 14–15). Thus, when Leroi-Gourhan talks about "evolution," we should not understand this simply as a metaphor, but as the statement that paleoanthropological and zoological lessons must be used in ethnology: weapons, baskets, and houses all have to obey mechanical laws in order to be efficient. Thus, roofs are peaked and arrows must be balanced at one-third of their length, the same way gastropods have spiral-shaped shells.

The *trend*, however, does not provide us with the details of a technology. Indeed, it is important not to confuse *trend* and *fact*. A trend can be anticipated. If we know the goal of an object, we can deduce what shapes it can have because the number of technical solutions to a given technical issue are limited. This is why we see the same technical object appearing in different, un-related places: "[…] one can say: men solve the wood issue by adze, the iron one by forge, the thread one by spindle" (Leroi-Gourhan, 1973 [1945], p. 336). The adze – a little axe whose edge is perpendicular to the handle rather than parallel – makes it possible to work easily with wood by, for example, taking bark off. Given the nature of matter and the biomechanically limited movements of prehension, rotation, and translation of the human hand and arm, there are very few possible shapes for an adze. In fact, the number of solutions is so small that the adze is the one shape which has been invented "all around earth since Neolithic" (Leroi-Gourhan, 1971 [1943], p. 320). In the same way as there is a process of convergence in biology, there is a process of convergence in ethnology (Leroi-Gourhan, 1973 [1945], p. 338) such that there is no need to assume that there was contact between people who have the same technology.

---

[1] Using the expressions "technical determinism", Leroi-Gourhan only means there are physical constraints operating upon the shape of our tools and does not imply there is technological predetermined path of development. Yet, as we will see, Leroi-Gourhan sometimes uses this determinism not only as an heuristic tool, but also as an historical one.

The technical *fact* relates to the details and the specificities of the technical object. The shape of a sabre, its matter, its curve, its ornamentation, or its symbolic and social meaning are different in different societies. The *fact* is the trend added with all the details; it is what appears once the *trend* goes through what Leroi-Gourhan calls the "inside environment" which is made up of all the mental traditions of a human group, the same way a sunbeam goes through bodies with different properties and is modified (p. 361). Thus, the *fact* appears at the junction point between the inside environment and the physical laws of the external environment. The better it is designed, the more efficient it will be, which is the "general movement" (p. 338) of the *trend*.

To fully understand the definition of the *fact* as "material witness" (p. 339) of the trend, it is important to understand that according to Leroi-Gourhan technological classification should not be organized by the tools themselves because they are not the kernel of technical activity. Tools are linked to gestures (1965, p. 35) and to matter. This is why Leroi-Gourhan defines the object of technology as "strength + matter = tool" (1971 [1943], p. 319), or more precisely: "elementary mean/matter = tool and product" (p. 320). The classification in his two-part book *Évolution et techniques* is organized by the criteria of means and matter. Consequently, the object of technology is the *operating gesture* and tools must be considered from the point of view of human beings' adaptation to the environment, of efficiency, and of improvement of efficiency along the line given by the *trend* and its prescribed goal.

However, Leroi-Gourhan does not state there is some kind of Platonic idea of a sabre or of any other technical object; he does not argue in favour of some objective and transcendent idea which would be only imperfectly embodied in our tools. Departing from the example of the sabre and studying the roofs of houses, Leroi-Gourhan underlines that physical conditions, such as wind or snow, do not make one shape necessary: human beings choose a relevant shape. That is what Leroi-Gourhan means when he claims that a *trend* does not come from the external environment, but comes from the internal one. Technical objects, *facts*, appear where the general physical conditions meet culturally specific ones. No mechanistic explanation of human adaptation to an environment is required: external and natural environments do not causally explain the development of societies. Thus, the theory of Leroi-Gourhan can be understood as a kind of possibilism, a geographical concept created by Lucien Febvre to name the approach of Paul Vidal de la Blache. Possibilism opposes geographical determinism and argues that the environment offers possibilities that can be used differently by different human groups in function of their culture. The distinction between *trend* and *fact* echoes possibilism because we cannot anticipate the shape and the details of a technical object. *Trend* is a range of potentialities which will exist only if one of them meets societies with the right technical knowledge, an appropriate cultural conception, and a suitable economic field. The environment offers a frame made of the matter of the technical object, the matter on which it must apply, and of human anatomical structure. Nature proposes and humans dispose.

Therefore determinism, as Leroi-Gourhan named his approach, should not be seen as being relevant to all levels of analysis. It is useful to notice that the shape of adze is natural in some way, but this approach cannot be used further. Details must be explained differently. There is no autonomy of technology, and determinism is a heuristic way of classifying our knowledge about technical objects.

12.3   Tension Between Technological Analysis and Historical Discourse in the Work of Leroi-Gourhan

Yet, ambiguities remain in the work of Leroi-Gourhan and it is not clear how the concept of evolution can be used to describe the history of technological objects without having prejudicial consequences on human freedom. In other words, Leroi-Gourhan does not use it only in a heuristic way. The way each technology gives birth to new one is displayed as a process that takes a long time, without any interruption by *ex nihilo* inventions:

> The internal combustion engine came from hydraulic machines of the seventeenth century, from the spinning wheel, from the Papin's digester (here again we can analyse endlessly: it is an additional proof of technical environment continuity). (1973 [1945], p. 366)

The problem is that what was supposed to be only a logical order tries to be as close as real historical order as possible: each shape of a tool has an ancestor, the same way each living being has one. So Leroi-Gourhan states: "tools are bound on a time scale in an order which appears, broadly speaking, as logical and chronological at the same time" (1971 [1943], p. 24).

If the concept of evolution has chronological ambitions, how can we not come back to notions of progress and deterministic history? Is it possible to escape from the conclusion that human beings are not in control of technological development when we underline the biological anchorage of technology? We need to clarify the relationship between evolution and autonomy in order to find out if the idea of biological anchorage does not condemn us to understand technical development as a process oriented by its origin. Just like Alfred Espinas, Leroi-Gourhan conceptualizes technical evolution only as the externalization of human capacities, as an extension of an organic body. Following Dominique Bourg, one can consider this the reason why Leroi-Gourhan predicts an apocalyptic end of human history in which all of us will become a, "de-personalised cell in an organism that admirably expands world-wide" (Leroi-Gourhan, 1965, p. 60). Reading the history of technologies using only the relation between human bodies and tools from a starting point of the human body "as mainly productive" (Bourg, 1996, p. 187) does not allow a full account of the diversity of technical entities and of their relationships. Considering this all together, ambiguities in Leroi-Gourhan's statements are the result of undue expansion of technological discourse to statements about human history: "Human would not be human if technology did not escape from its control since the beginning" (Leroi-Gourhan, 1983, p. 91). With the notion of *trend* being biologically marked by the reference to the externalisation movement – a process which would drive us to externalize our physical and intellectual capacities through our technical inventions – Leroi-Gourhan reaches a problematic reading of the direction taken by societies, an orientation that appears as planned, universal and non-sensitive to cultural variations. This technical history is the history of a complete externalisation of our physical and intellectual abilities until we will be completely externalized and obsolescent. Here lies the problem in the work of Leroi-Gourhan: humanity is thought of as a biological species with individuals being mainly *homo faber* and culture pushed into the background. Thus, here again, we find a cumulative and problematic conception of progress.

By locating the physical conditions which apply to the conception of a technology, Leroi-Gourhan allows us to understand what parts of technology are due to nature and what parts are due to culture. What I call "possibilism" refers to the contingency in the design of technologies: nature is a field of possibilities, a theme which could lead to the Marcusian idea of potentialities. However, Leroi-Gourhan does not entirely get rid of a pre-determined conception of the evolution of technologies, which path is designed by our body. On the contrary, Gilbert Simondon focuses on the internal functioning of each technology, its history and its potentialities through the notions of concretization and invention.

12.4 Differences Between Technical Nature and Utility

The terminology used by Gilbert Simondon is very similar to that employed by Leroi-Gourhan: changes of technical objects are described with concepts such as "evolution," "convergence", or "internal necessity". It should not come as a surprise, then, that both authors follow the same intention with the use of this terminology. Indeed, Simondon wants to study technical objects on their own, regardless of their use, stating that he is continuing the work of the anthropologist about pre-industrial civilisations by working on industrial ones (Simondon, 2014, p. 33). Yet, Simondon does not apply exactly the same method as Leroi-Gourhan, nor does he use the same classification framework and some noteworthy differences must be underlined for they radicalise the work of his predecessor. First, Simondon based his analysis on the technical object itself, and not on the combination of gesture and matter; on the technical object as distinct, and not on its type, its line, or its trend. Such a move is the result of Simondon's choice to study industrial civilisations. Indeed, studying pre-industrial technical objects requires reference to functional mediation. The fact that industrial technical objects are more or less free from human operatory gesture is an opportunity to focus on the technical dimension of objects, namely the internal consistency of their functioning.

Initially, to really focus only on technical objects, Simondon excludes moral and political discourses about good or bad uses of technologies, the history of technologies within civilisations, or the question of the essence of technology. We can call his method *épochè* for it consists of putting aside everything apart from the object itself. Thus, *On the Mode of Existence of Technical Objects* (*Du mode d'existence des objects technique*)[2] offers a kind of phenomenology from the point of view of the object, the goal of which is to describe the functioning and the evolution of a given object (Chap. 1) and its relationships with its environment (Chap. 2). In a way which can seem paradoxical, Simondon starts his book by regretting that technical objects are distinct from human culture, and yet, to understand how technical objects are cultural facts, one must first study it through what is, in it, the most technical and the least human.

Therefore, we cannot rely on an analysis of its use when we want to understand what a technical object is. It remains technical even if it is not used and, conversely, most of the things in the world can be used without being technical objects. This means that utility is not strictly speaking a technical category. Because it reduces technical object to "*utilia*, utensils whose only dimension is to answer a practical goal to a human need" (Simondon, 2014, p. 74), utility is not relevant to identify the specificity of an object as a technical object. Therefore, it is not the technical object as a mediation between humans and nature that must be studied. Such a mistake rises from confusion between *work* and *technical nature* (2017, p. 247 *sqq.*), because the model of work leads us to think of technical objects through their utility. Yet, this dimension is not the main one since a technical object does not necessarily aim at producing something. The category of technical nature is wider than the one of work: it is work which should be examined in relation to technology, and not the opposite. Against Bergson's conception of technical activity as an answer to human needs, the end of the *MEOT* seeks to establish that the right category to work on technical objects is *operatory functioning*, with a clear distinction between operatory and practical. Indeed, technical reality is not only a question of the "manipulation of solids" (2017, p. 259), in other words, of the primacy of materiality; a technical object is not useful at first, but must develop itself inside the framework of the physical laws of our world. Technologies are objective in the way they must first respect the laws of physics in order to function. Even if objects are useful, the objectivity of the functioning comes first. Another way of understanding this point is to add a remark from Jean-Yves Château (2005, p. 54): Simondon wants to differentiate his approach from the structuralist one because the technical object is not, in the first place, a meaningful object. It must work and, to do so, it must have an autonomy from human language and communication. Analyses based on languages cannot take account of the fact that technical objects are required to embody physical laws.

A distinction between use and *function* or *functionality* on one side and functioning on the other side is essential to assert the right definition of a technical object. One must not be blinded by what the objects mean for humans instead of what they are in their technical nature. Engines functioning with steam, gas, springs, or weight can be unified under the one category of "what puts something in motion" only if we consider the physical laws being used that allow them to be technical objects. Identical uses should not lead us to wrongly identify objects, no more than different

---

[2] From here on, I will use the French acronym: MEOT.

uses should hide a unity of functioning. Technically speaking, an engine based on spring-action is closer to a bow or a crossbow than to a gas engine, the same way an engine that uses weights and leverage is closer to a winch (Simondon, 2017, p. 25). Thus, on one side, technologies are a way of solving difficulties arising from the relation between living beings and their environment and what is important is their ability to provide effects. Understood as instrumental and functional mediation, they combine the way a technical object functions with the way a human being functions (if we consider human technologies) and this functioning of the object is the starting point of the technological study. On the other side, technology can, and must, be included in a "psychological and reflexive study at the highest level" (Simondon, 2005, p. 84) – to quote the sometimes-misleading terminology of Simondon – which targets the functioning of technical reality and is the only one to make possible the definition of the "type of consistency" (p. 84) that characterizes the essence of technology. This reflexive study requires a *synchronic* examination of "functioning regimes and structures" (p. 84), of the parts of the object, their relations, or their lack of relations. Yet, a static examination of technical functioning is not enough because an engine cannot be defined only by its structure. Indeed, technical structures are subject to permanent modifications. This second part of analysis gives birth, in turn, to two axis, a synchronic one and a *diachronic* one that applies to the *genesis* of technical objects.

12.5 Simondonian Concepts: Genesis and Concretization

Simondon's genetical method rests on his assessment that a technical object is unfinished when it first appears. We thus have to deal with its genesis because a technical object faces trials and errors and failures and improvements because there is distance between thought and reality or, to use terminology we will define later, between the abstract and the concrete, between science and technology. The way of being of a technical object is a temporal one (Simondon, 2017, p. 26). If we want to talk about a gas engine, we cannot talk about a specific engine, neither about a particular step among all its changes, but about a functioning which has known several transformations over time, the same way a "phylogenetic lineage" (p. 26) does. Thus, to study the nature of technical objects is difficult because objects are at the same time individual and multiple and have similarities that appear to be due to relationships of causality or generation. How can we cut a technical object out of its reality and abstract it from its context? To solve this difficulty, Simondon suggests that we "reverse the problem" (pp. 25–26) to clarify what the specificity of a technical object is from a clear starting point, something which is not an accidental feature, but a main aspect of its essence: change. A technologist or "mechanologist" must apply a *genetical method* and pay attention to the evolution of objects. In this way, a technical object is not different from a living being or an aesthetic piece of work: it is not something given, but a result. This conclusion reminds us of statements found in Simondon's *L'Individuation à la lumière des notions de forme et d'information* (Simondon, 1964): hylomorphism and theories of substance are wrong to start from what must be explained, that is to say, the individual and how it comes to exist.

Yet, to follow "phylogenetic lineage", kinship between objects does not mean that we can simply uncoil real temporal series in which we can observe variations of a framework of functioning. Indeed, chronology is not hierarchy. We have to be careful regarding improvements which would only affect uses or be just a fad. Hierarchical organization of objects regarding their technical nature must obey the criterion of concretization. Here we meet a very famous notion in Simondon's thought: the abstract shape of a technical object is its primitive one, when each part is like a closed system, a unity. In this situation, "the pieces of the engine are like people who work together, each in their own turn, but who do not know one another" (Simondon, 2017, p. 27). When an engine is abstract, each part has a role in one specific time of the cycle and we cannot really say that the elements function together. Thus, in order to make the engine function, structures must be added. For example,

> …the cylinder head of the thermal combustion engine bristles with cooling fins…In the first engines these cooling fins are as if added from the outside to the theoretical cylinder and cylinder head, which are geometrically cylindrical; they serve only one function, that of cooling. (p. 27)

A more concrete engine shape exploits as much benefit as possible from each element. Consequently, cooling fins do not only play one part anymore as if they were juxtaposed to some pre-existing system; they have a second task, a mechanic one, in which they oppose the deformation that the gas thrust can cause on cylinder head. Concretization can thus be defined as the implementation of a synergetic relation between parts of a technical object. Its main way of action is to unify several functions in the same structure. It is not a compromise, but a convergence towards unity.[3]

12.6 Concretization as Criterion of Technical Progress

Concretization is a way to save matter, energy, or equipment wear. Simondon situates concretization as the process that makes it possible to assess technical progress. But why is it the central criterion to assess a technical object? Due to "internal necessity" (p. 29), a technical object tends towards unification of its parts, like living beings do – and we must point out that "to tend towards" is not "to be." Therefore, in a technical object, what it will be and what it should be match: the more the object is abstract, the more it is imperfect, consumes resources, demands work, and, ultimately, is fragile (p. 30). A cooling system that is autonomous from the engine can stop and the engine will still work but risks being damaged; whereas, if the cooling system relies on the engine functioning, the engine is necessarily cooled when

---

[3] For a similar reading of Simondon, see for example Feenberg (2017, chap. 3).

working. From this, Simondon concludes that a car's air-cooling system is more integrated than a water-cooling one. Being more concrete, it is progress from the point of view of its own internal necessity because it is more likely to last and to persevere in its being, a goal which, for technical objects, as for all things, is primary.

We cannot overstate the importance of the notion of concretization to understand what is, in its specificity, a technical object. Concretization defines a technical object and allows us to distinguish it from other human productions. Textiles are not technical objects for there is not – at least, not yet – structures or functions in the fabric whose job is to allow the textile to last as long as possible. In other words, there are not any reciprocal causal relations, any *self-correlation*. Through self-correlation, the object is more efficient when facing its destruction. Such a correlation can be found even in the simplest technical objects, such as an adze or even an electrical pole. To design a handle is a difficult task of internal logic for many tools and so it is often a weak point: it has to fit the human hand and arm movements and, at the same time, has to stay attached to its ending (blade, pike, etc.) despite the impacts. Similarly, an electric pole has to be able to support some bending and to come back to its position.

Yet, such integration, such processes of convergence, should not be mixed up with the one we met in Leroi-Gourhan's work. Both meanings encompass the idea of efficiency within the framework of physical laws, but Leroi-Gourhan is interested in the function of the object and the way it is an instrumental mediation between humans and nature, whereas Simondon uses convergence to describe relationships between the elements that make up technical objects. The more complex they are, the more the criterion of self-correlation is relevant, for they need less and less human gestures.

12.7   The Part Played by Invention

Unlike in Leroi-Gourhan's work, technical determinism is not a problem in Simondon's thought. Indeed, concretization does not have historical and chronological ambitions. It is a prescriptive concept: it tells us what a technical object should be and which logic should drive its development and the direction of its evolution; not what it is or what really and historically happened. Leroi-Gourhan's work allows for steps in the evolution of technologies to be made backwards, but for him the development of technical objects always leads to increasing efficiency. On the contrary, Simondon notices – and sometimes regrets – how concretization is subject to interference and obstacles. All the work Simondon did on different intrinsic causes and what he calls a "halo of sociality" aims at accounting for the evolution of objects as it historically happens, meaning that the evolution of technical objects does not always follow an internal necessity. None of this requires economic or social explanations, that is to say, external causes are not real causes. But, before being accepted and used, a technical object must be physically viable. From the point of view of a technological analysis, it is only the technical object that matters, as long as the goal of the analysis is to understand the ontogenetical process of its development.

Yet, even if Simondon acknowledges the permeability of technical objects with cultural, social, or economic conditions, are we not forced by his theory to admit that, ultimately, the internal logic of the object will always prevail after it has overcome interferences and difficulties? Here again, an important difference between Leroi-Gourhan's and Simondon's work must be pointed out: the place given to invention in the thought of the latter. It is crucial here to explain the genesis of technical objects, giving the same attention that Simondon pays to it. The *MEOT* states the existence of an "absolute beginning" (p. 44) of technical lineages, and with it the possibility to introduce ruptures within the continuity of evolutions. Such a new genesis is linked to an act of invention which is both objective and subjective. In a way, indeed, the new technical object comes from previous ones: "the gas engine follows, in a sense, from the steam engine" (p. 46), something we know because we can recognize a similar distribution of functions of elements, which are yet different. So, in a way, it is true that "a gas engine follows the steam engine" because, in order to appear, a technology needs a society which possesses enough technical knowledge to understand it. But a new technical object is also an appearance and, because of that, is a result of human subjectivity. Inventors add the "new phenomenon" (p. 46), which allows design to go from steam engine to gas engine, a concrete reality which has to appear first as an abstract idea in human minds. Solving problems requires human intelligence.

Thus, the ontogenetical process of a technical object is at the same time a story of human freedom and the constraints of necessity, with a part to play by random steps, for there is no method to invent, no *ars inveniendi*. This is why the notion of invention is so important in Simondon's thought: it is "one of the main aspects of an acting freedom" (2005, p. 151), the place where we can realize that only intelligence can at the same time know existing technologies and remain free enough to try to find a new synthetic organization when nothing tells how to find it. Thus, invention is a condition of potentialities, and the theory of technology that Simondon offers is a genuine history of technical objects, a history that can deal with technical realities in their specificity without excluding the subjective aspect of human intention. From this, it is possible to understand that it is not the technical object "engine" that evolves, but the idea of an engine; and to understand how the filiations relate to the creators as much as to objects, because the latter are tangible traces for the former and feed the capacities of invention. Simondon proposes a history of technical objects which is neither a biological nor a human history nor a vague description of an evolving process. If there is an autonomy of technical objects in Simondon's thought, it is purely a methodological one in order to identify the specificity of technical nature. The *épochè* of psychosocial, political, and economic dimensions does not mean they do not play a role in this history, but only that we can put them aside for a while.

12.8   Evaluations of Technologies Through Simondon

Andrew Feenberg has found in Simondon's work the idea of concretization (Feenberg, 2017). Concretization allows us to assess technologies on a technical level by considering how integrated the different aspects of a technology are. This consideration, although purely technical, is not blind to the relation the technology has with its natural and social environment. On the contrary, the notion of associated *milieu* is absolutely crucial to this evaluation. Thus, Feenberg obtains a framework which combines the assets of Critical Theory (normative assessment) and constructivism (contingency). For a given technology, a diversity of concretizations exist that depend on diversity of actors. The notion of concretization gives us a normative perspective because it shows how values can be integrated into the functioning of a technology. It provides us with a tool to think what is now a well-known fact in technology studies: technical rationality is not pure, facts and values are mixed, and so is rationality and ideology. Values are in the technical fact and concretization tells us if they are well integrated in the way a technology works.

Obviously, Simondon's concretization is an interesting concept through which to assess the environmental consequences of a technology. The more a technology is concrete, the more it efficiently uses its natural environment and the less it consumes resources. For example, an air-cooling system uses ambient air to cool, let's say, a car, instead of consuming fuel to produce fresh air. Moreover, concretization allows us to do more than just to assess technologies. It also gives a frame to assess the system by which technologies are utilized, in particular the organization of work. Indeed, the process of concretization is linked to the capacities of individuals to understand a technology and to modify it in order to improve it. The human being is the "witness of machines," the one able to consider the components of a machine and the totality of a system and to improve their correlation, and technical nature lies in both these aspects. Thus, we understand that there is no reason for the hierarchical organization of industry. The only reason why those responsible of the totality are more valuable than those responsible for the components is that we give them chief prerogatives.

Whereas Simondon offers a very powerful tool to think about and assess technologies, he does not always seem to use it in his own assessment of technology. One aspect is striking: Simondon does not take in consideration the fact that the human being is an incarnate being. He tells us about the tasks of watching and invention, but the complementarity between our body and our tools, the pleasure we can have directly transforming matter is never mentioned and, through the way Simondon dismisses handicrafts, it is even excluded. He contends that handicraft is an abstract stage of technology, an insignificant state in which the knowledge is unconscious. All the technicity lies in the human beings and none in the technology because there is no spare part to convey it. Yet, recent historiography denies these considerations (Hilaire-Pérez, 2000; Jarrige, 2016).

Despite his consciousness of the problems that technologies can raise, Simondon is far too confident in technology. He judges that a technical problem has a technical solution. Simondon thinks there is a law of technical development which is that each deepening of technology allows us to face the damages created by its previous stage. For example, regarding nuclear power and waste treatment by a technology which is still not available today, Simondon expects a rapid development and speaks of "redemption", of "dialectical operatory recuperation" (Simondon, 2014, pp. 340–341). He is confident in the capacity of research to solve the problem of nuclear waste. And he shows no consideration about how democratic the appropriation of the required knowledge about nuclear plants can be (technical specifications are not only complex, but also secret).

The problem is that he does not consider seriously that there are multiple paths of concretization for a given object and that each path can express different values. Yet, the path that is followed depends on actors with different interests and goals, as constructivism has taught us. These paths can be regarded as the "potentialities" that Marcuse mentions (Marcuse, 1964): instead of having a relation of domination with nature, we could develop different technologies to favour ethical and aesthetic relationships. Yet, a reference to nature could easily be conservative and exclude the cultural aspect of technologies. Indeed, what is natural is universal and so applies in the same way all over the world. What is natural has not to be discussed and cannot be discussed. That is how the concept of nature is used in conservative positions to condemn the so-called "unnatural" behaviours. How to avoid this pitfall of a naturalism that would deprive us of the possibility to discuss normativity?

12.9   Vital Normativity, a Way to Join Nature and Culture

To help us with our search, we can look at the work of Georges Canguilhem, an author who deeply influenced Herbert Marcuse and one of his students who was... Gilbert Simondon. As with André Leroi-Gourhan and Henri Bergson, Canguilhem considers technology an activity of living beings. Thus, technology has a biological anchorage. But this does not mean that we are doomed to a naturalism which would prevent us from assessing technologies by saying that it is all natural and thus ineluctable. Indeed, the critical assessment of technology should not be organized around the couple "natural-artificial." Everything which is artificial is in a way natural, at least for it respects the laws of the universe, and there are not many natural things which did not undergo some kind of artificialisation. So, we face a continuum which makes it very difficult to make a clear distinction between "natural" and "artificial". More promising is the coupling of nature and (artificial) culture, and so the question is not if there is a continuity or an opposition, but how the mediation between a human being and its environment takes shape (Larrère & Larrère, 2015, pp. 167–171). A thorough reading of Canguilhem's paper, "*Machine et organisme*" (Canguilhem, 1992), allows for a complexification: technical activity is a biological fact, "*a fact of nature*" (p. 120), which comes before any knowledge about this activity itself. But this fact of nature embodies itself only in "*a fact of culture.*" Artificialization is a biological fact, but it is subjected to the mediation of culture and thus becomes also a social fact. There is no more purely natural activity than there is a purely natural technics of body, as Marcel Mauss already underlined in 1934 (Mauss, 1936). Technics is a fact of evolution for a species which uses culture as a way of development. Against a monism – of nature – which would ruin any normative statement, beyond a problematic distinction between natural and artificial, Canguilhem proposes a three-part relation, using nature, artifice, and culture. Therefore, Canguilhem is able to falsify the idea that the relationship between human beings and the environment is similar to the one between an organism and its environment. This idea would lead to the conclusion that actual environmental troubles are due to "a biological rupture of balance" (Canguilhem, 2000, p. 187), whereas we should incriminate social and economic conditions.

If the topic of the biological anchorage is mediated by the place of culture, the return to the primary pulsion that explains technology does not imply the neutralisation of normative discourse. On the contrary, to remind us of how technical activity belongs to living activity makes possible to conceptualize technology as a cause of disequilibrium of the livings and the environment. To consider technology as a direct application of sciences is to assume a lack of limits, an indefinite progress. Then, we are disconcerted when continuous improvements of the human condition are not achieved, whereas we should be able to understand that human action on the environment has limits for we act in a "finite world" (p. 190). Here we meet Bergson (2012) again, when he mentions, in *Les Deux sources de la morale et de la religion*, the distortion, "the emptiness" (p. 379) between the human being created by evolution and industrial technics. Life always faces the possibility of the dead end of matter, its immobility, its lack of value creation. Matter is one aspect of life, the one it takes when the *élan vital* drops and freezes. Thus, life itself is ambivalent, and the human being is the paroxysmal expression of that fact, a being able to follow up to the top the impetus of creation or to proceed to destruction of life itself.

These limits are not only those of the quantity of transformations the environment can tolerate to stay liveable. When denouncing an intellectualist understanding of technics and its conceptual relation to a mechanical idea of the living beings, Canguilhem notices that technologies must respect *vital normativity* for it is an activity of the living beings. He does not worry, as Bergson does, about the disappearance of life as the disappearance of the process of creation of values, but of the ability of technical activity as an activity which creates values to subdue vital normativity up to a point where it would drain it. Industrial mechanisation exhausts the vital normativity of human beings who have to follow the pace of production because this pace is not adapted to their biological one, the same way industrial breeding is a violence towards animal's biological normativity. From this point of view, nuclear power can be understood as the risk of exhausting vital normativity; procedures of enhancement of the human body and mind can be assessed from the question of the associated milieu of a human who has co-evolved in a definite environment; handicrafts can be assessed as an expression of vital normativity of our body. Simondon himself provides a useful example of relationships between technology and normativity of living beings through his analysis of the case of a young farmer who has to sell an eye to pay for surgery on his endocrine glands. The "purely operatory gesture" of the surgery is "infra-technical" because "it does not have a normativity fitting for the global reality" which would be subject to surgery and, so, will "alienate…the functional whole" (Simondon, 2014, p. 124) of the body.

Normativity is the way through which technical action should join living beings as a whole of internal relationships and external relationships with their environment and other living beings within an environment which is organized and polarised by living values, instead of considering living beings as an addition of chemical-physical processes. That is why this notion allows us to restore the point of view of culture, as a whole, far from a reduction of human being to *homo faber* and from its analysis simply as a biological species. To fully understand that point, it must be noticed that biological normativity does not represent, for Canguilhem, a natural datum which would be the last word of any assessment. As mentioned above, a biological fact is always transformed into a fact of culture, and this is true for both human and animal normativity: some situations are considered good or permissible depending on the value system of a culture. Violence to a human body does not have the same meaning in different cultures, depending on their nature of initiation rites, treatments, punishments or work conditions. It is the same in animal vital normativity: veganism, vegetarianism, hunting, "traditional" or industrial "breeding" give different meanings to it.

Culture is what chooses the possibilities that should be explored and those that should not. Artificiality, through experimentation, makes it possible to know reality in new ways. It is also what makes possible the creation of new possibilities for natural phenomena which are able to endure changes. Technical actions can slip themselves in interstices of the laws of nature because there is contingency in nature and because living beings can tolerate self-triggered modifications but also external modifications. Therefore, artificial and natural are not opposed. Their distribution is cultural and historical, and so are the causes that explain why we choose to undertake a specific exploration of this possibility and not of that other one. Thus, Xavier Guchet suggests combining vital normativity with the genesis of technical objects (Guchet, 2018). This allows, for example, to find a way to establish and to assess the difference between traditional genetic selection, cross-breeding, and contemporary modifications of genome. The first one does not necessarily imply that animals will be treated with respect. But it uses the potentialities already given in the genome of the animals. The second one creates potentialities that did not exist before, creating a new vital normativity, introducing for example a silk gene in a goat genome.

As Andrew Feenberg noticed, Marcuse's thesis of potentialities allows us to imagine different paths in the design of our technologies. Yet, referring to nature confronts us with a difficulty: how can nature be useful in our assessment of technology? Indeed, nature is often seen as a conservative notion, determining once and for all the essence of things. But it would be a mistake to limit it to this conservative notion and its caricatural Darwinian conception of life as a restrictive and economical process of adaptation. Life is profuse, abundant and generous. Leroi-Gourhan helped us to better understand the link between nature and technology and to locate the role of cultural and the contingency. It appeared that nature holds potentialities. Yet, we needed to go further for Leroi-Gourhan has a too restrictive conception of technology and of its path of development. Using Simondon's work, we found a way to focus on technology without considering it as an externalisation of the human abilities, a reading that would deeply reduce the scope of potentialities. Concretization and invention are two major notions to give content to the idea of potentiality.

As a final step, we then have to determine how to deal with the reference to the nature and with the idea of technology as an activity of the living without restricting ourselves in our normative judgments. Nature, not as opposition to what is artificial, but as that which opens up different paths to living beings, can give us ways to deepen our assessment of technology. The law of Gabor does not hold: not everything that could possibly be explored will be explored. Technical reality is always different from the ramifications of each technical realization. The coupling of the artificial and the natural cannot help us in the choices we have to make because it does not provide a helpful reference point in technology assessment. In contrast, the notion of life, understood as vital normativity, gives us such a reference point for questions to assess our practices because the living beings have needs and express preferences. And, among these living beings, the human being can use its reason to select among possibilities.

References

Bergson, H. (2012). *Les Deux sources de la morale et de la religion*. Flammarion.

Bourg, D. (1996). *L'Homme artifice: le sens de la technique*. Gallimard.

Canguilhem, G. (1992). Machine et organisme. In G. Canguilhem (Ed.), *La connaissance de la vie (révisée et augmentée)*. Vrin.

Canguilhem, G. (2000). La question de l'écologie (révisée et augmentée). In F. Dagognet (Ed.), *Considérations sur l'idée de nature. [Suivi de] La question de l'écologie* (pp. 183–191). Vrin.

Château, J.-Y. (2005). L'invention dans les techniques selon Gilbert Simondon. In G. Simondon (Ed.), *L'invention dans les techniques. Cours et conférences*. Seuil.

Feenberg, A. (2017). *Technosystem. The social life of reason*. Harvard University Press.

Guchet, X. (2018). Toward an object-oriented philosophy of technology. In S. Loeve Sacha, X. Guchet, & B. Bensaude-Vincent (Eds.), *French philosophy of technology. Classical readings and contemporary approaches* (pp. 237–256). Springer.

Hilaire-Pérez, L. (2000). *L'Invention technique au siècle des Lumières*. Albin Michel.

Jarrige, F. (Ed.). (2016). *Dompter Prométhée. Technologies et socialismes à l'âge romantique (1820–1870)*. Presses universitaires de Franche-Comté.

Larrère, C., & Larrère, R. (2015). *Penser et agir avec la nature. Une enquête philosophique*. La Découverte.

Leroi-Gourhan, A. (1971 [1943]). *Évolution et techniques (I). L'homme et la matière*. Albin Michel.

Leroi-Gourhan, A. (1973 [1945]). *Évolution et techniques (II). Milieu et techniques*. Albin Michel.

Leroi-Gourhan, A. (1965). *Le Geste et la parole (II). La mémoire et les rythmes*. Albin Michel.

Leroi-Gourhan, A. (1983). *Le Fil du temps*. Fayard.

Marcuse, H. (1964). *One-dimensional man: Studies in the ideology of advanced industrial society*. Beacon Press.

Mauss, M. (1936). Les techniques du corps. *Journal de Psychologie, XXXII*(3–4): https://doi.org/10.1522/cla.mam.tec

Simondon, G. (1964). *L'Individuation à la lumière des notions de forme et d'information*. Presses universitaires de France.

Simondon, G. (2005). *L'Invention dans les techniques. Cours et conférences*. Seuil.

Simondon, G. (2012). *Du mode d'existences des objets techniques*. Aubier. English edition: Simondon, G. (2017). *On the mode of existence of technical objects* (C. Malaspina, & J. Rogove, Trans.). Univocal.

Simondon, G. (2014). *Sur la technique (1953–1983)*. Presses universitaires de France.

Chapter 13
Is Critical Constructivism Critical Enough? Towards an Agonistic Philosophy of Technology

Alberto Romele

Abstract In this chapter, we discuss the value of Feenberg's critical constructivism for overcoming the limitations of the dominant empirical and ethical approaches in the field of philosophy of technology. In the first section, we show the advantages of critical constructivism. From an ontological point of view, it suggests that technologies are always more than the sum of their material parts. In fact, technologies are entangled with specific forms of life and worldviews. From an ethical-political perspective, critical constructivism suggests that these forms of life or worldviews are often crystallizations of forms of domination. In the second section, we discuss the limitations of critical constructivism, which lie not so much in its theoretical elements as in its practical propositions. In particular, we discuss the residue of Habermasian rationalism in the way Feenberg proposes to implement technological democracy. In the third section, we proposed two "exit strategies," namely, Bourdieu's sociology and Mouffe's agonistic approach. The first has the merit of renouncing any form of rationality in the behaviors of social groups; however, he recovers it, in a scientist and elitist manner, from the side of the social scientist. The second has the merit of making the struggle between social groups and classes a real resource for democracy. It is precisely this resource that we propose to apply to the field of the philosophy of technology.

Keywords Critical constructivism · Andrew Feenberg · Technology · Jacques Rancière · Jurgen Habermas · Democratization · Pierre Bourdieu · Chantal Mouffe · Rationality · Actor-network theory · Postphenomenology

A. Romele (✉)
IZEW (International Center for Ethics in the Sciences and Humanities),
University of Tübingen, Tübingen, Germany
e-mail: alberto.romele@uni-tuebingen.de

D. Cressman (ed.), *The Necessity of Critique*, Philosophy of Engineering and Technology 41, https://doi.org/10.1007/978-3-031-07877-4_13

13.1 Introduction

This chapter builds on two hypotheses. The first one, discussed in the first section, is that Andrew Feenberg's critical constructivism represents a valid alternative to the empirical and ethically oriented attitude that dominates the field of philosophy of technology nowadays. Philosophy of technology needs to go beyond the empirical attitude that has characterized it in recent years. Yet, we think that the overcoming of this attitude cannot happen through an "ethical turn," but through a political turn. According to Jacques Rancière, ethics has as its ultimate goal consensus. And consensus, says the French philosopher, "properly means a mode of symbolic structuring of the community that excludes what makes up the heart of politics, namely dissensus" (Rancière, 2009, p. 115). Put differently, ethics tends to cancel differences, not by indifference (this would in fact be an attitude not only un-ethical, but anti-ethical) but by absorption, inclusion, acceptance, and acceptability. Much of the work in ethics of technology consists in fact not in problematizing technology as such, but in making it acceptable, either by means of rhetorical actions, or by means of design actions. In both cases, however, what is excluded is the radical possibility of critique and dissensus, that which can come to trouble the very meaning of the technological process of innovation. Feenberg's critical constructivism has this double merit. From an ontological point of view, he defends the idea that technologies are always more than the sum of their material parts. From an ethical-political point of view, he suggests that technologies, as well as technological mediations or assemblages of human and non-human actors (to borrow, respectively, the language of postphenomenology and Actor-Network Theory [ANT]) are always taken within social dynamics of power, which most of the time are visible in terms of domination and exclusion.

Our second hypothesis, discussed in the second section, is that Feenberg's critical constructivism is nevertheless not critical enough. Although excluded from its theoretical premises, consensus is re-admitted in the practical applications of critical constructivism. We propose to translate this hypothesis in the following way: profoundly anti-Habermasian (Habermas, we will see, is the philosopher of consensus par excellence, and the philosopher par excellence on whom an ethics of the technique and technology of consensus has been built) in his theoretical premises, Feenberg turns out to be Habermasian in the practical outcomes of his thought. In particular, we are referring to the way he has proposed to put into practice the idea of "democratizing technology" (Feenberg, 2001). For him, the democratization of technology depends on re-distributing power between "lay and expert, object and subject of technological action" (p. 193). We argue that this is the continuation of the Habermasian ideal of transparent communication among equal and rational social actors or groups with other means.

In order to overcome this residue of Habermasian communicative rationality, in the third section we propose two "exit strategies." The first is that, already suggested elsewhere (Romele, 2020), of a re-appropriation of Pierre Bourdieu's sociology within the philosophy of technology. The second, in our view even more radical, consists in the use of Chantal Mouffe's (2013) agonistic approach – which not coincidentally closely resembles Rancière's philosophical perspective. Our ultimate goal, of course, is not to eliminate critical constructivism, but, if anything, to complement it by offering a theory of practice that is ultimately coherent with its theoretical intentions.

13.2 The Philosophy of Technology Beyond the Empirical and Ethical Turn

In many ways, we could say that critical constructivism represents a "return to Marcuse," to which elements from science and technology studies (STS), in particular social constructivism of technology (SCOT), are attached. Feenberg has always preferred Marcuse to Habermas. In *The Theory of Communicative Action*, Habermas famously distinguished between three worlds: the objective world of things, the social world of people, and the subjective world of feelings. Each of these worlds has its basic attitude, namely: an objectivating attitude which treats things; a norm-conformative attitude which views them in terms of moral obligation; and an expressive attitude which approaches them emotively (Feenberg, 1996, p. 50).

For Habermas, only the first attitude has been developed in capitalistic societies. Moreover, this attitude has expanded beyond its proper world and colonized both the social world of people and the subjective world of feelings. Habermas' efforts consist in criticizing this colonization and promoting an attitude proper to the social world of people in terms of communicative rationality. In separating between two worlds, realms, or spheres (the subjective sphere of sentiments is not considered in this context), he has also recognized their autonomy from each other. This is particularly helpful when it comes to defending the social world from the colonization of scientific-technical rationality and proposing a rationality proper to the social world; it becomes completely ineffective when one wants to problematize the way the scientific-technical rationality operates in what is supposed to be its proper domain, namely the objective world of things. The ideal-typical distinction between two worlds and two rationalities brings to neglect the continuous overlapping between them, at least from the side of the scientific-technological rationality.

Marcuse represents a valid alternative to solve this issue. Indeed, for him, the idea of the neutrality of technology is an ideological illusion. In the article "Industrialization and Capitalism in the Work of Max Weber" (1991 [1965]) he admits that technical principles are formulated in abstraction from any content, interest, or ideology. He has also highlighted, however, that these are just abstractions, and that as soon as they enter reality, these principles take on a socially and historically specific content (Feenberg, 1996, p. 51). In other words, scientific-technological rationality is not neutral at all. This might sound like bad news, because the social and historical content of rationality in which the abstractions of technical principles are concretized are those of class society. However, this is potentially good news, because it means that another rationality, or another way of understanding and practicing the same scientific-technological reason, can be put into place.

Feenberg has not limited himself to re-habilitating Marcuse. He has also proposed to go a step forward, a step that has implied the insertion of reflections borrowed from STS. There is indeed in Marcuse's perspective a certain amount of romanticism, but also ignorance towards concrete technologies. In his reflections on STS, Feenberg mobilizes both ANT and SCOT, generally favoring the latter. The problem with ANT, at least in its Latourian version, is indeed that to give full credit to technologies, it has partially emptied humans of their intentions. Latour's ANT reduces humans and nonhumans to their actualities. This flat perspective is clearly in contradiction with Marcuse's bi-dimensional anthropology. In SCOT, another principle of symmetry applies, which is not between humans and nonhumans, but rather among concerned social groups, that is, social groups that for some reason are or should be involved in a specific process of technological innovation.

The assumption of this first section is that critical constructivism represents a valid alternative to the empirical and ethically oriented attitudes that dominate the field of the philosophy of technology. Let us consider postphenomenology, one of the most influential research programs in the field. Postphenomenology tends to be as "flat" and descriptive as ANT, perhaps even more so. Indeed, while ANT is engaged in descriptions of the multiple and two-dimensional relations among humans and nonhumans, postphenomenology has often limited its interest to the linear and one-dimensional I-technology-world relations. Don Ihde (1990) has famously distinguished among four kinds of these relations (embodied, hermeneutic, alterity, and background), a list that has been considerably expanded by scholars since. In this sense, we might say that Feenberg's critical constructivism paves the way to a third dimension to explore, one of the human intentions, expectations, desires, and generally potentialities.

It would be unfair to say that postphenomenology is entirely descriptive. An ethics of technology has become one of the privileged domains of application of postphenomenology, especially after the work of Verbeek (2011). However, there is an important difference between the "surplus of meaning" characterizing Verbeek and other postphenomenologists' ethics of technology, and the one of Feenberg's critical constructivism. While particularly attentive to the ways in which ethics might orient some technical choices, postphenomenologists, of course with some important exceptions,[1] tend to ignore what lies 'upstream' and 'downstream' of these choices. Postphenomenology, like aspects of Husserl's phenomenology, has given too much credit to the gesture of *epoche*. In fact, the *epoche* ends up rendering inoperative any consideration of the conditions of possibility and the modes of existence of a phenomenon such as an object or a technological mediation. In the words of Cressman (2020), "while postphenomenology provides a philosophical framework that encourages careful consideration of the consequences and affordances of living

---

[1] See, in particular, the work of Robert Rosenberger (2017).

with technology, an important aspect of living with technology is to ask *why* we have the technical artifacts and socio-technical relations that we do" (p. 13).

An effective metaphor is that of the game, which is made up of rules and players, but also social and cultural conditions that allow for its existence and success. Some games, such as certain violent sports and games of chance, are accepted and played only to the extent that our societies and cultures accept a certain amount of harm to the individual and the community – physical violence to people or animals, psychological and emotional dependence, economic losses, etc. – which it would be important to be able to discuss to the point of questioning their *raison d'être*.

A more concrete example is that of the current status of the ethics of AI. This field has been for a long time dominated by the quest for general principles – beneficence, non-maleficence, transparency, trustworthiness, explicability, privacy, etc. – that can be implemented either "in the head" of the engineers or in the machines. Jobin et al. (2019) have observed that while among the several reports and guidelines about ethical AI a sort-of understanding about some of these principles is in place and disagreement arises when it comes to contextualizing them and putting them into practice. In general, one could say that there is still a lack of interest in the broader context in which a potentially unbiased, transparent, and trustworthy AI is implemented. These things are, of course, quickly evolving, but interestingly enough it is not much from the side of philosophy of technology, but rather from that of other disciplines like digital sociology and critical media studies that this is happening.[2] Among the many examples available, one can refer to Crawford and Plagen (2020) who analyze the way Image sets for machine learning training such as ImageNet reproduce classic social and cultural biases, such as the marginalization or exclusion of certain social categories or classes. However, one might wonder whether these studies really go far enough, that is, whether they are able to challenge the very reasons why we have such technologies and why our societies increasingly rely on them.[3]

The philosophy of AI and philosophy of technology *tout court* are less in need of an ethical turn than of a political turn. The ethical concern that characterizes many current discussions in the philosophy of technology has made of it a sort of *ancilla technologiae*. In part, this is the responsibility of the tech companies, the "hard" sciences, and all the public and private institutions that profit from the chronic lack of funding in humanities, and philosophy in particular. But this is also our fault, as philosophy scholars that have assumed a naïve positivistic attitude before our colleagues, incidentally in a historical moment in which the "positive" sciences had already started to question their "positive" nature. Ethics of technology is our version of voluntary servitude. In turning away from transcendental perspectives like

---

[2] Again, there are of course exceptions in the philosophical sphere, such as attempts to apply to AI the reflections that have long been initiated in the field of intercultural information ethics (Capurro, 2004).

[3] Interestingly, many highly regarded critics of AI work for companies like Google and Microsoft, leading to the paradox that those who have the means to fund critical research are those that can most use these insights, but will probably ignore it.

that of Heidegger and Marcuse, the philosophy of technology has also turned away from the possibility of engaging in "thick descriptions" of the conditions of possibility, be they social, political, metaphysical, technical, etc., in which technological artifacts and mediations are always entangled.

Finally, the ethics of technology has, if not as an aim, at least as a result, that of favoring consensus towards a technical innovation that can only be perceived according to a TINA (There is no Alternative) logic. This is what Rancière (1999) says about consensus democracy:

> According to the reigning idyll, consensus democracy is a reasonable agreement between individuals and social groups who have understood that knowing what is possible and negotiating between partners is a way for each party to obtain the optimal share that the objective givens of the situation allow them to hope for and which is preferable to conflict. But for parties to opt for discussion rather than a fight, they must first exist as parties who then have to choose between two ways of obtaining their share (p. 102).

What happens, however, to all those groups or social classes that exclude themselves or are in principle excluded from the possibility of participation in dialogue and reason? This question is even more urgent when technological innovation is at stake, because it is precisely with the excuse that technological innovation has a reason of its own that a large part of the concerned groups is excluded from these processes. Feenberg's critical constructivism represents a valid alternative. Feenberg himself has demonstrated that he is conscious of the differences between his work and postphenomenology. For instance, he has highlighted that despite several resemblances with his work, Ihde's postphenomenology has focused on individuals rather than on society. Moreover, postphenomenology has excessively concentrated on technological mediations and their role in approaching the world rather than on the social, economic, and political conditions of possibility of these relations: "[i]nstruments make modern science possible and influence our interpretation of nature, even our interpretation of our own sense experience. [...] But unfortunately, our science-influenced perceptual culture has also been influenced by commercialism and masculinist ideology" (Feenberg, 2015, Kindle Edition). This is the reason why in a more recent occasion, he has said: "were it [critical constructivism] to be schematized as a human-technology-world relation it would look like this: Humans $\rightarrow$ (world-technology)" (Feenberg, 2020, p. 29).

13.3   A Critique of Critical Constructivism

As noted above, our second assumption is that Feenberg's critical constuctivism is not critical enough. In this context, the term "critique" must be understood as the possibility of radically questioning any preconception that might be a source of domination or exclusion. To be precise, our idea is that Feenberg's critical constructivism contains a residue of Habermasian trust in communicative reason, not so much in its theoretical assumptions, but in its practical applications, in particular, the practice of "democratizing technology." Certainly, such an idea is *per se* coherent with the theoretical premises of critical constructivism. According to Feenberg (2001, p. 179), "the worst problem with the social contract is [...] theoretical. The contract is an imaginary agreement between individuals abstracted from all concrete social relations." One cannot but think of Habermas' transparent communication – which is not only an ideal but also a condition for democratic deliberation – as an heir of the social contractualism. Rather, the problem concerns the way Feenberg has proposed ways to apply these theoretical assumptions.

In Feenberg (1999), he presents three sorts of practices that might lead, if opportunely promoted, to a democratization of technology: controversies, innovative dialogue, and creative appropriations. Let us focus on creative appropriations and, in particular, on one of Feenberg's favorite examples, the French Minitel. By distributing terminals that could access a nationwide electronic directory of telephone and address information, the French national telephone and post company (PPT) hoped to increase the use of the country's 23 million phone lines and reduce the costs of printing phone books and employing directory assistant personnel. According to Feenberg (1999, p. 126) the Minitel quickly underwent a reconfiguration, since the users started to resort to it "for anonymous on-line chatting with other users in the search for amusement, companionship, and sex. Users 'hacked' the network in which they were inserted and altered its functioning introducing human communication where only the centralized distribution of information had been planned." We contend that there is a naïve optimism here.

First, it should be considered that there are several appropriations of techniques and technologies by consumers that are certainly not for the better. Think of the appropriation of Fentanyl patches in the US. Second, it should be considered that if users and consumers are not passive before the designed uses and ways of consumption, producers are also not passive before the multiple reappropriations. Sometimes there is a sort of dialogue between these two macro-subjects, but more often strategists have the means to absorb tactics.[4] For Boltanski and Chiapello (1999), the new spirit of capitalism has consisted, from the mid-1970s on, in abandoning the hierarchical Fordist work structure and developing a new network-based form of organization founded on employee initiative and autonomy in the workplace. In other words, strategists have absorbed the demands and tactics of those who criticized them and have thus transformed that criticism into an additional force. Think of Facebook, which in 2014, after much criticism, launched fifty gender options for the users – which allowed it to not only to make peace with the LGBT+ community but also to collect more basic data on the users of the platform.

The Minitel case itself is less a case of "hacking," as Feenberg has put it, than of negotiation and absorption. When users' behaviors began to differ from those initially thought of by the French telephone and post company, the company began to adapt to them to eventually make them economically beneficial. The Minitel soon became an infrastructure for other companies to provide service – from which the

---

[4] On the distinction between strategies and tactics, see De Certeau (1984). Feenberg refers to De Certeau in Feenberg (2002, pp. 83–85).

company made a constant profit. Amusement, companionship, sex, and many other things became paid services, just as we now pay, often with our data, for friendship on Facebook, public recognition on Twitter, and love and sex on Tinder. Consider, for example, that *messageries roses* ("pink messages," adult chat services hosted by operators purporting to be receptive women) were very popular. Widespread street advertising marketed services such as "3615 Sextel," "Jane," "kiss," "3615 penthouse," and "men," embarrassing government officials who preferred to discuss growing business usage of messaging.[5]

Feenberg believes that citizens' tactics can pave the way to broader participation in technology and that more participation would bring more democracy. But public participation in technology is not in itself a guarantee of democratization. Nothing omits that the same logic of exclusion that often characterizes, for instance, the technical personnel, becomes the common practice of the formerly excluded groups. Feenberg is aware of this problem and has resorted to "collegiality," a concept he has borrowed from Habermas, in a context in which he has also espoused Habermas' idea of participatory administration: "Refined and generalized, collegiality might be part of a strategy for reducing the operational autonomy of management and creating systematic openings for democratic rationalizations. The recovery of collegial forms would be a significant step toward democratizing modern technically based societies" (Feenberg, 1999, p. 145). We argue that this collegiality implies the same belief in transparent communication among rational beings that characterizes Habermas' perspective. "Collegiality" derives from the Latin *collegium*, which means "a partnership among equals sharing power." Habermas, who was expulsed from Feenberg's theoretical reflections, has been somehow readmitted in his practical conclusions.

On another occasion, Feenberg (1992) has spoken of "subversive rationalization." In our societies, he says, rationalization is strictly related to a specific definition of technology as a means to the goal of profit and power. However, "[a] broader understanding of technology suggests a very different notion of rationalization, based on responsibility for the human and natural context of technical action" (p. 320). For sure, this a very different form of rationalization of the technological milieu, but still is a form of rationality. A rationality in which every concerned group listens to the others and is listened to by the others. A rationality in which every concerned group sacrifice or at least partially renounces to its interests before

---

[5] In Feenberg (2017, pp. 89–111), he speaks of the Internet and arrives at similar optimistic conclusions. For him, the future of the Internet depends on which actors determine its technical code. In particular, he has opposed two models of the Internet, the community and the business-oriented, and praised for the former over the latter. However, Feenberg seems in this way to forget, first, the capacity, already largely in place, of the business-oriented Internet to profit from the communities online and their interests. Social interactions are constantly transformed into data structures, and these data structures represent the base of the business-oriented Internet. The more social relations, the more money for the big tech companies. Second, and most importantly, there is no evidence that an Internet of the communities (by the way, all communities, one community, some communities, which communities and communities' rules? Such questioning is not without consequences) would be necessarily better than the Internet we have today.

those of the others, and in which all the concerned groups renounce to their interests before those of nature. In this respect, one should not forget that the notion of responsibility has to do with "responsiveness," and that being "responsive" implies both the capacity and the will to listen, to welcome one's interests and intentions, to make room for them. Rationality, be it subversive or not, implies a certain form of transparency. "Subversive rationality" is a contradictory concept that implies, first, a confrontation among well-intended groups, that is, a certain dose of rational politeness; second, it suggests that any form of conflict represents just a historical phase towards a rationalized technological action, in which dominants and dominee are equally involved in the name of superior rationality, namely the one that should orient our behaviors towards nature.

13.4   Philosophy of Technology: An Agonistic Approach

The question now arises if there is a way to both maintain the theoretical and ontological advantages of Feenberg's critical constructivism while avoiding the residuum of a naïve communicative rationality that haunts his practical conclusions. In this third section, we propose two "exit strategies."

The first strategy consists in importing Pierre Bourdieu's social theory to the philosophy of technology. Like critical constructivism, a Bourdieusian philosophy of technology refuses to reduce technology to its most immediate and material parts. In particular, a Bourdieusian philosophy of technology is concerned by the social conditions of possibility in which technological artifacts and mediations are entangled both in their design and possible uses. We have fully developed this point elsewhere (Romele, 2020).

The second interesting aspect of a Bourdieusian philosophy of technology is that it refuses to rely on any form of naïve rationalization. Like Feenberg, Bourdieu has denounced the fallacy of social contractualism. And yet, the way Bourdieu has renounced the rationalization of social reality and its dynamics is, in our opinion, more radical and more critical than Feenberg's critical constructivism. Bourdieu has heavily criticized Habermas in the *Pascalian Meditations* (2000). For him, Habermas' theoretical universalization has brought to fictitious universalism that forgets the social and economic conditions of possibility to access the universal. In particular, Habermas has neglected the social and economic conditions of possibility that must be fulfilled "in order to allow the public deliberation capable of leading to a rational consensus" (p. 65).

On the same page, Bourdieu asks himself: "How indeed can it be ignored that [...] cognitive interests are rooted in strategic or instrumental interests, that the force of arguments counts for little against the argument of force [...], and that domination is never absent from social relations of communications?" Several arguments are contained in this sentence: (1) First, the idea that language, and hence communication, is not autonomous, but always depends on social dynamics of recognition, domination, and exclusion; (2) Second, this means that the force of the argument, for instance when it comes to "democratic" debate to deliberation, does not exclusively rely on the argument itself, but also, and in particular, on the social status of those who advance that specific argument; (3) Third, those who have a stronger argument are probably also those who designed the rules of the argumentation, the accepted way of talking, debating, etc. Inclusion and exclusion from the possibility of participating in debate and deliberation is already embedded in these rules; (4) Fourth, those who are at the margins or excluded from debate and deliberation are also probably those who are less interested in participating; (5) Fifth, there is no reason to believe that the inclusion of the excluded ones would coincide with a more democratic debate and deliberation. It seems more legitimate to suppose that, if the excluded were included, other strategic and instrumental interests would take the upper hand in determining the rules of acceptable debate and deliberation.

Our hypothesis is that if points 1–3 unite Feenberg and Bourdieu, points 4–5 represent the step forward of a Bourdieusian philosophy of technology from critical constructivism. The problem of Habermas is that "he throws the political back onto the terrain of ethics. He reduces political power relations to relations of [ethical] communication" (Bourdieu, 2000, p. 66). His communicative ethics is nothing but the re-actualization of the Kantian principle of the universalization of the moral judgment. These reflections strongly recall what we have said in the first section about the present status of the philosophy of technology that understands itself in terms of empirically oriented ethics of technology.

Feenberg has certainly understood this major issue, and his critical constructivism can be legitimately considered a strong response to it. Bourdieu and Feenberg agree on the fact that there is no contradiction in "fighting *at the same time* against the mystificatory hypocrisy of abstract universalism *and for* universal access to the conditions of access to the universal […]" (Bourdieu, 2000, p. 71). However, they differ in the way that they practically propose to implement such universal access. In Feenberg, this passes through a certain degree of rationalization of social reality, in particular when it comes to communication among concerned groups and deliberation. The moment of subversion exists, but soon or later must be domesticated. In Bourdieu, the universal access to the conditions of access to the universal passes through a total renunciation of rationalizing the social dynamics of prestige, recognition, inclusion, and exclusion. In other words, universal access depends on the universal recognition of the legitimacy of the non-universal interests that orient the intentions and behaviors of the social actors or groups.

From Bourdieu's perspective, we can therefore appreciate above all the unromantic way of describing social reality. Social groups and classes, whether dominant or dominated, are driven by *habitus* and particular interests. The only difference between dominant and dominated is that the former has managed to impose their worldview on the latter.

Yet, we realize today that there is a non-negligible limitation in the Bourdieusian perspective. At the core of the previous reflections has been the idea that while Feenberg has, in the end, rationalized society, Bourdieu has rather accepted its interests-oriented nature. This is however only partially true. Indeed, while for

Bourdieu social reality is not rational, it can be rationalized. The rationalization of the irrational society is delegated to the scientific work of sociologists – see, in particular, Bourdieu (2004), where he discusses the capacity for self-reflexivity amongst social scientists. Bourdieu's sociology thus turns out to be scientistic and, one could say, even elitist: that rationality that is refused to social actors is instead attributed to the social scientist. The social scientist would then make a gift (in the sense of *kharis* and *gratia*) of rational analyses to social groups and institutions. This perspective is then not very different from the reliance on engineers and scientists that characterizes our societies, as well as much contemporary work in the philosophy of technology.

In the rest of this section, we would like to discuss another "exit strategy" that is more radical and therefore more critical. This is the agonistic approach in political philosophy theorized by Chantal Mouffe (2013) that we propose importing to the philosophy of technology. It is interesting to note that the starting point of her reflections is a critique of the contractualism and rationalism that characterize many modern theories of political philosophy. Habermas is, for that matter, like for Bourdieu and Feenberg, one of Mouffe's favorite targets. However, her basic idea consists not only in the acceptance of the interest-oriented nature of society but also in her positive view of it. In Bourdieu's perspective, society is a battlefield from which only the sociologist, precisely by abstaining or making an abstraction from it, can emerge victoriously. For Mouffe, this battlefield that is society is not to be abandoned or transcended, but to be accepted and promoted.

What Mouffe accepts is the conflictual nature of social reality. It is a matter for her to make of this nature (or at least to see in it) not a destructive but rather a creative element. This naturally has a cost, which consists of being able to see and eventually transform antagonism into agonism. There is a difference between antagonism, which does not recognize the other and whose purpose is precisely to annihilate the other, and agonism, in which the other is a contender who enjoys the same rights to participate in the competition or game. Agonistic democracy is one in which these rights are accorded to all contenders, whatever their methods and approaches to the contending matter are. The only rule of the game is to recognize the other as a legitimate contender. The rationality of the social is certainly not eliminated. After all, agreeing on the legitimacy of the dispute, that is, on the agonistic and not the antagonistic nature of participation, already implies postulating a horizon of comprehensibility. Yet, this horizon is minimized and so the room for critique is maximized. We could also say that Mouffe touches, in this way, the limit of all possible critique, that also in its radicality must not be criticism at all costs, otherwise it risks to fall in the pure negativity or, still worse, in mere neutrality.[6]

---

[6] On this point, see Esposito (2018), who distinguishes (1) the Negativity of German Philosophy, in particular the Frankfurt School; (2) the Neutrality of the French Theory, in particular deconstruction, and (3) the Affirmativeness of the Italian Thought, in particular a tradition that goes from Machiavelli to Negri and Agamben. There is no room of a detailed reflection on this here, but it can be said that Mouffe is theoretically close to (3).

   The democracy of Mouffe is not a poor democracy, but rather one that is much richer than that proposed in theories that end up, willingly or unwillingly, domesticating and therefore submitting its participants to the rationality of the dominant participants. Of particular interest is how Mouffe's theories have already been applied to the fields of science and technology. For example, Rey (2017) has criticized the deliberative reading of scientific controversies, resorting to Mouffe's antagonism. According to her, "the deliberative analysis [dominant] model [...], should be replaced by a model which would allow [scientific] controversies to be conceived of beyond the scope of an inevitable collapse back into rational consensus" (p. 47). This would pave the way to an "'adversarial-model' in which pluralism supposes the legitimate coexistence of divergences without presupposing their resolution, while framing these divergences within a common symbolic space where their conflict is played out" (Ibid.).

   In the field of technology, a similar position has recently been proposed by Popa et al. (2021). According to them, the dominant approaches to the issue of technological conflict are oriented towards establishing (or re-establishing) consensus, either in the form of a resolution of the conflict or in the form of an "agree-to-disagree" standstill between the stakeholders. The authors distinguish between two dominant perspectives. The first is the conciliatory approach, which sees conflict as a danger and even a disease or deviant activity. The second is the constructive approach. The merit of this approach is that it considers conflict as necessary. Yet, according to the authors, the practices set up by those who follow this approach often seem to be mere theoretical exercises – for example listening to the narratives of the other – without any guarantee that these exercises will have any effect in practice – a change in the actions. In response to these two approaches, Popa et al. (2021, np.) propose importing the agonistic approach in the debate on technological conflicts:

> From an agonistic perspective, conflict is to be sought and agonistic respect must take the form of being responsive to the other party's ethical demand. Concretely, this will mean that Greenpeace and Shell [this is one of the examples they resort to in the article] must seek conflict with each other not because of the possibility of an ideal agreement, or a comfortable 'agree-to-disagree' standstill, but rather to continually restate and reinterpret each other's ethical demands and translate these demands into responsive behavior. A resolution or a standstill, in agonistic thinking at least, would be bad news for both organizations.

The agonistic approach in philosophy of technology certainly poses some problems with regard to the *description* of the dynamics of technological innovation. For example, one could ask if the agonism in the field of technological innovation, where the clash is often between big companies and small concerned groups – sometimes supported by national and international institutions – is not in fact a disguised form of antagonism. This antagonism would be similar to a slow and positional war in which the final winner is the one who manages to exhaust the other's resources – and we know well that the resources of some large companies are practically infinite. One might also wonder, in the context of technological innovation, how long agonism can really last. After all, is it not right at the heart of technological innovation that something must finally be created and produced? And is it not true that this product can satisfy some but will inevitably end up excluding others?

However, the agonistic approach may prove fruitful in *prescriptions* related to technological innovation. If we take the first example above, we could say that an agonistic perspective (an agonistic technological democracy, to join Feenberg and Mouffe) would consist in the effort to guarantee equal resources to all contenders. If we take the second example, we could say that an agonistic perspective would consist in ensuring that technological innovation processes are represented not by one, but by multiple cycles. These are elements that we find in embryo in critical constructivism already but that, in our opinion, are not translated into a theory of practice.

13.5 Conclusion

In this chapter, we discussed the value of Feenberg's critical constructivism for overcoming the limitations of the dominant empirical and ethical approaches in the field of philosophy of technology. In the first section, we have shown the advantages of critical constructivism. From an ontological point of view, it suggests that technologies are always more than the sum of their material parts. In fact, technologies are entangled with (and help reconfigure) specific forms of life and worldviews. From an ethical-political perspective, critical constructivism suggests that these forms of life or worldviews are often crystallizations of forms of domination. In the second section, we discussed the limitations of critical constructivism, which lie not so much in its theoretical elements as in its practical propositions. In particular, we discussed the residue of Habermasian rationalism in the way Feenberg proposes to implement technological democracy. In the third section, we proposed two "exit strategies," namely, Bourdieu's sociology and Mouffe's agonistic approach. The first has the merit of renouncing any form of rationality in the behaviors of social groups; however, he recovers it, in a scientist and elitist manner, from the side of the social scientist. The second has the merit of making the struggle between social groups and classes a real resource for democracy – on the condition, minimal but necessary, that antagonism be transformed into agonism. It is precisely this resource that we propose to apply to the field of the philosophy of technology. The philosophies of technology of the empirical turn want to be extremely concrete; yet, the way their ethical applications are mostly based on rationalizing perspectives à la Habermas, they prove to be paradoxically abstract. The opposite is true of an agonistic approach that, however abstract, or extreme and radical, it may seem, has as its starting point the concreteness of social reality, which is not always as "smooth" as reason. In this context, we are not yet able to go beyond a mere theory of the antagonistic practice. We can however, in conclusion, suggest a brief example.

Take the case of AI ethics. We have said that consensus is, in the end, the disappearance of politics in favor of what Rancière (1999) calls the police, "an order of bodies that defines the allocation of ways of doing, ways of being, and ways of saying [...]; it is an order of the visible and sayable that sees that a particular activity is visible and another is not, that this speech is understood as discourse, and another is noise" (p. 29). The police establish and maintain a *status quo* that for Rancière has to do with the "distribution of the sensible," that is, with the allocation of spaces and places, the attribution (or interdiction) of the right to speak, to see and to be seen. Politics is the radical possibility of subverting this *status quo*. The police annihilate any possibility of disagreement, and, with it, one could say of true agonism. Not only the classical universalism in AI ethics, but also current efforts to include marginalized individuals in the processes of innovation can be understood as forms of domestication of the true agonism. Indeed, they mostly lead to an account, but with no definite results, of the voices being heard, or to an absorption of the critique itself into a process of technological innovation that is not critiqued as such. An agonistic ethics of AI would then be a perspective in which the purpose of AI, or at least its horizon, would be neither that of a general consensus on norms and values, nor that of an equally consensual promotion of certain virtues. Its purpose would be to promote, rather than anesthetize, agonism. An agonistic AI politics (Coeckelbergh 2022) would be more radical, however, insofar as it the very processes of AI-related technological innovation on a global scale would be openly questioned.

References

Boltanski, L., & Chiapello, E. (1999). *The new spirit of capitalism*. Verso.

Bourdieu, P. (2000). *Pascalian Meditations*. Stanford University Press.

Bourdieu, P. (2004). *Science of science and reflexivity*. The University of Chicago Press.

Capurro, R. (2004). *Intercultural information ethics*. http://www.capurro.de/iie.html. Accessed May 1st 2021.

Coeckelbergh, M. (2022). The political philosophy of AI: An introduction. Polity.

Crawford, K., & Plagen, T. (2020). *Excavating AI: The politics of images in machine learning training sets*. https://excavating.ai/. Accessed May 1st 2021.

Cressman, D. (2020). Contingency and potential: Reconsidering a dialectical philosophy of technology. *Techné: Research in Philosophy and Technology, 24*(1), 138–158.

De Certeau, M. (1984). *The practice of everyday life*. University of California Press.

Esposito, R. (2018). *A philosophy for Europe: From the outside*. Polity.

Feenberg, A. (1992). Subversive rationalization: Technology, power, and democracy. *Inquiry: An Interdisciplinary Journal of Philosophy, 35*, 301–322.

Feenberg, A. (1996). Marcuse or Habermas: Two critiques of technology. *Inquiry: An Interdisciplinary Journal of Philosophy, 39*(1), 45–70.

Feenberg, A. (1999). *Questioning technology*. Routledge.

Feenberg, A. (2001). Democratizing technology: Interests, codes and rights. *The Journal of Ethics, 5*(2), 177–195.

Feenberg, A. (2002). *Transforming technology: A critical theory revised*. Oxford University Press.

Feenberg, A. (2015). Making the gestalt switch. In R. Rosenberger & P.-P. Verbeek (Eds.), *Postphenomenological investigations: Essays on human-technology relations* (Kindle ed.). Rowman & Littlefield.

Feenberg, A. (2017). *Technosystem: The social life of reason*. Harvard University Press.

Feenberg, A. (2020). Critical constructivism, postphenomenology, and the politics of technology. *Techné: Research in Philosophy and Technology, 24*(1), 27–40.

Ihde, D. (1990). *Technology and the lifeworld: From garden to earth*. Indiana University Press.

Jobin, A., Ienca, M., & Vayena, E. (2019). Artificial intelligence: The global landscape of ethics guidelines. *Nature Machine Intelligence, 9*(2), 389–399.

Marcuse, H. (1991). Industrialization and capitalism in the work of Max Weber. In P. Hamilton (Ed.), *Max Weber: Critical assessment 2* (pp. 123–136). Routledge.

Mouffe, C. (2013). *Agonistics: Thinking the world politically*. Verso.

Popa, E. O., Blok, V., & Wessenlik, R. (2021). An agonistic approach to technological conflict. *Philosophy & Technology, 34*, 717–737.

Rancière, J. (1999). *Disagreement: Politics and philosophy*. Minnesota University Press.

Rancière, J. (2009). *Aesthetics and its discontents*. Polity.

Rey, A.-L. (2017). Agonistic and epistemic pluralism: A new interpretation of the dispute between Emilie du Châtelet and Dortous de Mairan. *Paragraph, 40*(1), 43–60.

Romele, A. (2020). Technological capital: Bourdieu, postphenomenology, and the philosophy of technology beyond the empirical turn. *Philosophy & Technology, 34*, 483–505.

Rosenberger, R. (2017). *Callous objects: Design against the homeless*. Minnesota University Press.

Verbeek, P.-P. (2011). *Moralizing technology: Understanding and designing the morality of things*. University of Chicago Press.

Chapter 14
Geschichtlichkeit, Life, and Technicity: From *Heideggerian Marxism* to the Critical Theory of Technology

Taila Picchi

Abstract From a historico-philosophical perspective, this study traces *à rebours* Marcuse's Philosophy of Technology through Feenberg's reading. It focuses on three notions – historicity, life, and technicity – in Marcuse's early writings that enable a genealogy of his thinking about technology. Historicity (*Geschichtlichkeit*) represents the attempt to make compatible historical materialism and existential phenomenology. Life is a pivotal concept for a dialectical thinking which evolves from a dialectics of life of the so-called Marcuse's Heideggerian Marxism to a dialectics of technology in his mature production. Finally, technicity seems to be the key notion of the technological shift by the quitting of the Heideggerian legacy alongside the joining of Frankfurt School. The representation of technology in *One-dimensional man* rearticulates these notions of historicity, life and technicity and shapes a particular kind of objectivity that makes technology a multilayered and ambivalent field of human historical action. Thus, Marcuse's account of technology can be integrated with Simondon's conception of concretization towards a redefinition of reification that allows for an open – although critical – idea of progress, as Feenberg assumes.

Keywords Herbert Marcuse · Critical theory of technology · Ontology · Technology · Historicity · Karl Marx · Martin Heidegger · Heideggerian Marxism · Gilbert Simondon · Andrew Feenberg

T. Picchi (✉)
Terricciola, Pisa, Italy
e-mail: taila.picchi@unifi.it

14.1 Introduction

The definition of technology as multilayered makes it a particular ambivalent object, as we find in one of Andrew Feenberg's main influences, Herbert Marcuse. Marcuse's mature critical theory of technology is rooted in his early works that focus on an ontology of Life, which is, at the same time, a historical ontology. Ontological implications constitute possibilities for historical transformations because ontology and politics are linked through the movement of historicity. The notion of historicity in Marcuse's early writings is pivotal in the rethinking of the relation between essence and existence. At the same time, historicity means rethinking a homogeneous and one-dimensional *Dasein* as human and historical through the perspective of the productive class which represents a contradictory form of existence and introduces the existential difference of human *Dasein*.

Around the notion of historicity and Life it is thus possible to chronicle, *à rebours*, Marcuse's critical theory of technology through Feenberg's reading. According to Feenberg, the ontology of Life in Marcuse's early writings undergoes a progressive translation into the dialectics of technology of *One-dimensional Man* through the passage of Life in *eros*, in *Eros and Civilization*, which expands the pleasure drive as the great force unifying and preserving all life. In other words, we can find in Marcuse's *Heideggerian Marxism* the germinal elements of his critical theory of technology. Like Feenberg, I agree that the idea of Life in Marcuse's early works continues through his later works. In addition to this, I argue that we can notice in Marcuse's production the shift from an ontology of Life based on the notion of historicity to a critical theory of technology based on a notion of technicity that is influenced by the work of Gilbert Simondon to better explain the essence of technique.

Historicity and Life are significant for understanding the essence of technique, which represents the same ontological bi-dimensionality of historicity and Life. For this reason, Marcuse's definition of technology is based on the reform of essence, expanding the Heideggerian definition with a socio-political dimension. Through the notion of technicity Marcuse provides technology with a concrete objectivity different from the objective rationalization and standardization of technocracy. In my account, the reform of the objectivity of technology in a social and political perspective derives from the reform of essence found in Marcuse's early writings. This represents the speculative attempt for reading Marcuse's critical theory of technology through Feenberg's review.

14.2 *Geschichtlichkeit*

Between 1928 and 1933, under the supervision of Martin Heidegger, Marcuse developed his personal interpretation of existential philosophy. This early period is characterized by Marcuse's political goals. Radicalizing Heidegger's concept of being through Marx's historical materialism culminated in Marcuse's first work on Hegel (*Hegel's Ontology and the Theory of Historicity*). Moreover, the publication of Marx's *Economic and Philosophic Manuscripts of 1844* in 1932 played a significant role in the ontological foundation of history upon which Marcuse's anthropology was developed, as outlined in his *Contributions to a Phenomenology of Historical Materialism* (1932).[1]

Marcuse dedicates *Contributions to a Phenomenology of Historical Materialism* to answering the question, "can human Dasein in general be thought of as primarily historical?" (Marcuse, 2005, p. 10). In light of a conciliation between Heidegger's ontology and Marx's materialism, Marcuse's attempt to ground the Heideggerian *Dasein* in history is led by a specification of *Dasein*'s primary Care (*Sorge*) and a differentiation between care as practical-active and the motility of history as historical-constitutive of *Dasein* itself. Historicity (*Geschichtlichkeit*)[2] is the happening of *Dasein* that, according to Marcuse, is a historical happening. Since the material content (*materiale Bestand*) of historicity is "a concrete Dasein in a concrete world" (Marcuse, 2005, p. 24), "the motility of history is the happening of human existence" (Marcuse, 2005, p.33).

The early notion of historicity enables one to think of human *Dasein* as happening. At the same time, the historical understanding of *Dasein* as historicity means the shift from the ontological perspective to the societal one. Indeed, the historical understanding of *Dasein* is linked with the prerogative of a concrete philosophy that Marcuse tries to build up in these early writings and presents as a theory of historicity through his study on Hegel's ontology. According to Alfred Schmidt, the writings of the so-called *Heideggerian Marxism* primarily deal with the question of history since history is the domain where existential ontology collides with materialistic trends.[3] Indeed, in *Contributions to a Phenomenology of Historical Materialism*, Marcuse tries to gather existential ontology and historical materialism into the project of a dialectical phenomenology.

Marcuse takes into account the two determinations of Dasein as being-in-the-world and being-with (*Mitsein*), both of which are based on the notion of historicity. On the one side, human *Dasein* as historical concreteness represents a particular social fact linked with a generic totality; on the other, philosophical thought does

---

[1] The writings between 1928 and 1932 refer to the period called Heideggerian Marxism and are collected in Marcuse's early writings *Heideggerian Marxism*, edited by R. Wolin and J. Abromeit (2005). In this study we will focus mainly on *Beiträge zu einer Phänomenologie des Historischen Materialismus* (1928. *Contributions to a Phenomenology of Historical Materialism*, pp. 1–33); *Über konkrete Philosophie* (1929. *On Concrete Philosophy*, pp. 34–52); *Neue Quellen zur Grundlegung des Historischen Materialismus* (1932. *New Sources ion the Foundation of Historical Materialism*, pp. 86–121).

[2] Marcuse employs the term *Geschichtlichkeit* ("happeningness") deriving from *geschehen* ("to happen") as the term History (*Geschichte*) derives from it as well (and in opposition to *Historie*). This use reveals Marcuse's Heideggerian influence but also the necessity to historicize Heidegger's *Dasein* in the form of human and historical happening, i.e. in a materialistic perspective. See also the glossary of *Heideggerian Marxism* edited by J. Abromeit (Marcuse, 2005, pp. 183–184).

[3] See Schmidt (1968).

not subordinate to Life as the motility of *Dasein* but becomes a kind of knowledge which grasps the movement of being as the motility of history. According to Marcuse, historicity is linked with being-in-the-world and the being-with of *Dasein*. Being-in-the-world implicates the biological determination of species and their generation. Consequently, the being-in-the-world of *Dasein* is influenced by past generations that pass on not only practical-active conduct, but also their historical-constitutive fundament of the provision of life-space. As Marcuse writes, "Dasein's original mode of conduct is practical-active, as production and reproduction, on which the domains of cultural, spiritual, and intellectual are founded" (Marcuse, 2005, p. 27). While Dasein's *Mitsein* (being-with) implicates the unity of historicity in a society,

> Historical society constitutes itself in the mode of production corresponding to its thrown-ness, in the modes in which provisions its life-space in accordance with its existential needs. Only then, when a society truly provisions its life-space in a unified way *as* a society, is it a historical unity, the bearer of historical movement. In the moment when this unity is torn asunder, when the whole society no longer exists in making-provision for its existential needs, and where a *division of labor* is sufficiently advanced that the provision for life-space is no longer regulated through the voluntary act of the whole society, but is rather distributed by means of various coercive measures (such that now the hardest work appears as the activity of the subordinated and inferior) – in this moment the existential needs, too, grow out of and differentiate themselves from this division of labor within a society that was once unified. (Marcuse, 2005, p. 29)

The unity of the world of *Dasein* is fragmented and the unity of society as the historical motility of *Dasein* is broken by the introduction of the capitalist division of labour. The two determinations of *Dasein* presented by Marcuse, to-be-in-the-world and being-with, can be understood as biosocial needs expressed through the primary care of *Dasein*, and the necessity of labour that introduces the fragmentation of the historical social unity through the division of labour. In this early writing, we already find the germinal elaboration of the struggle between the pleasure and the reality principle, since the specification and differentiation of biological and social needs refer to the pleasure drive, while the break of the unity of *Mitsein*, by introducing the division of labour, seems to also introduce a principle of reality that authorizes the specification and differentiation of biological and social needs. Indeed, unified society (as a further interpretation of the myth of the origins, or an ideal type, of the original unity of human societies) means the absence of conflict and the identification of needs and desires: it is a homogeneous ensemble, where each individual produces and reproduces themselves in the same way.[4] The introduction of the division of labour as a historical necessity is also a historical motor because, "the mode

---

[4] This unified society seems to represent a sort-of state of nature incompatible with the historical progress. Thus, in Marcuse's early writings we can recognize the germinal idea expressed through the twofold thesis of *One-Dimensional Man* according to which human progress makes possible the improvement of the conditions of existence but does not produce the historical subject of societal transformation. This twofold thesis is based on the fact that society becomes more and more complex, crossed by conflict and differences, where not only needs but also desires influence and are influenced by the mode of production.

of production of such a society necessarily comes into contradiction with its forms of existence, and the authentically productive class must, on the strength of its sheer existence, break through the reification and sublate the contradiction" (Marcuse, 2005, p. 32).

The division of labour fragments the social unity of historicity and produces a contradictory form of existence, i.e. the productive class. The "mode of production" is the historical determination of the historical *Dasein*, different from the practical-active behavior of Care (*Sorge*), characterized by a historical constitutive movement manifesting itself as production of the environment and *Mitsein* (being-with-others) as society, but a society crossed by contradictory forms of existence. The mode of production, as the ontological and historical unity of society, produces a homogeneous being-in-the-world and being-with that comes into contradiction with forms of existence produced by the mode of production itself. Consequently, the unity of historicity implicates different forms of existence.

> Social production and reproduction are truly the most originary and the ultimate constituent of every historical unity because they, without exception, affect its pure existence, and the essential distinction and characterization of human Dasein can only be derived from an existential difference. (Marcuse, 2005, p. 28)

The primary sense of historicity as care of *Dasein* draws the individual's self-conservation as provision of her or his life-space; nevertheless, the division of labour collapses the social unity of society and leads from historical unity to class conflict. Human *Dasein* differs existentially in its biological and social production and reproduction, but Marcuse is saying that social production and reproduction grounds the existential difference of human *Dasein*. In other words, when individuals come into existence, they come into a determined world, an environment determined by social relations. This world is not given once for all, but can be transformed materially and spiritually, as a world of objects and meaningful entities. Moreover, it is not a unique world, but changes from one individual perspective to another, from the productive class to the exploiting élite.

In this plurality of the historical *Dasein*, where the fragmented unity can be recomposed through the revolutionary act by the form of existence that comes into contradiction with the given world of capitalism, technology constitutes a layer of objectivity. The objectiveness of technology is the topic of Marcuse's mature work, but it is linked *ontologically* with historicity, i.e. the ontological sense of the historical happening. Since history is a mode of being, Marcuse's early ontology of Life – especially in his *Hegel's Ontology and the Theory of Historicity* – is based on this ontological sense of what is historical. Life is always historical even if it is placed in the Hegelian Idea, because, according to Marcuse, the motility of historicity as motility of Life is this very movement of existence taking its own self-consciousness and thus producing knowledge as self-consciousness of the historical subject.

14.3   Dialectics of Life

In *One-dimensional Man*, the dialectic of technology takes into account the sensuous element of life and desire as the utopic possibility for the transformation of bidimensional society, while the early dialectics of Life articulates the bi-dimensionality of being as essence and existence. Through a reform of essence, Life represents both a transcendental subject and the negativity of existence. Thus, it translates the movement of historicity in a historical dialectics. Life similarly to historicity moves from the ontological perspective to the societal and political one.

Since his early writings, Marcuse has attempted to combine historical materialism and Heideggerian phenomenology. As he wrote in *Über konkrete Philosophie* (1929), "Dasein does not 'make' history as its product, it does not live in history as if history were its more or less coincidental space or element; rather, the concrete existing of *Dasein* 'is' happening [*geschehen*] that is understood as 'history' [*Geschichte*]" (Marcuse, 2005, p. 38). The need for a concrete philosophy and the materialistic conception of existence characterizes Marcuse's *Bildung* with Heidegger. Within the project of a dialectical phenomenology as a concrete philosophy, Life, as fragmented unity and permanent totality is a dialectical concept. Whether we call it human *Dasein* or Life, it is always the negative moment of the historical movement, the moment where the break of the historical and social unity – as we read in *Contributions to a Phenomenology of Historical Materialism* – or the fragmented unity of self-consciousness (*Selbstbewusstsein*) – as we read in *Hegel's Ontology and the Theory of Historicity* – must be dialectically recomposed.

The dialectics of Life in the early writings allows for the articulation of Life as the self-conscious subject of its own development with the historicity of Life as historical and socially effective reality. As Marcuse shows in *Hegel's Ontology*, essence presupposes the negativity of existence and Life is the concept of a negative object. In other words, Life is both a transcendental subject and an empirical object, it is the essence of the historical motility (*Bewegtheit*)[5] but, at the same time, is the contradictory existence of the historical subject. Fragmented unity between essence and existence, universal medium and liquid substance, bi-dimensionality applied to Life makes it an empirical concept as well as transcendental one. As Feenberg reminds us,

> Marcuse is famous for having written a book entitled *One-Dimensional Man*. What is less well known is that the notion of dimensions plays a role in this early Hegel thesis where it signifies the relation of essence to immediate or determinate being. [...] The two dimensions, essence and existence, coexist necessarily in the self-manifesting of essence in existence. They cannot be thought independently of each other and their unity is the motility of being. (Feenberg, 2005, p. 59)

---

[5] For the notion of motility (*Bewegtheit*) see the glossary in *Heideggerian Marxism* (Marcuse, 2005, pp. 177–178). As the notion of *Geschichtlichkeit*, it represents Marcuse's appropriation of Heidegger's terminology. For the relation between motility and historicity see also Pippin's "Marcuse on Hegel and historicity" (Pippin, 1988, p. 73).

Thus, according to Feenberg, bi-dimensionality means the coexistence of the historical-material dimension and the ontological-transcendental one.

Feenberg argues that Marcuse "turned to Hegel's concept of life for a radically future-oriented ontology that owes as much to Heidegger as to Hegel" (Feenberg, 2005, p. 18). Life belongs to the negative and the first definition of bi-dimensionality appears precisely in *Hegel's Ontology*, even if it presents some differences with the definition found in *One-Dimensional Man*. Essence and existence are two dimensions of being, of the motility of being as historicity. The ambivalence of Life as object and concept, existence and essence, grounds the ontology of Life as a theory of historicity.

> Marcuse argues that the earlier concept of life can support a true "fundamental ontology" oriented toward history. This is because life implies world, world in Heidegger's sense. The unity of the living thing and its world can be conceived phenomenologically as an ultimate context embracing all levels of being. Even the nature of the natural sciences can be founded in the world, if not reduced to it. All forms of knowledge can take their place within life. Cognition understood as a life function necessarily involves a people and a history. Thus, historicity lies at the center of the theory so understood. (Feenberg, 2005, p. 64)

Marcuse's dialectics of life is a historicization of human *Dasein* through Dilthey's *Lebensphilosophie* and Hegel's idea of life as "a process of movement, negating and accommodating an environment" (Feenberg, 2005, p. 19). The fundamental ontology Marcuse presents concerns reforming the notion of essence through historicity. His idea of a concrete philosophy means precisely the historicization of the concept of Life, since Life is not only a transcendental category but also an empirical one, linked with human existence and its contingencies. Concerning the reform of the notion of essence, Feenberg explains that "essence is both the internal relation of the thing in itself to its determinations, and the external relation of the thing to the other things with which it necessarily coexists. Essence is relatedness and the development that proceeds out of relatedness" (Feenberg, 2005, p. 61).

Life is a dialectical concept in constant contradiction between broken unity and permanent totality.[6] As Pippin writes, "what it means to consider this recollected essence of a totality as a historical notion depends for its full defense on Marcuse's insistence that the essence of some historical form of life can be understood as the interconnected unity of a whole series of historical appearances, that a culture's religion, mode of production, philosophy, law, etc., can be accounted for as a whole, each illuminating what the other is, by all being aspects of human agency, and that this whole, and not the fragmentary self-understanding of any part, is what this form of life essentially is" (Pippin, 1988, p. 79). Marcuse's reform of essence through the

---

[6] "The choice of life as a fundamental ontological theme makes sense of the emphasis on interconnectedness and process in the dialectics of development. The life process has a direction: life seeks to preserve and further itself. Yet it is not confined by a predetermined end but invents its future as it moves. This is of course eminently true of modern human beings and their society. There is no longer a prior essence that defines what it is to be human. Human beings now must make themselves. In Marcuse's reading, Hegelian life, like Heideggerian Dasein, discovers its meaning ahead of itself as a conditioned choice, an appropriation, not behind as a determining cause. It is negative, not positive" (Feenberg, 2005, p. 19).

concept of Life occurs through the negativity of existence and the possibility of societal and political transformations led by this contradictory form of existence. As Feenberg argues, Hegel's dialectic has two patterns: "in one pattern a synthesis leaves behind its contradictory origin by creating a third entity. [...] But there is another pattern in which a continuing mediation between contradictory entities transforms them without dissolving them into a third. This pattern is exemplified in Hegel's notion of 'being at home in the other as such.' It is the pattern of life, which objectifies itself in its environment, transforming object and subject without abolishing their difference" (Feenberg, 2014, p. 123). Following this second pattern, according to Marcuse, the paradigm of activity is grounded in the concept of Life.

> We attain thereby a new and deeper determination of *Being* itself, which emerges out of the full disclosure of the dimension of "having-been" *(Gewesenheit)*, namely, "existence" as an essential being, proceeding from essence. [...] Here we want to add only that Hegel characterizes the motility of essence throughout as "deed" *(Tun)*, as "activity" *(Tatigkeit)*. This provides two extremely important hints: first, "deed" and "activity" signify an increased intensity of the motility of beings; in fact, this intensity leads toward the character of beings qua subject. Activity is a form of self-contained, self-incited, and self-relating movement. It is not an unmediated and flowing process, as in the sphere of being-there, but a mediated movement, reflected-into-self and remaining-by-itself. Second, it is no accident that with the expressions "deed" and "activity" one hears the Greek *poiesis,* as an ontological category which defines Being as a product, as fabricated, and as "prepared." This certainly does not imply something produced by another, being as prepared by humans; it means rather that Being is produced by and through itself. (Marcuse, 1987 [1932], p. 79)

The determination of being as activity is first, the motility of historicity, and second, *poiesis*, i.e. a technical activity. Here, Marcuse links the Hegelian notion of Life with Marx's concept of alienation that he finds in *Economic and Philosophic Manuscripts of 1844*. Indeed, in the *New Sources on the Foundation of Historical Materialism*, he presents the definition of work activity as an ontological category. Thus, within the concept of Life, Marcuse conceives of the historical production and reproduction of human *Dasein* as a historical subject. As Marcuse argues, *Economic and Philosophic Manuscripts of 1844* provides the "economic-philosophical basis of the theory of revolution through a quite particular, philosophical interpretation of human essence *(Wesen)* and its historical realization" (Marcuse, 2005, p. 87). In the frame of the question of the *Gattungswesen* (species-being), Marcuse analyses the conflict between human essence and work activity, or the fact that "objectification can become reification and that externalization can become alienation" (Marcuse, 2005, p. 98).

According to Marcuse, the basic level of objectification is sensuousness and "in Marx it is this concept of sensuousness (as objectification) that leads to the decisive turn from classical German philosophy to the theory of revolution, for he inserts the basic traits of practical and social being into his determination of man's essential being. As objectivity, man's sensuousness is essentially practical objectification, and because it is practical it is essentially a social objectification" (Marcuse, 2005, p. 100). Human essence is primarily characterized by sensuousness in which human subjectivity and technical activity are linked in a process of objectification. For this reason, Marcuse concludes that sensuousness is an ontological category, "within the determination of man's essence and that it comes before any materialism or sensualism" (Marcuse, 2005, p. 98).

Therefore, the reform of essence through the concept of Life means that human essence is historical since historicity makes human existence negative and contingent, but it also means that the technical basis of work activity is essentially linked with life as sensuousness, "in labor the specifically human universality is realized" (Marcuse, 2005, p. 102). Marcuse's ontology of life takes into account the negativity of human existence as historicity and material activity. For this reason, essence is reformed through the negativity of existence that makes human work activity and sensible qualities essential categories. In this sense, to-be-in-the-world and being-with as biological, environmental, and societal specifications of *Dasein*, are related to sensuousness as objectification in work activity and as an expression of bio-social needs.

14.4 The Technological Shift

The account of the notions of historicity and Life as correlative notions is significant in understanding the question of the essence of technique in Marcuse's thought. Indeed, the essence of technique represents the same ontological bi-dimensionality of historicity and Life. Therefore, it is necessary to follow Marcuse's reform of essence concerning his definition of technology, which expands the Heideggerian definition with a socio-political dimension.

The demand for a concrete philosophy and a materialistic conception of existence characterized Marcuse's *Heideggerian Marxism*. The frame of an ontology of Life that he traced during those years in Freiburg seems to have disappeared after joining the *Institut fur Social Forschung* in Geneva and then in New York. Nevertheless, a residual romanticism within Marcuse's dialectics of technology reveals a deep anchorage to his previous *Heideggerian Marxism*. Thus, we can expand the ontological and existential frame of Life and historicity to Marcuse's representation of technology.

In his early writings, technical activity appears as a specification of the primary care of *Dasein*. It is a provision of life-space, material production, and objectification. It is also a complex element linked to the introduction of the division of labour that fragments social unity. This fragmented unity implies technical possibilities, breaks the socio-historical unity, and opens a differentiated and conflictual world where forms of existence come into contradiction with the mode of production. Technology is thus a socio-natural attitude, as a provision of life-space, and a socio-political perspective characterized by the openness of historicity to the historical subject. The repressive function of technology arises in Marcuse's work first in 1941 and then reappears in *Eros and Civilization* (1955) as the reality drive limiting and domesticating Life.

Marcuse's first essay dedicated to technology is "Some Social Implication of Modern Technology" from 1941. Here, Marcuse distinguishes between technics and technology. He argues that "technics proper (that is, the technical apparatus of industry, transportation, communication) is but a partial factor" (Marcuse, 1998, p. 41). In contrast, technology is a social process that has a twofold meaning: it is both mode of production and ideology. Mainly, as he writes, "as a mode of production, as the totality of instruments, devices and contrivances which characterize the machine age is thus at the same time a mode of organization and perpetuating (or changing) social relationships, a manifestation of prevalent thought and behaviour patterns, an instrument for control and domination" (Marcuse, 1998, p. 41). Therefore, as mode of production, technology materializes the Marxist idea as instrumentalization according to which the means of production are subsumed to capital. While as ideology, technology recalls Heidegger's definition, a world of ready-to-end things as meaningful entities. In this definition, technology is both material and cultural production and the reproduction of capitalism, as it is presented in *One-Dimensional Man*, where the technological *a priori* is the ideology of the capitalistic mode of production based on quantification of sensible qualities and work activity: in a word, the quantification of Life. In contrast with this definition of technology, in the 1941 article, technics allows for liberation as well as domination. This means that technics keeps open a de-reified realization of technology as mode of production and, in this sense, technics seems to be the ontological foundation of technology, as a historical and determined mode of production.

A further notion appears in a 1958 conference held in Paris titled "From Ontology to Technology".[7] Here, Marcuse argues that "technology has replaced ontology" (Marcuse, 1990, p. 121) and "being assumes the ontological character of *instrumentality*" (Marcuse, 1990, p. 123). Furthermore, he uses the notion of technicity in contrast with technology, as the inner sense of the technical system, as human life as the final technological cause. Indeed, he says, "if we consider the existential character of technicity, one can speak of a *final technological cause* and the repression of this cause through the social development of technology" (Marcuse, 1990, p. 123). This means that technology, replacing ontology, replaces as well as a teleology based on human life with its needs and desires, a biosocial life. Technology appears as a transcendental subject no more able to take into account the negativity of existence except in its repressive form. For this reason, Marcuse introduces the idea that he will explain further in the last chapter of *One-Dimensional Man* of the necessary suppression of technological finality in a freed society.

From Ontology to Technology witnesses the passage from the ontology of Life to the dialectics of technology of *One-Dimensional Man*. This shift is mediated by *Eros and Civilization* (1955), in which Marcuse translates the existential dialectics

---

[7] Between 1958 and 1959 Marcuse gave six lectures at EPHE (*École Pratiques des Hautes Études*). Only the second lecture has been published, in 1960, as "De l'ontologie à la technologie" in *Arguments* – a review directed at that time by Kostas Axelos (see Marcuse, 1960). The English translation – "From Ontology to Technology" – appeared in H. Marcuse. *Critical Theory and Society: A Reader* (1989). It can also be found also in H. Marcuse. *Philosophy, Psychoanalysis and Emancipation, Collected Papers of Herbert Marcuse, Volume Five* (2011). The full text of the six lectures has been published by R. Laudani and is available in Italian (see Marcuse, 2008).

of Life and historicity through the Freudian theory of drives, where the concept of Life is translated into an enlarged definition of *eros*. *Eros* is conceived by Marcuse not only as a pleasure drive but also as the great force unifying and preserving all life. Marx's negative anthropology of *The Economic and Philosophic Manuscripts* of human essence as sensuousness and passivity is transformed into a positive anthropology based on the libidinal economy of Life as *eros* and desire.

From Ontology to Technology is a key text through which Marcuse gives up Heidegger's ontological perspective. Indeed, here he translates the ontology of instrumental action into another ontology where a *final technological cause* is required in opposition to the not specified and unpolitical orientation of the *Care* of *Dasein*. Referring to *From Ontology to Technology*, Feenberg comments,

> Compressed in these few lines is the move Marcuse made in the early 1930s from Heidegger to Marxism via Hegel and Marx's *Economic and Philosophical Manuscripts* of 1844. In the *Manuscripts* Marx describes the ontological unity of man and nature in terms of need and labor. Translated into Heideggerian terms, this would be equivalent to being-in-the-world as the ontological condition realized in everyday instrumental action. But Marx's notion has a normative character, Heidegger's does not. The fulfillment of rich and complex human needs through the application of human capacities and powers in labor contrasts with the impoverishment and alienation of capitalism. In Heidegger's case there is, to be sure, a "final technological cause," but it is left completely vague, relative to the contingent world of *Dasein*. If Marcuse retained this curious parallel despite the difference, it is no doubt because he needed the concept of transcendental project to ground the opposition of capitalism and socialism in a historicized theory of the preconditions of experience. (Feenberg, 2013, p. 608)

According to Marcuse, the final technological cause is not human Life but capitalism, through a twofold abstraction – abstraction of sensible qualities and abstraction of labour – that separates a system of means from a system of ends.

In the light of *From Ontology to Technology*, it is possible to understand the argument of chapter 6 of *One-Dimensional Man*[8] on the relation between scientific rationality and political domination. Here, Marcuse establishes a parallel between technological rationality and the historical *a priori* of domination by presenting a progressive process of abstraction. First, scientific rationality suppresses secondary qualities (i.e. sensible qualities) from nature in order to account for objective causes. Then, by a second kind of abstraction that follows the abstraction of labour from human activity by the introduction of the economic concept of labour-time. Thanks to these abstractions, scientific rationality considers humans and machines as quantifiable things. Moreover, chapter 6 *of One-Dimensional Man* articulates four important sources in the economy of Marcuse's critical theory of technology: Heideggerian "technology," Adorno and Horkheimer's "dialectic of Enlightenment," Lukács' "reification," and Husserl's "*Lebenswelt* (lifeworld)," as Feenberg has highlighted (Feenberg, 2013). However, along with these sources Marcuse focuses on a quite unfamiliar French philosopher, Gilbert Simondon. Here, Simondon is quoted concerning his technocratic argument on the "autocratic philosophy of technics"

---

[8] See "From Negative to Positive Thinking: Technological Rationality and the Logic of Domination" (Marcuse, 2002, pp. 147–173.)

(Simondon, 1958, pp. 126–127).[9] Therefore Chap. 6 (of *One-Dimensional Man*) focuses on technocracy, based on the identification of scientific rationality with the logos of domination, which is properly the technological rationality that determines a new historical *a priori*: a technological one.

How is technological rationality a historical *a priori*? As we read in Marcuse's definitions of technology, a second technological nature seems to suppress the traditional concept of nature (from Hegel to Lukács). Marcuse stresses a repressive sense of the term, while technique in itself is neither bad nor good, rather it is open to its possible future historical realizations. Thus, an ontological shift seems to occur between technique and technology. Another ontology – a "second nature" characterized by a lack of negative dimension as the antagonistic opposition to the repressive capitalistic order – grounds the technological *a priori* and the technocratic tendency based on the identification of scientific rationality with the logos of domination. Nevertheless, the inner relation between historicity, life, and technicity grounds a specific kind of objectivity that does not coincide with Marcuse's repressive definition of technology.

14.5   Technicity Between Life and Historicity

Marcuse's definition of technology articulates a positive representation based on the essence of technique and a negative one based on technocracy. The ontological perspective of technology is still connected with Life and historicity and it properly refers to the objective field where human Life and activity take place. On the contrary, the historical realization of technology presents a kind of objectivity which means standardization and social homogeneity. However, a non-repressive representation of technology rises through the notion of technicity that Marcuse seems to borrow from Simondon.

The objective dimension of technology in shaping a quantifiable world of commodities is not the only one. The essence of technics represents the reform of essence from Marcuse's early writings and it concerns its historical dimension. The historicity of human essence is at the same time historicity of its own activity as *poiesis* and the notion of essence contains the negativity of existence as contingency of Life and alienation of work activity.[10] As Marcuse writes in *New Sources on the*

---

[9] Gilbert Simondon published the complementary thesis of his Ph.D. dissertation in 1958 as *Du mode d'existence des objets techniques* (2017. *On the Mode of Existence of Technical Objects*. Translated by C. Malaspina and J. Rogove. Minneapolis: Univocal Publishing) and Marcuse was one of his first readers outside of the French context.

[10] As Feenberg argues, "Life resembles Heidegger's Dasein in seeking its unity and wholeness through a future-oriented construction of its own potentialities. It does not have a prior essence but must create itself under the given conditions. In this sense it is "historical," a being that relates its past and future. Yet Marcuse's concept of life differs from Heidegger's Dasein in that the master–slave dialectic introduces social division and labor into its motility. The expression of its "care" in work and world leads to objectification and mutual recognition, themes entirely absent from

Foundation of Historical Materialism, a particular relation between humans and work activity occurs with the advent of capitalism, the fact that "labor produces itself and the worker as a commodity" (Marcuse, 2005, p. 92). It is a specific capitalistic relation modifying human life and activity in terms of "alienated man, estranged labor, estranged life, estranged man" (Marcuse, 2005, p. 90). Subsumed by capitalism, technology presents itself ambivalently: as an ensemble of the means of production and as a repressive ideology of capitalistic society. Technology is thus the objective and concrete reality able to structure the technological environment and our experience of it, but also the more abstract technical and scientific rationality of administration and domination.

In *From Ontology to Technology*, Marcuse introduces the problem of teleology through the notion of technicity. Following the argument of the 1958 conference, technicity as the essence of technique is a key concept for the replacement of an ontology of instrumental action, such as we find in Heidegger, to a critical theory of technology that problematizes the definition of technology as material production and ideology of capitalist society. The notion of technicity introduces the problem of teleology – the final technological cause – in a modified one-dimensional ontology. This teleology is distorted because it replaces human Life as its final cause with performance, rationalization and optimization of production, and the bureaucratic administration of life in performing needs and desires.

Marcuse translates the ontological conception into a historical and concrete representation of technology. Simondon plays a very significant role in this rethinking of technology since he provides a notion of technicity split off from the idea of utility and focused on its essential relation with human free activity. Indeed, according to Simondon, technicity defines the relation between objects and humans in a concrete existential sense, as a process of concretization. The machine's mode of existence, like the instrument or the technical ensemble, consists of communication between material elements and human relations. Thus, technicity in Simondon is an essential relation between humans and technical reality and, at the same time, a concrete relation. As Simondon argues, according to Marcuse, technicity should realize the project of a qualitatively different existence.

In *One-Dimensional Man*, the concept of technicity disappears while the phenomenological interpretation of technology articulates the relation between reason and experience through the technological rationality of methodological domination and a technological *a priori* that structures experience. Nevertheless, I argue that Marcuse sees in Simondon's technicity the potential fulfilment of technological progress and he exploits Simondon's thinking for its liberating potential. Indeed, in chapter 9 of *One-Dimensional Man*, Marcuse redefines the idea of technological progress in terms of Simondon's "technical treatment of finality problems".

Heidegger's existential analytic. Marcuse conceives the notion of the human "essence" in Hegelian–Marxist terms, as self-conscious unity of self, community, and world, and on this basis he argues that it can only be realized through overcoming the alienation of the worker under capitalism" (Feenberg, 2005, p. 92).

Science itself has rendered it possible to make final causes the proper domain of science. Society, "through a raising and enlarging of the technical sphere, must treat *as technical* problems, questions of finality considered wrongly as ethical and sometimes religious. The *incompleteness* of technics makes a fetish of problems of finality and enslaves man to ends which he thinks of as absolutes." (Gilbert Simondon, op. cit. p. 151; my italics). Under this aspect, "neutral" scientific method and technology become the science and technology of a historical phase which is being surpassed by its own achievements – which has reached its determinate negation. (Marcuse, 2002, p. 237)

Marcuse presents, in another form, the argument concerning the relation between technicity and the technological final cause. Technicity is linked with life and experience rather than the capitalist imperative of efficiency and production. It is possible to conceive of Life as a final cause of technics. Contemporary industrial society can be experienced in a different way than as mercantile objectivity. Therefore, the inherent possibility of the social development of technology depends on a different understanding of technics, a new way to see the world, a new and qualitatively different objectivity. According to Marcuse, this would be realized in an aesthetic dimension different from capitalist domination and constitute the determinate negation of the efficiency of the technological system.

In Simondon, the essence of technics represents cultural and human values different from utility, and the inherent finality of technicity is the internal coherence of technical objects linked to humans as an external finality. Thus, technical finality is the relation established between internal and external ends, between internal operating and human utilization; it is recurrent causality. Marcuse misunderstands Simondon's technical notion of teleology because Life as final cause is the basis of his reform of technology, while in Simondon, anthropogenesis and technogenesis are the same process in which finality is replaced by recursive information in a technical, biological and societal sense. The perfecting of the technical is possible through the study of information because it would effectively represent the technical organization of technicity and solve the problem of teleology in a different way than a means-ends conception of technology.

Despite Simondon's rethinking of the cybernetic theory of information, which Marcuse does not take into account, Marcuse and Simondon agree on the potential of technicity to structure a new experience of the world. In the change of theoretical perspective from ontology to technology, Marcuse requires the historical determination of a concrete technicity. Between Heidegger's essentialist representation of technics and Marcuse's critical theory of technology, Simondon represents an intermediary step toward materializing the essence of technics as a human value. In this sense, Simondon's reading attempts to concretize Heidegger's idealistic conception of technology and provides Marcuse with a concrete normative idea of technical reality, although Simondon does not conceive of the historical relation between technology and political domination.

14.6 The Reform of Objectivity Through Technicity

Bi-dimensionality as the conflict between essence and existence in Marcuse's early writings also characterize his work on technology. Technicity is a feature of the negativity of existence: technical activity and work are rooted in the essence of technique as historical necessities, but their actualization does not necessarily mean a repressive realization, i.e. technocracy. The technological world is thus complex and multilayered and represents the objective field that proceeds alongside the subjectivity of life, as motility of historicity, fragmented unity, and a form of existence that comes into contradiction with the mode of production. Technology is an intermediary field between ontology and politics where the dialectic between being and the motility of history is still in action. Technology is properly the field of the historical realization of a freed society. It is not only a world – in a Heideggerian sense – but constitutes a layer of objectivity that comes between subject and object. Feenberg's critical theory of technology presents the conceptualization of a sociotechnical rationality as a kind of objectivity beyond the separation between subject and object. Within a dialectical frame, the objectivity of technology is something that exceeds the mere opposition of an external alterity and represents a subjective feature in shaping Life and societal attitudes.

Simondon provides Marcuse with a normative idea of technology in contrast to Heidegger's essentialism. In Simondon, technicity is a key concept in the genetic essence of technique, which is value-laden from a socio-symbolical perspective, and historical in orienting the evolution of technical objects. Like historicity, technicity is the ontological sense of what is technical, therefore it is historical, human, and social. As Feenberg shows in *Technosystem*, Simondon's concretization means elegance; as an object evolves it unifies more technical functions. This is the sense of becoming *concrete* in Simondon's thought. For this reason, Simondon's definition of the mode of existence of technical objects can refer to technical systems including machinery, digital devices, social networks, and constitute multiple layers of objective reality. The relative autonomy and normativity of the technical system expressed by the concretization process are close to Feenberg's contemporary definition of *technosystem* as the technically rational multilayered field of social operations, disciplines, and knowledge. Technological objectivity, in this sense, means communication between nature and society. It is normative and structures experience, environment, and social relations. In Simondon, an objective and concrete technology presupposes an evolutionary idea of technological progress and, according to Feenberg, critical constructivism is oriented towards the idea of open progress. Technical progress as an open process means at a basic level the disclosure of new possibilities for a better existence, but also unexpected outcomes and the invention of new forms of Life. Thus, it presupposes a multiple non-deterministic trajectory.

Moreover, Simondon distinguishes between an objective mode of existence and an objectal one. Objectivity is properly the fact to be produced by technical activity, to be an alterity that embodies human activity. Objectality, in contrast, means independence from the producer through which the object is free from technical and productive relations.[11] In other words, Simondon, through objectivity and objectality, expresses the same *aporia* that Marcuse finds in Marx's *Manuscripts*, namely, an essential difference between objectification and alienation. In Simondon, objectivity means objectification without alienation, it is a kind of "good reification" by which technological alterity is recognized as an objective field structuring environment and experience. Objectality is "bad reification" by which the technical world constitutes economic and social values independent from productive relations. Objectivity means an instrument of perception since it is a transparent reification of work activity, while objectality refers to commodities since it is a marketable value independent of sociotechnical activity and thus is not only reification but also alienation.

Similarly, according to Marcuse, technics is neither bad nor good, but technology by the neutralization of human values represents a means-end system which distorts the inner technological final cause, human Life (or even the historical *Dasein* of his early writings). There is a material and sensible issue in Marcuse's representation of technology that takes into account his reform of human essence through the notions of historicity and Life that introduce the negative and sensuous dimension of existence. Existential ontology progressively becomes a critical theory of technology in which two historical concretions – society and technique – represent the negative moment of a future and promised liberation: the liberation of Life and technology, or even of Life through technology. Thus, the reform of objectivity through a normative idea of technology is the ultimate sense of Marcuse's idea of a new sensibility rising up through the consumption or destruction of the technological apparatus.

To conclude, the notion of technicity appears as the new ontological structure of reality and allows for the shift from ontology to technology. Feenberg's reading of Simondon's "concretization" contributes to design the objectiveness of technology between ontology and politics, or even nature and society.[12] Simondon's representation of technology is still compatible with Marcuse's. Thus, Marcuse's shift from his dialectics of Life to the dialectics of technology articulates his early theory of historicity with the notion of technicity. A speculative understanding of Marcuse's reading of Simondon seems to be placed in the inner relation between historicity, Life, and technicity that grounds a new kind of objectivity: a technological one. The bi-dimensionality of the essence of technique replaces the conflict between essence and existence with the relation between repressive or emancipatory representation of technology. The ambivalence of technology between technocracy and the realization of a freed society is based on technicity, which presupposes the inclusion of the negativity of existence in the essence of technique. Nevertheless, this negativity as work activity and labour does not necessarily mean the reification and alienation of human essence. Rather it implies an evolutionary idea of human progress as well as

---

[11] See Simondon (2014).
[12] See Feenberg (2017).

a technological one toward the direction outlined by Feenberg: "we would then be at the beginning rather than the end of progress. The possibility cannot be excluded; if history is contingent, its contingencies may include a general transformation" (Feenberg, 2017, p. 203). Thus, technological objectivity implies multiple layers and directions integrating the contingency of history, while sociotechnical rationality includes the negative dimension of existence which Marcuse expressed in his early writings in terms of historicity and dialectics of Life and translating the question of being into the question of technology.

References

Feenberg, A. (2005). *Heidegger and Marcuse: The catastrophe and redemption of history*. Routledge.

Feenberg, A. (2013). Marcuse's phenomenology: Reading chapter six of one-dimensional man. *Constellations, 20*(4), 604–614.

Feenberg, A. (2014). *The philosophy of praxis: Marx, Lukacs and the Frankfurt School*. Verso Press.

Feenberg, A. (2017). *Technosystem: The social life of reason* (pp. 66–86). Harvard University Press.

Marcuse, H. (1987 [1932]). *Hegel's ontology and the theory of historicity*. MIT Press.

Marcuse, H. (1998 [1941]). Some implications of social change. In H. Marcuse (Ed.), *Technology, war and fascism. Collected papers of Herbert Marcuse: Volume One* (D. Kellner, Ed., pp. 39–65). Routledge.

Marcuse, H. (1990 [1958]). From ontology to technology. In H. Marcuse (Ed.), *Critical theory and society: A reader* (E. Bronner & D. Kellner, Eds., pp. 119–127). Routledge.

Marcuse, H. (1960). De l'ontologie à la technologie. Les tendances de la société industrielle. *Arguments, 18*, 54–59.

Marcuse, H. (2002 [1964]). *One-dimensional man*. Routledge.

Marcuse, H. (2005). *Heideggerian Marxism* (R. Wolin & J. Abromeit, Eds.). University of Nebraska Press.

Marcuse, H. (2008). Lezioni parigine del 1958. In R. Laudani (Ed.), *La società tecnologica avanzata* (Vol. 3, pp. 55–140). Manifestolibri.

Pippin, R. (1988). Marcuse on Hegel and historicity. In R. Pippin, A. Feenberg, & C. P. Webel (Eds.), *Marcuse: Critical theory and the promise of utopia* (pp. 68–94). Macmillan.

Schmidt, A. (1968). Existential-Ontologie und historischer Materialismus bei Herbert Marcuse. In *Antworten auf Herbert Marcuse* (pp. 17–49). Suhrkamp.

Simondon, G. (1958). *Du mode d'existence des objets technique*. Aubier.

Simondon, G. (2014). Psychosociologie de la technicité. In *Sur la technique* (pp. 27–129). PUF.

Index

© The Editor(s) (if applicable) and The Author(s), under exclusive license to Springer Nature Switzerland AG 2022
D. Cressman (ed.), *The Necessity of Critique*, Philosophy of Engineering and Technology 41, https://doi.org/10.1007/978-3-031-07877-4

Printed in Great Britain
by Amazon

44121336R00163